A Flora of

SOUTHERN ILLINOIS

Robert H. Mohlenbrock

AND

John W. Voigt

Southern Illinois University Press

CARBONDALE AND EDWARDSVILLE

Feffer & Simons, Inc.

LONDON AND AMSTERDAM

Library of Congress Cataloging in Publication Data

Mohlenbrock, Robert H.
    A flora of southern Illinois.

    (Arcturus books, AB120)
    Bibliography: p.
    1. Botany—Illinois.   I. Voigt, John W., joint
author.   II. Title.
[QK157.M6 1974]   582′.13′09773   73-12984
ISBN 0-8093-0662-X

ARCT
URUS
BOOKS ®

CONTENTS

LIST OF ILLUSTRATIONS

Figures

Plates

A Flora of Southern Illinois

Introduction

A *Flora of Southern Illinois* has been prepared for use by anyone having an interest in our native flora and who has had some introductory study of botany. The book will be of particular use to workers in the applied fields of biology, such as conservation, wildlife, forestry, and agriculture.

This book is still another effort to avoid or overcome the monotony of anonymous scenery. It will help the uninformed to the pleasure of recognition and to develop an appreciation for nature's forms and their botanical, geographical, and historical relationships.

Southern Illinois as treated in this work encompasses the southern twelve counties. The southern boundary of the area is the confluence of the Ohio and Mississippi rivers. These rivers also mark the eastern and western borders respectively. The northern boundary is a line running from northwestern Randolph County eastward along the 38th parallel.

The Southern Illinois area comprises 4,355 square miles which is about equal to that of the state of Connecticut (4,965 square miles). This twelve-county area has three neighboring states at its borders and is within one hundred miles of five states. This location is most favorable to floristic richness since some of these states are southern and the area is between or at the convergence of several important migration routes.

Climate and other conditions work together in setting the limits of where certain plants shall grow. Some plants grow only in groups of their own kind while other plants of many different kinds grow to-

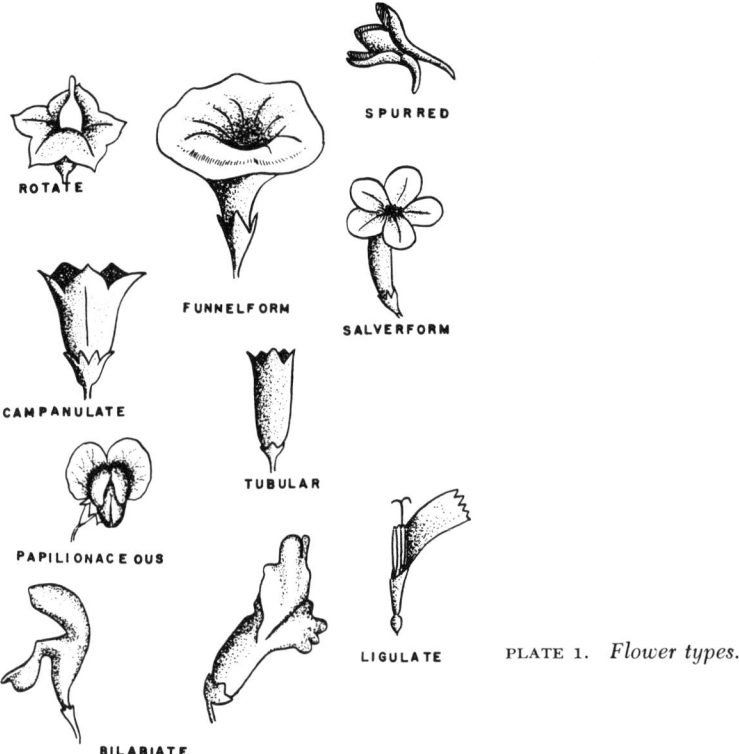

ROTATE

SPURRED

FUNNELFORM

SALVERFORM

CAMPANULATE

TUBULAR

PAPILIONACE OUS

LIGULATE   PLATE 1. *Flower types.*

BILABIATE

gether in various combinations. These groups of plants which are recognizable units are called communities.

Some of the more easily recognized plant communities of the Southern Illinois area are those of the following habitats: moist ravine forests; moist ditches; swamps; marshes; sink-hole ponds; stream beds and sandy river banks; hill prairies; and rocky slopes and ledges of variable aspect made of both limestone and sandstone. A few springs and caves occur, and numerous overhanging rock-ledges are known. A large number of introduced species have been recorded. These have for the most part been found in the larger cities, Carbondale and Murphysboro, and along railroads and highways.

MOIST RAVINES

The moist ravines may be considered the richest locations for plant communities. Dominant trees in the lowlands are Sugar Maple (*Acer saccharum*), Southern Sugar Maple (*Acer barbatum*), and American Elm (*Ulmus americana*). The layer of shrubs in communities like this is composed primarily of Spicebush (*Lindera benzoin*), American Bladdernut (*Staphylea trifolia*), and Running Strawberry (*Euonymus obovatus*). A host of colorful herbs present endless patterns of beauty from March through June.

MOIST DITCHES

The moist ditch flora is closely related to other aquatic habitats. Among the species of these ditches are Quillwort (*Isoetes melanopoda*), Spike Rush (*Eleocharis smallii*), Beak-rush (*Rhynchospora corniculata*), Sedge (*Carex annectens*), and Water Starwort (*Callitriche heterophylla*).

SWAMPS

There is considerable swampy ground in the southern counties. Dominant trees of shallow swamps or merely wet woods are Pin Oak (*Quercus palustris*) and Bur Oak (*Q. macrocarpa*). Deeper swamps may have Bald Cypress (*Taxodium distichum*), Swamp Red Maple (*Acer drummondii*), Swamp Cottonwood (*Populus heterophylla*), Deciduous Holly (*Ilex decidua*), and such characteristic herbaceous plants as *Carex muskingumensis, Carex grayii,* Purple Fringeless Orchid (*Habenaria peramoena*), and Swamp Milkweed (*Asclepias perennis*).

MARSHES

While Southern Illinois is not a paludal region, one tiny marsh occurs just north of Murphysboro. The area is very spongy underfoot

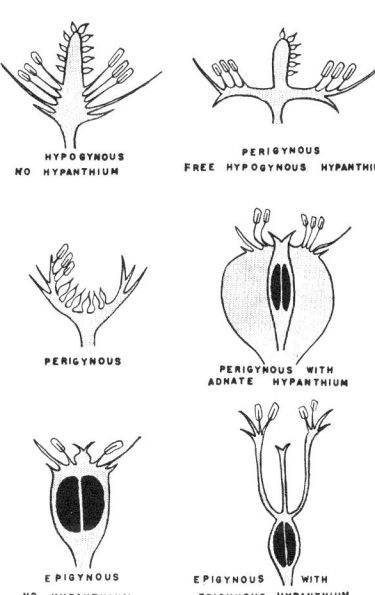

PLATE 2. *Ovary position—the fundamental types.*

for most of the year. It affords habitats for the following more interesting plants: sedges (*Carex comosa* and *Carex crinita*), Turtlehead
(*Chelone glabra*), Marsh Fern (*Dryopteris thelypteris*), Narrow-
leaved Buttercup (*Ranunculus pusillus*), Swamp Rose (*Rosa palustris*), and Viburnum (*Viburnum recognitum*).

SINK-HOLE PONDS

Sink-hole ponds are common in Monroe and Hardin counties, but
are rare in Jackson County. In Missouri these sinks reveal many relict
coastal plain elements. Only one such species, *Carex brachyglossa*,
has been found in our area. Further study of these ponds is needed.

STREAM BEDS

Along the banks of streams are such trees as Black Willow (*Salix
nigra*), Cottonwood (*Populus deltoides*), Sycamore (*Platanus occidentalis*), and River Birch (*Betula nigra*). Few shrubs are found.
Those present are Buttonbush (*Cephalanthus occidentalis*) and Sandbar Willow (*Salix interior*). In running streams a sedge (*Carex torta*)
is a characteristic though not too common species. Along sandy
shores two grasses commonly present are *Eragrostis hypnoides* and
Leptoloma filiformis. Winged Pigweed (*Cycloloma atriplicifolia*)
may also occasionally be found in these sandy areas. In the nearly
dry stream bed of intermittent streams the flowering season begins
with Clammy Hedge-hyssop (*Gratiola neglecta*) in May and ends in
July with Sharp-winged Monkey-flower (*Mimulus alatus*) and Lobe-
lias (*Lobelia* spp.).

HILL PRAIRIES

Atop some of the limestone river bluffs are isolated remnants of
prairie. These small relict areas are minute replicas of prairies to the
west. Characteristic species include Little Bluestem (*Andropogon
scoparius*), Big Bluestem (*Andropogon gerardi*), Side-oats Grama
(*Bouteloua curtipendula*), False Boneset (*Kuhnia eupatorioides*),
Blazing Star (*Liatris cylindracea*), Purple Prairie Clover (*Petalostemum purpureum*), and White Prairie Clover (*Petalostemum candidum*).

ROCKY UPLANDS

The dry woods community is a very extensive one in Southern
Illinois. It is dominated by Red Oak (*Quercus rubra*), Black Oak

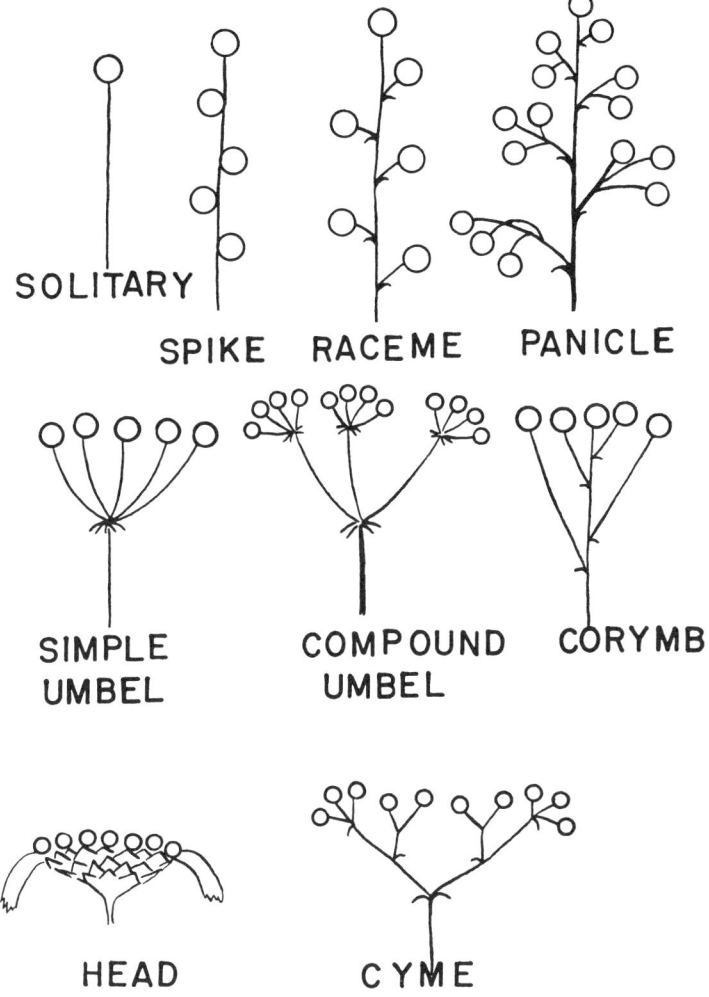

SOLITARY

SPIKE   RACEME   PANICLE

SIMPLE   COMPOUND   CORYMB
UMBEL   UMBEL

HEAD   CYME

PLATE 3. *Inflorescence types.*

(*Quercus velutina*), and species of hickory (*Carya* spp.). These woods are not without their share of shrubs, however, as New Jersey Tea (*Ceanothus americanus*) and Fragrant Sumac (*Rhus aromatica*) are abundant. Non-woody growth in this community is practically nonexistent until late May when Spiderwort (*Tradescantia virginiana*), Wild Comfrey (*Cynoglossum virginianum*), and others begin to flower.

DRY SANDSTONE BLUFFS

The thin soil atop the dry bluffs supports scrub oak vegetation of Post Oak (*Quercus stellata*) and Black Jack Oak (*Q. marilandica*). Red Cedar (*Juniperus virginiana*) and Winged Elm (*Ulmus alata*) are also found here. Patches of bare sandstone are plentiful and in some places are covered with a thin layer of soil. Herbs of crevices and thin soil areas or in mats of Black Moss (*Grimmia*), which has trapped fine sediments, include Rockrose (*Talinum parviflorum*), Pineweed (*Hypericum gentianoides*), Stonecrop (*Sedum pulchellum*), Prickly Pear Cactus (*Opuntia humifusa*), American Aloe (*Agave virginica*), Goat's-rue (*Tephrosia virginiana*), and Sundrops (*Oenothera linifolia*).

ROCKY LEDGES FACING NORTH

Rocky gorges are occasionally found and harbor many mesophytic and northern forms such as Clubmoss (*Lycopodium lucidulum* and its variety *occidentale, Lycopodium complanatum* var. *flabelliforme*), Maidenhair Spleenwort (*Asplenium trichomanes*), Bishop's-cap (*Mitella diphylla*), Saxifrage (*Saxifraga forbesii*), and Partridge-berry (*Mitchella repens*). Peat Moss (*Sphagnum*), a northern bog moss, is found in cushiony patches in a few localities and usually in the company of the clubmosses.

LIMESTONE RIVER BLUFFS

Most of the limestone bluffs occur along the Mississippi River. Several species are confined to these bluffs in Jackson County, some of which are found in small crevices and others are found on the talus so characteristic of limestone areas.

Numerous calcicolous ferns occur, the most characteristic being Walking-fern (*Camptosorus rhizophyllus*), Black Spleenwort (*Asplenium resiliens*), Slender Lip-fern (*Cheilanthes feei*), and Purple Cliffbrake (*Pellaea atropurpurea*).

Several vines are found clinging to exposed faces or on limestone bluffs. The more important of these include Virginia Creeper (*Parthenocissus quinquefolia*), Carolina Snailseed (*Cocculus carolina*), Moonvine (*Calycocarpum lyoni*), and Dutchman's Pipe (*Aristolochia tomentosa*).

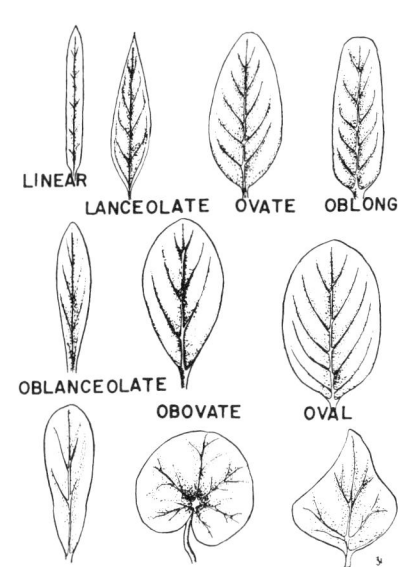

PLATE 4. *Leaf shapes.*

OVERHANGING ROCKS

Rock shelters are numerous in Southern Illinois and their openings may face almost any direction. These shelters, once formed by undercutting by streams, have usually a small stream near the sandstone overhang. The soils of this habitat are of fine sand and are moist in spring. Plants of usual occurrence in such areas are French's Shooting Star (*Dodecatheon frenchii*), Bitter-cress (*Cardamine pennsylvanica*), Pellitory (*Parietaria pennsylvanica*), Brookweed (*Samolus parviflorus*), and Enchanter's Nightshade (*Circaea latifolia*).

Summary of taxa in Southern Illinois and Jackson County

	Southern Illinois	Jackson County
Families	143	134
Genera	601	545
Species	1518	1295
Lower Taxa	81	68
Total Taxa (Species and lower)	1599	1363

A total of 1599 taxa exist in the southern twelve-county area. These plants represent over 60 per cent of the species listed as occurring throughout Illinois by Jones *et al.* (1955). Jackson County, the largest in the area, is represented by 1363 taxa.

A key is provided to all taxa which occur in the twelve-county area. A master key, divided into sections, enables one to ascertain families, while each family is provided with a key to its genera. Genera with more than three taxa in our area have keys to these taxa; if

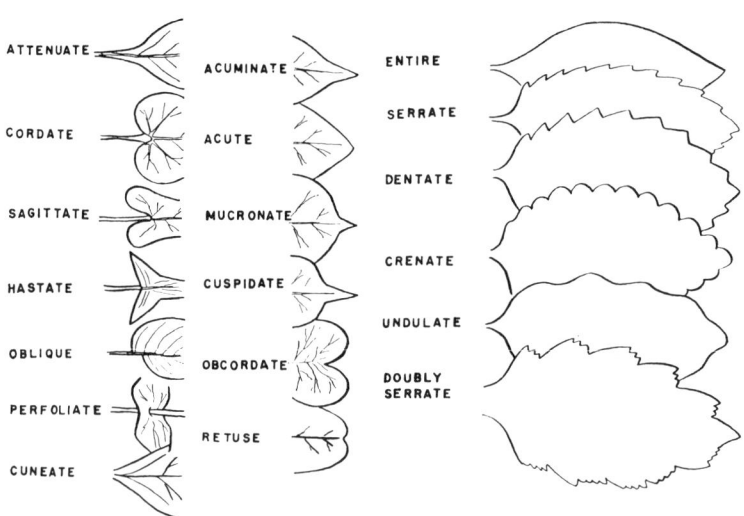

PLATE 5. *Leaf margins, leaf tips, and leaf bases.*

three entities or less occur, their distinguishing characteristics are incorporated in notes and no key is given. An asterisk by a species indicates that it has not been collected as yet in Jackson County.

A discussion of each species is presented which is concerned with nomenclatorial status, ecological notes, and distributional records. The specimens cited (following "SP. CIT.:" in the text) are only for Jackson County, except for the more rare species. All photographs of the more interesting plants and habitats are by J. W. Voigt unless otherwise indicated. For line drawings of each species, a reference is given to *The New Britton and Brown Illustrated Flora* (Gleason, 1952); e.g., *Allium stellatum* Ker. (1:415) means that a drawing of this species may be found in volume one, page 415. Synonyms are given in brackets if other names have been used frequently in more recent works.

In the specimens cited, abbreviations are used for the most frequent collectors; thus, BS—William M. Bailey and Julius R. Swayne,

Mc—John McCree, V—John W. Voigt, and M—Robert H. Mohlenbrock. For most species, our records for the earliest and latest flowering dates are given.

In completing this work the authors have collected extensively
throughout the area during the past several years and have, in addition to their own collections, cited many specimens from other excellent collections in the Southern Illinois University herbarium. The
historic French collection has provided records of plants no longer
found in the area. The McCree collection, the first intensive work
since the French, has provided other valuable records as have the
Bailey and Swayne collections of more recent years. A few plants
have been checked in the herbarium collections of the University of
Illinois and of the Illinois State Museum.

The treatment of certain more difficult groups follows recent mon-
ographic studies when there have been any available. The arrange-
ment of families follows Jones *et al.* (1955), a sequence based on the
Englerian system.

The authors wish to acknowledge financial support through grants
from the Research Council of the Southern Illinois University Grad-
uate School and a grant from The Illinois State Academy of Science.

PLATE 6. *Terms relating to fern structure.*

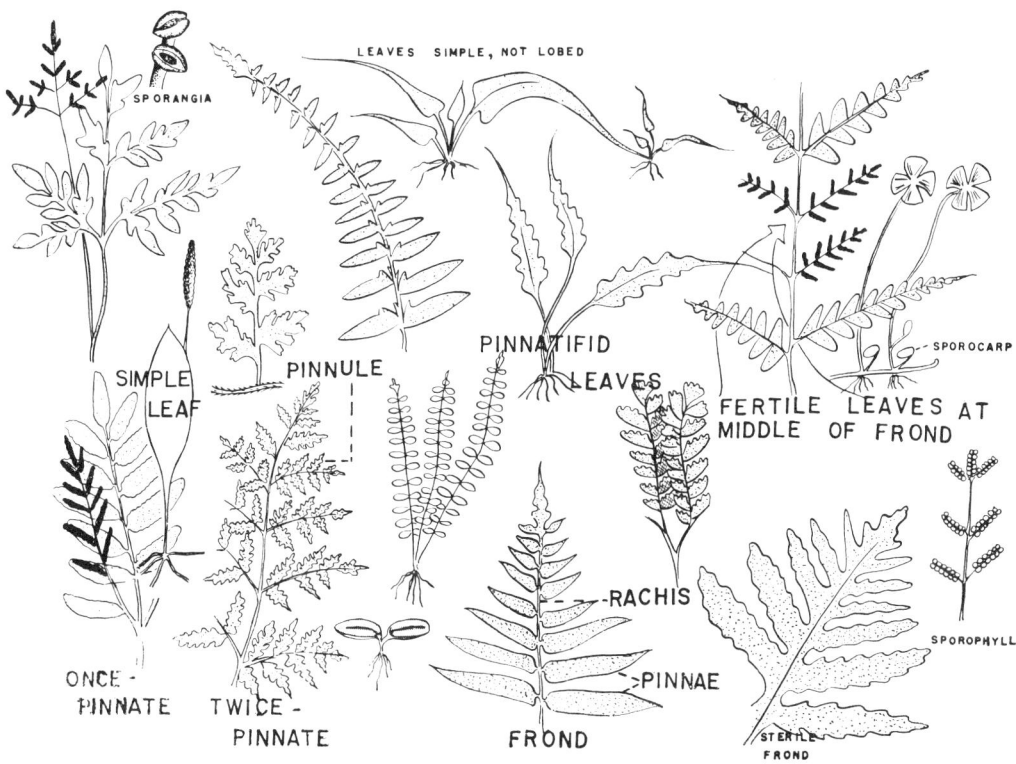

Key to Sections

Key to Families

Section 1:

NON-FLOWERING VASCULAR PLANTS: FERNS AND ALLIES

1. Sporangia borne in cones or in axils of leaves 2
  2. Cones with one kind of spore 3
    3. Stems leafy throughout; leaves narrow or oval, sessile and pointed . . . . 1. *Lycopodiaceae*
    3. Stems not leafy throughout; leaves, when present, whorled; stems grooved, jointed, and hollow . . . .
      . . . . 4. *Equisetaceae*
  2. Cones with two kinds of spores, these borne in different sporangia of the same cone; cones somewhat 4-angled; low growing plants . . . . . . 2. *Selaginellaceae*
1. Sporangia not as in 1 above 4
  4. Aquatics (in water or some phase of growth in water) . . 5
    5. Free floaters; leaves usually two-lobed; sporocarps in pairs beneath the stem . . . . 9. *Salviniaceae*
    5. Plants anchored, usually growing in shallow water or wet soil 6
      6. Leaves quadrifoliolate, long-stalked; sporocarps oval, hard coated, and containing two kinds of spores
        . . . . 10. *Marsileaceae*
      6. Leaves awl-like, tufted, from a dark corm-like stem; sporangia at base of leaves with two kinds of spores, each kind borne in a different sporangium . . .
        . . . . 3. *Isoetaceae*
  4. Terrestrial forms 7
    7. Plants of delicate nature, the fronds membranaceous-translucent, growing upon lichens and wet, shaded cave

walls or seeping rocky overhangs . *7. Hymenophyllaceae*

7. Plants of less delicate nature, except in winter rosettes and juvenile stages of some species   8

     8. Fertile leaves at the middle of the frond .   .   .

         .   .   *Osmunda claytoniana* in *6. Osmundaceae*

     8. Fertile leaves not at middle of frond   9

         9. Sporangia globular and in 2 rows   .   .   .   10

             10. Sporangia on a stalked terminal spike; one or more sterile blades present .   .   .

                 .   .   .   .   .   *5. Ophioglossaceae*

             10. Sporangia on a terminal upright contracted panicle   11

                 11. Rachis of sterile leaves winged .   .

                       .   .   *Onoclea* in *8. Polypodiaceae*

                 11. Rachis of sterile leaves not winged

                     .   *Botrychium* in *5. Ophioglossaceae*

         9. Sporangia not globular and not in 2 rows .   .   12

             12. Stipes of leaves winged at base; fertile apical pinnae contracted into an upright panicle   .   .   .   *6. Osmundaceae*

             12. Stipes not winged at base; sporangia in sori on underneath side of leaves or on the margins .   .   .   .   *8. Polypodiaceae*

Section 2:

CLIMBING OR TRAILING VINES OR PARASITES, SAPROPHYTES, OR CACTI

1. Plants parasitic, saprophytic or with thick or succulent stems with no leaves, or leaves reduced to spines   2

     2. Plants not parasitic or saprophytic, but with succulent jointed stems and no leaves, or leaves reduced to spines .   .   .

         .   .   .   .   *103. Cactaceae*

     2. Plants parasitic or saprophytic   3

         3. Saprophytic plants   4

             4. Plants whitish, waxy; ovary superior .   .   .   .

                 .   .   .   .   *113. Ericaceae*

             4. Plants not whitish; flowers irregular, ovary inferior

                 .   .   .   .   *31. Orchidaceae*

         3. Parasitic plants   5

             5. Parasitic upon roots or stem branches, not twining .   6

                 6. Plants parasitic upon roots   7

   7. Colorless plants; stamens 4, alternate with corolla lobes . . 133. *Orobanchaceae*

   7. Green plants; flowers creamy white; stamens 4, opposite corolla lobes . . .
     . . . *Comandra* in 41. *Santalaceae*

  6. Plants parasitic upon stem branches . . .
    . . . . 42. *Loranthaceae*

 5. Plants twining or climbing without tendrils . .
   . . . . . . 123. *Convolvulaceae*

1. Plants not parasitic, saprophytic or with thick or succulent stems with no leaves, or leaves reduced to spines 8

 8. Plants climbing by means of tendrils 9

  9. Plants with opposite or whorled leaves . . . .
  . . . . 134. *Bignoniaceae*

  9. Plants with alternate leaves 10

   10. Leaves simple 11

    11. Stems woody . . . 93. *Vitaceae*

    11. Stems herbaceous 12

     12. Throat of flower unfringed; fruit a pepo or berry . . . 105. *Cucurbitaceae*

     12. Throat of flower fringed; fruit a berry
    . . . . . 102. *Passifloraceae*

   10. Leaves compound 13

    13. Fruit an inflated capsule; leaves biternate .
    . . . . . 91. *Sapindaceae*

    13. Fruit a berry; leaves palmately compound .
    . . . . . 93. *Vitaceae*

 8. Plants climbing without tendrils 14

  14. Plants with compound leaves 15

   15. Leaves alternate 16

    16. Leaves trifoliolate 17

     17. Plants with woody stems; aerial roots; poison to the skin . 87. *Anacardiaceae*

     17. Plants with herbaceous stems; no aerial roots . . . 75. *Leguminosae*

    16. Leaves pinnately compound, with five leaflets
    . . . . . 75. *Leguminosae*

   15. Leaves opposite 18

    18. Leaves pinnately compound; plants climbing by aerial roots . . . . .
    . . . *Campsis* in 134. *Bignoniaceae*

    18. Leaves pinnately compound or ternate; numerous stamens; no aerial roots present .
    . . . *Clematis* in 55. *Ranunculaceae*

  14. Plants with simple leaves 19

   19. Leaves alternate 20

Section 3:

EVERGREEN OR CONE-BEARING PLANTS; GYMNOSPERMS BEARING NEEDLE-LIKE EVERGREEN OR DECIDUOUS LEAVES; CONES OR BERRY-LIKE FRUIT

Section 4:

WOODY PLANTS WITH OPPOSITE OR WHORLED SIMPLE LEAVES THAT ARE NOT LOBED OR DEEPLY DIVIDED

1. Leaves whorled; leaves large (6 to 15 inches long), cordate at the base; fruit a long pendent capsule . . . . .
. . . . *Catalpa* in 134. *Bignoniaceae*
1. Leaves opposite (uppermost may be whorled in *Cephalanthus*) .   2
   2. Leaves large (6 to 15 inches long), ovate, cordate; flowers purplish; fruit a short, thick (1½ to 2 inches) capsule . .
. . . . . . *Paulownia* in 134. *Bignoniaceae*
   2. Leaves smaller   3
      3. Leaves pellucid-punctate or with small black dots; flowers yellow, with stamens more numerous than petals . .
. . . . 96. *Hypericaceae*
      3. Leaves without small black dots; flowers not as above .   4
         4. Flowers in short, simple cymes or fascicled at the nodes   5
            5. Flowers in short simple cymes   6
               6. Ovary 2-loculed; calyx tube campanulate, its limbs minutely 4-toothed . 110. *Cornaceae*
               6. Ovary 3- to 5-loculed; sepals 4 to 5, spreading or recurved; seeds enclosed in an aril
. . . . . 85. *Celastraceae*
            5. Flowers in fascicles at the nodes; leaves silvery, stellate-scurfy underneath . 100. *Elaeagnaceae*
         4. Flowers in large cymes, corymbs, or globose heads .   7
            7. Flowers in globose heads; ovary inferior; upper leaves often in whorls . . . . . .
. . . *Cephalanthus* in 137. *Rubiaceae*
            7. Flowers not in globose heads   8
               8. Staminate flowers fascicled, pistillate flowers short-paniculate; shrubs or small trees . .
. . . . *Forestiera* in 118. *Oleaceae*
               8. Flowers, when perfect, in large (5 to 13 cm. broad) cymes or corymbs; marginal flowers often sterile; ovary inferior   9
                  9. Fruit drupaceous; all flowers fertile .
. . . . 138. *Caprifoliaceae*
                  9. Fruit a capsule; marginal flowers sterile.
. . . . 70. *Hydrangeaceae*

Section 5:

WOODY PLANTS WITH OPPOSITE, COMPOUND LEAVES, OR WITH OPPOSITE SIMPLE LEAVES THAT ARE LOBED OR DEEPLY DIVIDED

1. Leaves compound 2
   2. Leaves palmately compound or mostly all trifoliolate . . 3
      3. Leaves palmately compound (5 to 7 leaflets) . . .
      . . . . 90. *Hippocastanaceae*
      3. Leaves trifoliolate 4
         4. Leaflets finely serrate; fruit a bladdery capsule .
         . . . . 88. *Staphyleaceae*
         4. Leaflets coarsely toothed; fruits twin samaras; twigs
         green . . . *Acer negundo* in 89. *Aceraceae*
   2. Leaves pinnately compound (more than 3 leaflets) . . 5
      5. Leaf scars large, crescent-shaped; fruit a single samara;
      stamens 2; ovary superior; trees . . . . .
      . . . . . . *Fraxinus* in 118. *Oleaceae*
      5. Leaf scars not large, not crescent-shaped; fruit a berry;
      ovary inferior; weak shrub . . . . . .
      . . . . *Sambucus* in 138. *Caprifoliaceae*
1. Leaves simple, lobed, or deeply divided 6
   6. Lobes of leaf angled or toothed 7
      7. Fruit a pair of samaras; trees . . . 89. *Aceraceae*
      7. Fruit a one-seeded drupe; shrubs . . . . .
      . . . . *Viburnum* in 138. *Caprifoliaceae*
   6. Lobes not acutely angled, merely undulate or entire; ovary
   inferior . . *Symphoricarpos* in 138. *Caprifoliaceae*

Section 6:

WOODY PLANTS HAVING ALTERNATE SIMPLE LEAVES, NONE OF WHICH ARE LOBED OR DEEPLY DIVIDED

1. Leaves with entire margins 2
   2. Twigs or leaves with an aromatic odor; flowers appearing
   before the leaves . . . . . . 62. *Lauraceae*
   2. Twigs or leaves without an aromatic odor 3
      3. Leaves oblanceolate, large (4 to 12 inches long) . . 4

4. Naked buds with rusty, silky hairs; fruit a berry (banana-like); sepals 3, petals 6, stamens numerous . . . . 54. *Annonaceae*

4. Buds covered; aggregate accessory fruits; flowers with numerous stamens and pistils . . . . 53. *Magnoliaceae*

3. Leaves smaller 5

   5. Leaves prominently veined; flowers small, cymose, bracted . .*Cornus alternifolia* in 110. *Cornaceae*

   5. Leaves not prominently veined; flowers not as above 6

      6. Buds clustered at ends of branches; leaves with a bristle-tip . . .*Quercus imbricaria* and *Q. phellos* in 36. *Fagaceae*

      6. Buds not clustered at ends of branches . . 7

         7. Flowers perfect 8

           8. Flowers pink to red 9

              9. Leaves pinnately veined, 1 to 2 inches long, lanceolate-ovate; flowers in short 1-sided racemes . . . .*Gaylussacia* in 113. *Ericaceae*

              9. Leaves palmately veined, about as broad as long, heart-shaped; flowers in sessile umbel-like clusters . . .*Cercis* in 75. *Leguminosae*

           8. Flowers whitish-yellowish . . . 10

             10. Flowers in axillary clusters; fruit a berry . . 115. *Sapotaceae*

             10. Flowers not in axillary clusters; fruit a drupe 11

               11. Terminal bud absent; base of petiole hollow . . . *Dirca* in 99. *Thymelaeaceae*

               11. Terminal bud present; base of petiole not hollow . . *Vaccinium* in 113. *Ericaceae*

         7. Flowers imperfect 12

           12. Pith of twigs with cross-plates; bark separating into square plates; fruit a drupe . .*Nyssa* in 110. *Cornaceae*

           12. Pith chambered in part; bark separating into square plates; fruit an orange, globose, several-seeded berry (2.5 to 5 cm. in diameter) . . . .*Diospyros* in 116. *Ebenaceae*

1. Leaves serrate, dentate, or coarsely dentate .　.　.　.　.　13

　13. Leaf base oblique (uneven) .　.　.　.　.　.　.　14

　　　14. Leaves pinnately veined; fruit a samara, drupe, or
　　　　nut .　.　.　.　.　.　.　.　*37. Ulmaceae*

　　　14. Leaves palmately veined; fruit a bracted drupe .　.

　　　　.　.　.　.　.　.　.　.　.　*94. Tiliaceae*

　13. Leaf base symmetrical (even) .　.　.　.　.　.　15

　　　15. Leaves with stipules .　.　.　.　.　.　.　16

　　　　16. Inflorescence a cyme, corymb, or panicle .　.

　　　　　.　.　.　.　.　.　*92. Rhamnaceae*

　　　　16. Flowers in catkins; leaves usually narrow; twigs
　　　　　slender and flexuous; buds incurved .　.　.

　　　　　.　.　.　.　.　.　*33. Salicaceae*

　　　15. Leaves without stipules .　.　.　.　.　.　.　17

　　　　17. Flowers borne in catkins; fruit a 1-seeded nut
　　　　　or samara .　.　.　.　.　*35. Betulaceae*

　　　　17. Flowers not in catkins .　.　.　.　.　.　18

　　　　　18. Plants polygamous or dioecious .　.　.　19

　　　　　　19. Fruit a red, berry-like drupe .　.

　　　　　　　.　.　.*Ilex* in 86. *Aquifoliaceae*

　　　　　　19. Fruit a multiple type; leaves serrate
　　　　　　　or lobed on same plant .　.　.

　　　　　　　.　.　.　.　.　*38. Moraceae*

　　　　　18. Plants with perfect flowers .　.　.　.　20

　　　　　　20. Stamens 15 to 20; inflorescence cy-
　　　　　　　mose, corymbose, or umbellate .　.

　　　　　　　.　.　.*Prunus* in 74. *Rosaceae*

　　　　　　20. Stamens less than 15 .　.　.　.　21

　　　　　　　21. Inflorescence racemose .　.　22

　　　　　　　　22. Inflorescence a narrow ra-
　　　　　　　　　ceme; plants of swampy
　　　　　　　　　places .　.　.　.

　　　　　　　　　Itea in 69. *Escalloniaceae*

　　　　　　　　22. Inflorescence a short ra-
　　　　　　　　　ceme of large (up to 2.5
　　　　　　　　　cm. long) whitish flowers

　　　　　　　　　.　.　117. *Styracaceae*

　　　　　　　21. Inflorescence cymose; leaves
　　　　　　　　coarsely dentate .　.　.　.

　　　　　　　　Viburnum in 138. *Caprifoliaceae*

Section 7:

WOODY PLANTS WITH ALTERNATE, COMPOUND LEAVES OR ALTERNATE, SIMPLE LEAVES WHICH ARE LOBED OR DEEPLY DIVIDED

1. Leaves compound 2
   2. Leaves trifoliolate 3
      3. Stems prickly; flowers perigynous, or perigynous with free hypogynous hypanthium (Plate 2). 74. *Rosaceae*
      3. Stems not prickly 4
         4. Leaflets sessile; fruit a suborbicular samara . .
           . . . .*Ptelea* in 81. *Rutaceae*
         4. Leaflets or at least the terminal one petiolulate; fruit a drupe . . . . . . 87. *Anacardiaceae*
   2. Leaves pinnately compound 5
      5. Leaves once-pinnately compound 6
         6. Paired stipules at base of petiole 7
           7. Stipules as short spines at base of petiole . . 8
              8. Leaflets 3 to 11; flowers regular, plants polygamous or dioecious . . . .
                . .*Zanthoxylum* in 81. *Rutaceae*
              8. Leaflets 9 to 21; flowers perfect, irregular
                . . . *Robinia* in 75. *Leguminosae*
           7. Stipules as reduced leaflet-like appendages; ovary apocarpous, perigynous (Plate 2) . .
            . . . . 74. *Rosaceae*
         6. Paired stipular spines absent 9
           9. Leaves having paired glands at base of leaflets
            . . . . 82. *Simarubaceae*
           9. Leaves without paired glands at base of leaflets 10
             10. Flowers imperfect; staminate catkins drooping; fruit a drupe or nut; branches without prickles . . 34. *Juglandaceae*
             10. Flowers perfect (no catkins present) . 11
                11. Branches without prickles . . .
                . . *Rhus* in 87. *Anacardiaceae*
                11. Branches usually prickly . . .
                . . . . . 111. *Araliaceae*
      5. Leaves twice-pinnately compound; fruit a legume . . 12
         12. No spines present; pith large, chocolate-colored .
           . . . *Gymnocladus* in 75. *Leguminosae*

12. Branched spines present; pith not as above . .
. . . . .*Gleditsia* in 75. *Leguminosae*

1. Leaves simple, lobed, or divided 13

13. Leaves pinnately lobed 14

14. Buds clustered at ends of branches; leaves bristle-
tipped . . . . .*Quercus* in 36. *Fagaceae*

14. Buds not clustered at ends of branches; leaves 4-lobed;
stipule scars nearly encircling the twigs . . .
. . . . .*Liriodendron* in 53. *Magnoliaceae*

13. Leaves palmately lobed or lobes irregular 15

15. Leaves palmately lobed 16

16. Leaves star-shaped; twigs often corky-winged;
fruit a pendent, bur-like aggregate . . .
. . .*Liquidambar* in 72. *Hamamelidaceae*

16. Leaves not star-shaped 17

17. Twigs bearing spines 18

18. Trees; fruit a pome . . . .
. . .*Crataegus* in 74. *Rosaceae*

18. Shrubs; fruit a berry . . . .
. . . . 72. *Grossulariaceae*

17. Twigs without spines 19

19. Leaf scars encircling the bud; stipule
scars encircling the twig; trees . .
. . . . . 73. *Platanaceae*

19. Leaf scars not encircling the bud; no
stipule scars; shrubs . . . .
. . .*Hibiscus* in 95. *Malvaceae*

15. Leaves irregularly lobed 20

20. Leaves 3-lobed or 2-lobed as in a left-hand and
right-hand mitten, some entire; twigs green,
aromatic . . .*Sassafras* in 62. *Lauraceae*

20. Leaves not lobed as above; margins serrate; twigs
not green, not aromatic; juicy multiple fruit .
. . . . 38. *Moraceae*

Section 8:

HERBACEOUS AQUATIC FLOWERING PLANTS FLOATING ON OR UNDER WATER OR PARTLY EMERGENT

1. Monocotyledonous plants 2

2. Leaves elongate and narrow 3

3. Flowers borne in globose heads, staminate heads uppermost; stigmas 1 or 2; fruit an achene . 15. *Sparganiaceae*

3. Flowers not in globose heads 4

 4. Flowers large, 2 to 4 inches broad, with 3 sepals, 3 petals, 3 stamens, and a 3-carpellate ovary . . . . 30. *Iridaceae*

 4. Flowers smaller 5

  5. Leaves mostly submerged; flowers on long slender scapes . . . . *Vallisneria* in 19. *Hydrocharitaceae*

  5. Leaves emergent 6

   6. Flowers on a spadix, enclosed by a spathe; leaves and rhizomes aromatic . . . . *Acorus* in 22. *Araceae*

   6. Flowers in spikes; pistillate and staminate parts of spike contiguous or short-separated; pistillate part of spike thickest and lowermost; perianth of bristles; leaves and rootstocks not aromatic . . 14. *Typhaceae*

2. Leaves not elongate and narrow or no leaves present . . 7

 7. No leaves present (thallus 1 to 3 mm. broad), free-floating . . . . 23. *Lemnaceae*

 7. Leaves present, larger than above (over 2 cm. broad) 8

  8. Leaves floating and spongy thick, cordate at base . . . . *Limnobium* in 19. *Hydrocharitaceae*

  8. Leaves not floating 9

   9. Inflorescence paniculate; fruit an achene; leaves not glossy-green . . . 18. *Alismaceae*

   9. Inflorescence spicate; fruit a capsule; leaves glossy-green, cordate at base . . . . 25. *Pontederiaceae* °

1. Dicotyledonous plants 10

 10. Plants generally submerged 11

  11. Leaves alternate, 1- to 2-dichotomous, segments few, capillary, bladder-bearing . 132. *Lentibulariaceae*

  11. Leaves opposite, whorled, or verticillate . . . 12

   12. Leaves not dissected, the margins minutely serrate or spinulose; flowers axillary, solitary . . . . 17. *Naiadaceae*

   12. Leaves pinnately dissected, 2 to 3 times forked, or palmately dissected 13

    13. Submerged leaves opposite, palmately dissected; stems mucilaginous . . . . 56. *Cabombaceae*

    13. Leaves verticillate 14

    14. Leaves pinnately dissected, capillary
      . . . . 108. *Haloragidaceae*
    14. Leaves 2 to 3 times forked . . .
      . . . 59. *Ceratophyllaceae*
10. Plants floating on water or emersed (not submerged) . . 15
 15. Leaves all opposite 16
   16. Stems long, creeping on mud or floating in water,
     rooting at nodes; ovary inferior, 4-carpellate;
     flowers yellow . . . . 107. *Onagraceae*
   16. Stems shorter, not creeping 17
    17. Flowers without perianth; stamens 2 . .
     . . . . . 109. *Callitrichaceae*
    17. Flowers irregular, white; stamens 4 . .
     . . . *Bacopa* in 130. *Scrophulariaceae*
 15. Leaves alternate, basal, or fascicled on stolons, or uppermost leaves opposite 18
   18. Leaves floating on water 19
    19. Leaves large, simple, round or oval, petio-
     late, leathery 20
     20. Leaves with a sinus at base . .
      . . . . 58. *Nymphaeaceae*
     20. Leaves without a sinus at base . .
      . . . . 57. *Nelumbonaceae*
    19. Leaves smaller, not leathery . . . 21
     21. Flowers small, pinkish; nodes swollen,
      sheathed . . . 44. *Polygonaceae*
     21. Flowers on axillary peduncles, without
      a perianth; submerged leaves often lin-
      ear . . . 16. *Potamogetonaceae*
   18. Leaves, or at least some of them, emersed . . 22
    22. Leaves large, round, leathery; flowers large,
     yellowish, with stamens numerous . .
     . . . . . 57. *Nelumbonaceae*
    22. Leaves not large, round, or leathery . . 23
     23. Flowers white, sepals 4, petals 4, sta-
      mens 6 . . . . 65. *Cruciferae*
     23. Flowers white, calyx 5-parted, corolla
      tubular, 5-parted, stamens 5; stems
      hollow, jointed, and constricted at the
      joints . *Hottonia* in 114. *Primulaceae*

Section 9:

GRASSES AND GRASS-LIKE PLANTS (SEDGES AND RUSHES)

Leaves 3-ranked; flowers enclosed by scales; fruit an achene; stems solid, usually triangular in cross-section (some may be round, flattened or square) . . . . 21. *Cyperaceae*

Leaves 2-ranked; leaf sheath usually split; flowers enclosed by dry scales; fruit a caryopsis; stems generally round or oval in cross-section and hollow, jointed . . . . 20. *Gramineae*

Leaves grass-like, round in cross-section or channeled; leaf sheaths with free margins; stems round, solid; fruit a capsule containing 3 to many seeds; perianth of 6 glumaceous parts . . 26. *Juncaceae*

Section 10:

MONOCOTYLEDONOUS TERRESTRIAL HERBS (NOT GRASSES, SEDGES, OR RUSHES)

1. Flowers regular or nearly so 2
    2. Ovary superior 3
        3. Perianth undifferentiated (sepals and petals look alike) 4
            4. Stamens 6, all exserted; corolla essentially regular
                . . . . 27. *Liliaceae*
            4. Stamens 6, only 3 exserted; flowers somewhat 2-lipped . . . . . 25. *Pontederiaceae*
        3. Perianth differentiated (sepals are green) . . . 5
            5. Flowers subtended by a leaf-like bract; flowers usually bluish or roseate, rarely white. . . . .
                . . . . *Tradescantia* in 24. *Commelinaceae*
            5. Flowers not subtended by a leaf-like bract; 3 leaves are in a whorl at top of stems; flowers brownish red, white, or, rarely, yellow . . . . . .
                . . . . . . *Trillium* in 27. *Liliaceae*
    2. Ovary inferior 6
        6. Stamens 6; stems underground (bulbs) . . . .
            . . . . 29. *Amaryllidaceae*
        6. Stamens 3; stems underground (rhizomes) . . .
            . . . . 30. *Iridaceae*
1. Flower irregular 7
    7. Ovary inferior; flowers not subtended by a leaf-like bract;

leaves prominently veined in some plants, others with leaves
reduced to scales . . . . . . 31. *Orchidaceae*
7. Ovary superior; flowers subtended by a spathe-like bract
. . . . . .*Commelina* in 24. *Commelinaceae*

Section 11:

DICOTYLEDONOUS HERBS WITH OPPOSITE OR WHORLED OR VERTICILLATE SIMPLE LEAVES WHICH ARE NOT LOBED OR DISSECTED

1. Plants with verticillate leaves   2
   2. Plants with milky sap . . . . 122. *Asclepiadaceae*
   2. Plants without milky sap   3
      3. Plants with punctate-glandular leaves, flowers yellow, stamens numerous (few in *Hypericum*) . . . .
        . . . . 96. *Hypericaceae*
      3. Plants without punctate-glandular leaves, flowers not yellow or not commonly yellow   4
        4. Plants with swollen, sheathed nodes . . . .
        . . . . 44. *Polygonaceae*
        4. Plants with nodes not swollen and sheathed . .   5
          5. Flowers cream to greenish-white; usually 4 petals with nectariferous glands; plants 1 to 3 meters high. . . . .
            . . *Swertia carolinensis* in 120. *Gentianaceae*
          5. Flowers small; apetalous; nectariferous glands absent on petals; low growing plants . . .   6
            6. Plants with complete flowers; stems often square, occasionally retrorsely barbed . .
            . . . . *Galium* in 137. *Rubiaceae*
            6. Plants with apetalous flowers . . .   7
              7. Stems and branches forking; inflorescence cymose; calyx 5-parted . .
              . . . . . 49. *Illecebraceae*
              7. Stems and branches not forking; flowers axillary; calyx 5-parted; plants prostrate
              . . . *Mollugo* in 50. *Aizoaceae*
1. Plants with opposite leaves (upper leaves alternate in *Asclepias tuberosa*)   8
   8. Plants with milky sap (sap clear in *A. tuberosa*) . .   9
      9. Flowers apetalous; ovary 3-parted . 84. *Euphorbiaceae*

9. Flowers with petals and sepals 10
   10. Ovary 2-parted, forming a twin follicle fruit; flowers
       without a corona . . . . . 121. *Apocynaceae*
   10. Ovary 2-parted, forming a single several-seeded fol-
       licle; flowers with a corona . 122. *Asclepiadaceae*
8. Plants without milky sap 11
   11. Plants with square stems 12
       12. Ovary inferior 13
           13. Leaves 3- to 5-nerved, sessile, ciliate-serrulate;
               stamens twice the number of sepals or petals
               . . . .*Rhexia* in 106. *Melastomaceae*
           13. Leaves with one prominent (mid) vein, and
               not ciliate-serrulate; petals if present 2 to 6;
               stamens 4, 8, 10, 11, or 12 . 101. *Lythraceae*
       12. Ovary superior 14
           14. Flowers regular. 15
               15. Flowers complete, corolla of fused pet-
                   als, scarlet on outside, yellow on inside
                   . . .*Spigelia* in 119. *Loganiaceae*
               15. Flowers incomplete (petals absent); con-
                   spicuous bracts reddish-purple and sub-
                   tending the calyx; leaves broad, cordate
                   at the base, acuminate at the tip . .
                   . . . . . 48. *Nyctaginaceae*
           14. Flowers irregular 16
               16. Stems smooth or only slightly pubescent,
                   not ciliate; ovary not 4-lobed . . .
                   . . . . 130. *Scrophulariaceae*
               16. Stems and/or leaves mostly pubescent;
                   plants usually with aromatic or mint
                   odor; ovary usually 4-lobed . . . 17
                   17. Flowers 2-lipped; style 2-cleft at
                       apex; ovary lobed at flowering time,
                       splitting into 2 to 4 nutlets at ma-
                       turity . . . 128. *Labiatae*
                   17. Flowers only slightly irregular (fun-
                       nelform or somewhat rotate); style
                       1, entire; ovary not lobed at flower-
                       ing time, splitting into 2 to 4 nutlets
                       in fruit, or fruit a drupe . . .
                       . . . . 127. *Verbenaceae*
   11. Plants with round stems 18
       18. Ovary inferior 19
           19. Flowers irregular or nearly regular; inflores-
               cence small, or 2-flowered cymes; sepals 5;

petals 5, united; stamens 5; ovary 2-carpel-
late . . . . . 138. *Caprifoliaceae*

19.   Flowers regular, not as above . . . 20

20.   Flowers not in heads 21

21.   Flowers in corymbs or very short
paniculate cymes; stamens 1 to 4;
ovary 1- to 3-loculed . . .
Valerianella in 141. *Valerianaceae*

21.   Flowers not in corymbs or paniculate cymes 22

22.   Flowers in terminal and axillary cymes; stamens as many
as the petals; ovary 2-carpel-
late . 137. *Rubiaceae*

22.   Flowers not in terminal and
axillary cymes . . . 23

23.   Sepals and petals 4;
stamens twice as many
as the petals; ovary
4-carpellate . . .
. . 107. *Onagraceae*

23.   Sepals and petals 2;
petals white, notched
at apex; stamens 2;
fruit obovoid with
hooked bristly hairs .
. . . .*Circaea*
in 107. *Onagraceae*

20.   Flowers in flattened heads or flowers in
a subglobose or ovoid head . . . 24

24.   Flowers in flattened heads; calyx
reduced to a pappus; stamens 5
. . . . 143. *Compositae*

24.   Flowers in subglobose or ovoid
heads; calyx not reduced to a pappus; stamens 2 to 4 . . .
. . . . 142. *Dipsacaceae*

18.   Ovary superior 25

25.   Petals united 26

26.   Leaves fleshy; carpels 4 or 5, separate
. . . . . 67. *Crassulaceae*

26.   Leaves not fleshy 27

27.   Corolla irregular; stamens four   28

28.   Flowers funnelform or funnelform-campanulate . . 29

29. Flowers purple in short
terminal racemes; fruit
4 to 6 inches long,
curved, splitting at ma-
turity into 2 divergent
horns   .   .   .   .
.   135. *Martyniaceae*
29. Flowers lavender, terminal or axillary; fruit
not as above   .   .
.   131. *Acanthaceae*
28. Flowers distinctly 2-lipped,
the upper lip 2-cleft, long-
pointed, the lower lip shorter
and 3-cleft; flowers distant,
in slender elongate terminal
spikes   .   .   .   .   .
Phryma in 127. *Verbenaceae*
27. Corolla regular   30
30. Stamens opposite the corolla
lobes; ovules several; free
central placentation   .   .
.   .   .   114. *Primulaceae*
30. Stamens alternate with the
corolla lobes and attached to
the corolla   31
31. Corolla small, papery,
uncolored   .   .   .
.   136. *Plantaginaceae*
31. Corolla not small, papery, uncolored .   .   32
32. Flower parts reg-
ularly in fives;
ovary 3-carpellate
.   .   .   124.
Polemoniaceae
32. Flower parts in
either fours or
fives; calyx 3- to
12-lobed (2 se-
pals in *Obolaria*);
corolla 4- to 12-
lobed; stamens as
many as the co-
rolla lobes; ovary
mostly 2-carpel-

late   .   . 120.
Gentianaceae

25. Petals and sepals separate or with one or the
other absent 33

   33. Petals and sepals present   .   .   . 34

      34. Sepals 2; petals 5; ovary 2-carpel-
late; stamens 5 .   .   .   .
. *Claytonia* in 51. *Portulacaceae*

      34. Sepals more than 2   .   .   . 35

         35. Stamens numerous (5 to 12
in *H. mutilum*); leaves with
punctate dots   .   .   .
.   .   96. *Hypericaceae*

         35. Stamens 3 to 5; leaves with-
out punctate dots; calyx
5-parted; corolla 5-parted . 36

            36. Petals each deeply di-
vided; ovary 2- to
5-carpellate; leaves not
scurfy .   .   .   .
. 52. *Caryophyllaceae*

            36. Petals not deeply di-
vided; ovary 1-car-
pellate; leaves scurfy .
.   .   . *Crotonopsis*
in 84. *Euphorbiaceae*

   33. Petals absent; sepals present   .   . 37

      37. Leaves entire; plants woolly;
flowers perfect   .   .   .   .
Froelichia in 46. *Amaranthaceae*

      37. Leaves serrate, coarsely serrate or
dentate; plants not woolly .   . 38

         38. Leaves with stellate hairs;
stigmas 2 or 3; carpel 1;
fruit a capsule .   . *Crotonopsis* in 84. *Euphorbiaceae*

         38. Leaves without stellate hairs;
stigma 1; fruit an achene
.   .   . 40. *Urticaceae*

Section 12:

DICOTYLEDONOUS HERBS WITH OPPOSITE OR WHORLED LEAVES; SIMPLE AND LOBED OR DISSECTED OR COMPOUND LEAVES

1. Compound leaves 2
   2. Flowers in heads (disk, rays, or both) . 143. *Compositae*
   2. Flowers not in heads 3
      3. Trifoliolate leaves; perianth undifferentiated; numerous pistils and stamens; flowers not yellow; fruit with plumose hairs . . . *Clematis* in 55. *Ranunculaceae*
      3. Pinnately compound leaves; perianth differentiated; stamens 10; flowers yellow, peduncled, axillary; fruit 5-angled, spiny . . . *Tribulus* in 80. *Zygophyllaceae*
1. Simple leaves, lobed or dissected, or pinnatifid 4
   4. Leaves whorled, 3-parted nearly to the base, each division lanceolate and incisely toothed . *Dentaria* in 65. *Cruciferae*
   4. Leaves opposite (at least some of them). 5
      5. Leaves palmately veined, lobed, or dissected; ovary superior, 5-carpellate; stamens 10 . . 76. *Geraniaceae*
      5. Leaves pinnately divided . . . .
         . . . . . *Valeriana* in 141. *Valerianaceae*

Section 13:

DICOTYLEDONOUS HERBS WITH ALTERNATE, SIMPLE, ENTIRE, OR TOOTHED LEAVES

1. Perianth undifferentiated; stamens and carpels numerous . .
   . . . . 55. *Ranunculaceae*
1. Perianth differentiated (sepals and petals distinct if both are present) 2
   2. Perianth parts free from each other; ovary superior . . 3
      3. Carpels several, freely splitting in fruit; stamens numerous . . . . 95. *Malvaceae*
      3. Carpels few (under 5 or single) 4
         4. Carpels 3, splitting at maturity into 3 one-seeded segments; petals absent . . 84. *Euphorbiaceae*
         4. Carpels not as above; petals present 5
            5. Flowers irregular or nearly regular . . . 6
               6. Flowers spurred 7
                  7. Sepals and petals 5; ovary 1-loculed; stems not succulent . . 98. *Violaceae*

        7. Sepals 3; petals 3 or 5; stamens 5; stigmas 5-lobed; stems clear and succulent . . . . . 79. *Balsaminaceae*

    6. Flowers not spurred 8

        8. Stamens 10; pistil simple; flowers papilionaceous, or nearly regular (in *Cassia*) . . . . 75. *Leguminosae*

        8. Stamens 8; pistil not simple; sepals 5 (2 are larger than other 3 and usually colored); petals 3 or 5 . . . . 83. *Polygalaceae*

5. Flowers regular 9

    9. Petals in fives 10

        10. Sepals usually 2; leaves and stems somewhat succulent . . . . 51. *Portulacaceae*

        10. Sepals 5 or more; leaves and stems not succulent 11

            11. Leaves usually in a basal rosette 12

                12. Plants of rocky places; carpels 1 to several or 2 basally fused . . . . 68. *Saxifragaceae*

                12. Plants not of rocky places; carpels 5; free-central placentation; ovules numerous . . . 114. *Primulaceae*

            11. Leaves not usually forming a basal rosette . . 78. *Linaceae*

    9. Petals 3, 4, or 5, or petals absent . . . 13

        13. Petals 3; sepals 5; stamens 3 to 12; ovary 3-carpellate . . 97. *Cistaceae*

        13. Petals and sepals 4 to 5 or petals or sepals absent 14

            14. Petals and sepals 4 to 5 with stamens twice the number of petals, or stamens 6 15

                15. Plants with succulent leaves; carpels several and separate; stamens twice the number of petals . 67. *Crassulaceae*

                15. Plants not having succulent leaves; sepals 4; petals 4; stamens 6; carpels 2 . . . . . 65. *Cruciferae*

            14. Petals or sepals absent . . 16

16. Petals absent (sepals present) 17
    17. Leaves often mealy; flowers small, greenish; styles 1 to 3 .
      45. *Chenopodiaceae*
    17. Leaves not mealy . 18
      18. Stems with swollen joints, sheathed at the nodes; stamens 2 to 9; fruit 3-angled or 2-angled when flower is 2-carpellate .
        44. *Polygonaceae*
      18. Stems not swollen or sheathed at nodes . . 19
19. Leaves 3-nerved 20
    20. Flowers glomerate in axillary clusters; ovary simple; fruit an achene . . .
    Parietaria in 40. *Urticaceae*
    20. Flowers not glomerate in axillary clusters; ovary 3-parted; fruit a capsule . *Acalypha* in 84. *Euphorbiaceae*
19. Leaves not 3-nerved . . . 21
    21. Scarious bracts present; stems lined; leaves rough or harsh . . . . .
    . . 46. *Amaranthaceae*
    21. Scarious bracts not present; stems not lined; leaves not rough 22
      22. Ovary 3-parted . .
      . 84. *Euphorbiaceae*
      22. Ovary 10-loculed .
      . 47. *Phytolaccaceae*
    16. Sepals absent; stamens 6; plants of swampy places .
    . . . 32. *Saururaceae*

2. Petals fused 23
  23. Flowers irregular 24
    24. Flowers pipe-shaped, or calyx tube 3- to 6-lobed,

adnate to the ovary; petals none   .   .   .   .
.   .   .   .   .   .   43. *Aristolochiaceae*
24. Flowers not pipe-shaped   25
25. Flowers blue or purple   26
26. Corolla purple; stamens 4; fruit 10 to
15 cm. long, curved, splitting at maturity into 2 divergent horns .   .   .
.   .   .   .   135. *Martyniaceae*
26. Corolla blue; petals in corolla 2 plus 3;
sepals 5; locules 2   . 140. *Lobeliaceae*
25. Flowers not blue or purple; stamens 5 (one
is sterile); ovary 2-carpellate .   .   .   .
.   .   *Verbascum* in 130. *Scrophulariaceae*
23. Flower regular   27
27. Plants with milky sap; fruit a twin follicle; corolla
tubular, bluish   . *Amsonia* in 121. *Apocynaceae*
27. Plants with milky sap   28
28. Ovary superior; sepals, petals, and stamens 5   29
29. Corolla with a corona; fruit a single follicle; sap watery.   .   .   .   *Asclepias tuberosa* in 122. *Asclepiadaceae*
29. Corolla without a corona   .   .   .   30
30. Ovary separating into 4 nutlets at
maturity .   . 126. *Boraginaceae*
30. Ovary not as above; stigma capitate; ovary 2-carpellate   .   .
.   .   .   . 124. *Solanaceae*
28. Ovary inferior   31
31. Sepals and petals present; ovary more
than 1-loculed   32
32. Inflorescence an umbel; ovary
2-lobed   .   . 112. *Umbelliferae*
32. Inflorescence not an umbel .   .   33
33. Sepals 5; petals 5; flowers
solitary or racemose; ovary
3-carpellate; flowers blue   .
.   . 139. *Campanulaceae*
33. Sepals and petals 4; stamens
8; ovary of 4 carpels .   .
Ludwigia and *Jussiaea decurrens* in 107. *Onagraceae*
31. Petals absent; sepals 5, creamy white
or greenish; ovary 1-loculed; plants
parasitic upon roots of other plants   .
.   .   *Comandra* in 41. *Santalaceae*

Section 14:

DICOTYLEDONOUS HERBS WITH ALTERNATE COMPOUND LEAVES OR SIMPLE LEAVES WHICH ARE LOBED OR DISSECTED

              14. Fruit a follicle or dehiscent capsule
                    .   .   .   .   . *Spiraeoideae*
                of some authors in 74. *Rosaceae*

              14. Fruit an achene or drupelets
                (achenes sometimes enclosed by
                hypanthium) .   .   .*Rosoideae*
                of some authors in 74. *Rosaceae*

          13. Flowers epigynous (ovary inferior);
             fruit dry, carpels separate at maturity;
             carpels 2; inflorescence an umbel . .
             .   .   .   .   . 112. *Umbelliferae*

      12. Petals united; flowers hypogynous (ovary
         superior) 15

          15. Leaflets entire .   .   .   .   .
           .   *Polemonium* in 124. *Polemoniaceae*

          15. Leaflets toothed . 125. *Hydrophyllaceae*

2. Palmately compound leaves 16

      16. Flowers hypogynous (ovary superior), solitary or
         racemose .   .   .   .   .   . 66. *Capparidaceae*

      16. Flowers not as above 17

         17. Flowers epigynous, no hypanthium (Plate 2) . 18

            18. Fruit a drupe or berry; carpels 2 to 5 .   .
               .   .   .   .*Panax* in 111. *Araliaceae*

            18. Fruit dry, carpels separate at maturity;
               carpels 2 .   .   .   . 112. *Umbelliferae*

         17. Flowers perigynous; free hypogynous hypanthium
         (Plate 2) .   .   .   .   . 74. *Rosaceae*

1. Alternate simple leaves which are lobed or dissected .   .   . 19

  19. Leaves toothed, dissected, or pinnatifid 20

      20. Flowers perfect 21

         21. Flowers having a superior ovary 22

           22. Plants with colored sap (greenish-yellow or
              orange-red) .   .   . 63. *Papaveraceae*

           22. Plants without colored sap 23

             23. Petals free from each other .   .   . 24

               24. Flowers irregular and spurred .
                 .   .   .   . 98. *Violaceae*

             24. Flowers regular 25

               25. Flowers with numerous
                 stamens and pistils .   .
                 .   . 55. *Ranunculaceae*

               25. Flowers with 6 stamens, 4
                 sepals, 4 petals and a 2-
                 carpellate ovary .   .   .
                 .   .   . 65. *Cruciferae*

23. Petals united, lobes 5; calyx lobes 5; stamens 5 26

    26. Stigma capitate . . . .
    . . . . 129. *Solanaceae*

    26. Stigma deeply bifid . . .
    . . 125. *Hydrophyllaceae*

21. Flowers having an inferior ovary 27

    27. Umbellate inflorescence . . . .
    . . . . . 112. *Umbelliferae*

    27. Axillary or cymose inflorescence; stems and leaves viscid pubescent . . . .
    . . . *Mentzelia* in 104. *Loasaceae*

20. Flowers unisexual; plants dioecious . . . .
. . . . 39. *Cannabinaceae*

19. Leaves entire or with rounded lobes 28

28. Leaves mostly basal, round-lobed, crenate; plants of rocky places . . . . . 68. *Saxifragaceae*

28. Leaves not basal or as above 29

    29. Flowers with an undifferentiated perianth (petals and sepals look alike) 30

    30. Flowers with numerous pistils and stamens
    . . . . . 55. *Ranunculaceae*

    30. Flowers with a single pistil, the style wanting; leaves large, with 4 to 7 rounded lobes . . . . 60. *Berberidaceae*

    29. Flowers with a differentiated perianth (petals and sepals distinct or petals may be absent) . 31

    31. Plants with milky sap; ovary 3-parted . .
    . . . . . 84. *Euphorbiaceae*

    31. Plants without milky sap; ovary 3-parted . 32

    32. Stems with swollen nodes, the nodes sheathed; stamens 2 to 9 . . .
    . . . . 44. *Polygonaceae*

    32. Stems without swollen nodes . . 33

    33. Flowers small, greenish, perfect; plants polygamous, dioecious or monoecious; ovary 1-loculed; stamens usually 5 . . .
    . . . 45. *Chenopodiaceae*

    33. Flowers with 1 pistil, 5 to several carpels, these freely splitting in fruit; stamens many .
    . . . . 95. *Malvaceae*

Division Pteridophyta

1. *Lycopodiaceae* – Clubmoss Family

Lycopodium L.

1. Stems erect; sporangia not in terminal cones 2
   2. Leaves entire . . . . .*Lycopodium lucidulum* *
   2. Leaves minutely toothed . . . .
      . . . . *Lycopodium lucidulum* var. *occidentale*
1. Stems creeping; sporophylls in terminal cones . . . .
     . . . . *Lycopodium complanatum* var. *flabelliforme* *

Lycopodium lucidulum Michx. (1:3). The typical species is known from a moist sandstone ravine at Jackson Hollow in Pope County. Variety *occidentale* (Clute) L. R. Wilson is known only from one station in Illinois, a wet north-facing bluff with slightly acid soil, in Little Grand Canyon, Jackson County. *Lycopodium complanatum* L. var. *flabelliforme* Fern. (Fig. 1) has been found once in Illinois (*J. R. Swayne 2328* in 1952). It occurs along a rugged sandstone bluff over-

FIG. 1. *Trailing Ground-Pine* (Lycopodium complanatum *var.* flabelliforme), *a distinctly northern species, occurs at Indian Kitchen on Lusk Creek. It is near the southern-most limit of its range.*

FIG. 2. *Lusk Creek has cut deeply through Pennsylvania sandstone in Pope County. The stream is viewed from about one hundred feet above the water at a point known locally as Indian Kitchen.*

looking Lusk Creek in Pope County (Fig. 2). SP. CIT.: (*L. lucidulum* var. *occidentale*), Little Grand Canyon, November 22, 1949, *Hatcher.*

2. *Selaginellaceae* – Small Clubmoss Family

Selaginella   BEAUV.

1. Leaves delicate, not bristle-tipped; plants of moist situations
   .   .   .   . *Selaginella apoda* °
1. Leaves firm, bristle-tipped; plants of dry exposed places   .   .
   .   .   .   . *Selaginella rupestris* °

Although no species of *Selaginella* has been collected in Jackson County, two are to be looked for. *Selaginella apoda* (L.) Spreng. (1:9) occurs along streams or in rich woods and has been collected in Union and Pope counties. *Selaginella rupestris* (L.) Spreng. (1:9) occurs on bare sandstone blufftops in Pope County. Jones *et al.* (1955) attribute *Selaginella apoda* to Jackson County, but we have seen no specimens. The report may be based on a collection by Hatcher near Alto Pass, which is actually in Union County.

3. *Isoetaceae* – Quillwort Family

1. Leaves usually about 15 in number, rarely more than 20 cm. tall
   .   .   .   . *Isoetes butleri*
1. Leaves usually more than 15, sometimes up to 90 in number,
   often up to 30 cm. tall   .   .   .   . *Isoetes melanopoda*

Isoetes butleri Engelm. (1:11). This rare species was first discovered in Illinois in March, 1953, at Giant City State Park (Jackson County), *Voigt 1320.* It has since been collected in Pope County (Belle Smith Springs) and at Dixon Springs by the authors. Unlike other species of *Isoetes* of northeastern North America, this species grows in moist depressions on ledges of bluffs. The leaves appear in early March, the spores in mid-May, and the plants die down completely by early June. Our plants, which seldom have more than fifteen erect, slender leaves, are difficult to detect, since they grow mixed with *Allium cernuum, Nothoscordum bivalve,* and various grasses and sedges. Pfeiffer (1922) gives the habitat for *Isoetes butleri* as "limestone ledges" or "limestone barrens," but the stations in Southern Illinois are on sandstone. Our specimens have been identified by Dr. Rolla M. Tryon. A previous report of this species from Illinois by Hill (1912) is an error (see Jones, 1947). SP. CIT. Giant City State Park, April 30, 1953, *V 1320;* April 30, 1954, *M 1904.*

Isoetes melanopoda Gay and Durieu (1:11). This species is larger than the preceding in all respects. Some specimens have corms which measure nearly one inch in diameter. This species is the most recent addition of a pteridophyte to Southern Illinois. It was found growing abundantly in frequently inundated roadside ditches along Illinois highway 3 between the junction with Illinois highway 144 and Fountain Bluff. It is associated at these stations with *Eleocharis smallii, Callitriche heterophylla,* and other amphibious species. Specimens with pale leaf-bases which occur with the species have been segregated as forma *pallida* (Engelm.) Fern. The nearest station of *Isoetes melanopoda* to the Jackson County one is in St. Clair County, about one hundred miles to the north (Neill, 1950). SP. CIT.: typical: along Illinois highway 3, two miles south of junction of Illinois highways 3 and 144, June 13, 1955, *M 5386.*

4. *Equisetaceae* – Horsetail Family

Equisetum L.

Equisetum arvense L. (1:15). Common Horsetail, our most abundant *Equisetum,* is found along almost every railroad embankment, where it is often classed as a weed. This is a perennial species with rough, branched stems and with teeth of the sheaths persistent. SP. CIT.: Carbon Lake, July 29, 1949, *Hatcher;* Southern Illinois University campus, May 23, 1940, *McCree 65.*

Equisetum hyemale L. (1:17). This plant may be found growing with Winged Monkey-flower (*Mimulus alatus*), False Pimpernels (*Lindernia anagallidea* and *L. dubia*), Blue Lobelia (*Lobelia siphilitica*), and *Leucospora multifida.* It is our only evergreen *Equisetum,* and is further distinguished by its roughened, unbranched stems, and by the teeth of its sheaths being usually persistent. SP. CIT.: along Stonefort Creek, Giant City State Park, May 25, 1953, *M 559;* Kinkaid Creek, June 27, 1951, *BS 1949;* Grand Tower, November 5, 1881, *French & Seymour.*

Equisetum laevigatum A. Br. (1:17). This is an annual species with smooth unbranched stems, and with teeth of the sheaths soon falling.

The only station in the county is a moist clay railroad embankment. It is reported in Southern Illinois from Perry County by Jones (1947). SP. CIT.: along the Gulf, Mobile and Ohio Railroad, about three miles north of Etherton, May 12, 1954, *M 1901*.

5. *Ophioglossaceae* – Adder's-tongue Family

1. Leaves simple, entire   .   .   .   .   .   . *Ophioglossum*
1. Leaves lobed or compound .   .   .   .   . *Botrychium*

Ophioglossum   L.

Ophioglossum vulgatum L. (1:21). The curious Adder's-tongue Fern is infrequently collected in Southern Illinois, perhaps because

FIG. 3.   *Adder's-tongue Fern* (Ophioglossum vulgatum) *growing among oak leaves at Giant City State Park. Photo taken April 15, 1956.*

of its inconspicuousness (Fig. 3). The plants usually die down in July. This species has been collected three times in Jackson County, all in low woods or marshy areas. Records from Union, Saline, Gallatin, and Alexander counties are known. The principal veins of the leaf form smaller areoles which do not enclose secondary areoles. Our specimens belong to variety *pycnostichum* Fern. SP. CIT.: one-half mile south of Murphysboro, July 26, 1949, *Stewart 781;* Giant City State Park, April 30, 1953, *Stewart.*

Another Adder's-tongue, *Ophioglossum engelmannii* Prantl (1:21), has been collected even less frequently with three Illinois stations known. It was first collected on limestone bluffs in Hardin County in 1919 by E. J. Palmer and subsequently in the same county north of

Cave-in-Rock by Winterringer in 1950. It was discovered in 1955 under cedar trees on limestone bluffs north of Chester in Randolph County. It has one station in Jersey County. The principal veins of the leaves form large primary areoles which enclose several secondary areoles.

Botrychium sw.

1. Leaves persisting throughout the year, turning bronze in the fall; stipe of sterile leaf much shorter than the leaf   2
    2. Blades often 4-pinnate, the pinnae much dissected .   .
        .   .   .   . Botrychium dissectum var. dissectum
    2. Blades 2- to 3-pinnate, the pinnae entire or at most only toothed .   .   . .Botrychium dissectum var. obliquum
1. Leaves deciduous, dying down in the fall; stipe of sterile leaf longer than the leaf .   .   .   .   .Botrychium virginianum

Botrychium dissectum Spreng. var. dissectum (1:19). The typical form of this species happens to be the rarer one in our area. It has been collected four or five times in Jackson County, but nowhere is it abundant. It occurs usually in moist woodlands. Much more common is var. obliquum (Muhl.) Clute which generally occupies drier habitats. It is commonly called Bronze Fern because the blades' color turns bronze in winter. SP. CIT.: var. dissectum: Giant City State Park, March 7, 1953, Weber 491; Little Grand Canyon, Brewer 1. Var. obliquum: Hatcher & Stewart 231:2; Mc 1194; M 4791; M 4794. An immature specimen collected in a dry woodland at Giant City State Park by Hatcher (No. 1156) and determined by Dr. Rolla M. Tryon doubtfully as B. multifidum (Gmel.) Rupr. should probably be considered as B. dissectum var. obliquum.

Botrychium virginianum (L.) Swartz (1:21). This fern, popularly known as Rattlesnake Fern, is the most common Botrychium in Southern Illinois. It is abundant in mesic woodlands, but may be found in dry woods as well. SP. CIT.: Steagall 16; Mc 716; BS 29.

6. Osmundaceae – Royal Fern Family

Osmunda L.

1. Fronds twice pinnate; pinnules mostly entire . Osmunda regalis
1. Fronds once pinnate; pinnules pinnatifid   2
    2. Petioles with a dense cinnamon-colored tomentum; fertile

leaves not at middle of the frond . *Osmunda cinnamomea*
2. Petioles only loosely tomentose when young; fertile pinnules
    at middle of the frond .   .   .   .*Osmunda claytoniana*

Osmunda regalis L. var. *spectabilis* (Willd.) A. Gray (1:26). Royal Fern, one of our most beautiful ferns, is known in Jackson County only from a low swampy woods at Little Grand Canyon. Elsewhere in Southern Illinois it has been recorded from Union, Alexander, Pulaski, Williamson, Saline, and Pope counties. At Jackson Hollow in the latter county, it occurs along a ledge near the top of a sandstone bluff. Variety *spectabilis* may be distinguished from the species by its lower stature (0.5 to 1.8 meters), relatively broader fronds, more slender panicle, and rachises which do not bear numerous and persistent black, hair-like scales. SP. CIT.: Little Grand Canyon, May 18, 1954, *M 1902*.

Two other species of *Osmunda* have been found in Southern Illinois where they are exceedingly rare. Cinnamon Fern, *O. cinnamomea* L. (1:26) has been collected on a sandstone bluff along Lusk Creek in Pope County (*V 1193*). Interrupted Fern, *O. claytoniana* L. (1:26) is reported by Jones *et al.* (1955) from Union County.

7. *Hymenophyllaceae* – Filmy Fern Family

Trichomanes L.

The rare Filmy Fern, *Trichomanes boschianum* Sturm (1:24), has not been found in Jackson County, but has been collected in Jackson Hollow, Pope County (Fig. 4) where it was first located by Dr. Mary Steagall (No. 37, August 2, 1923) at the base of a wet, dripping, sandstone bluff.

FIG. 4. *Filmy Fern* (Trichomanes boschianum) *growing under a darkened cave-like ledge which drips with moisture at Jackson Hollow, the only Illinois station for this rare plant.*

8. *Polypodiaceae* – Fern Family

1. Blades simple, not lobed or toothed . . . . *Camptosorus*
1. Blades either toothed, shallowly lobed, or deeply dissected . . 2
   2. Stipe forking near apex; pinnae kidney-shaped and with low round lobes on the upper margin . . . . *Adiantum*
   2. Stipe simple; pinnae usually not kidney-shaped . . . 3
      3. Each pinna, including the lowest, connected by the winged rachis to the pinna above or below it . . . 4
         4. Leaves with veins forming a closed network . . . . *Onoclea*
         4. Leaves with veins open and not connected to each other 5
            5. Leaf segments with irregular margins . . . 6
               6. Blades of leaves glabrous . . *Asplenium*
               6. Blades of leaves pubescent at least below . . . . *Dryopteris*
            5. Lobes of leaf entire . . . . *Polypodium*
      3. At least the lowest pinnae distinct from ones above them 7
         7. Leaves with veins forming a closed network . . . . *Onoclea*
         7. Leaves with veins open and not connected to each other 8
            8. Blades once-pinnate 9
               9. Pinnae with short spine-tipped teeth . . . . *Polystichum*
               9. Pinnae either entire or with teeth which lack spiny tips 10
                  10. Stipes purplish; margins of leaf segments curled back over sporangia . . . . *Pellaea*
                  10. Stipes not purplish; margins of leaf not as above 11
                     11. Some pinnae over 4 cm. long . . . . *Athyrium*
                     11. All pinnae less than 4 cm. long . . . . . *Asplenium*
            8. Blades more than once-pinnate 12
               12. Rachis brown about two-thirds its length, becoming green and flattened near its apex . . . . *Asplenium*
               12. Rachis not as above 13
                  13. Margins of fertile pinnule segments curled back over sporangia . . 14

Camptosorus LINK

Camptosorus rhizophyllus (L.) Link (1:41). Walking Fern, so named from its ability to take root at tips of slender fronds, is an indicator of slightly basic conditions, although it is often found on sandstone rocks. It grows in moist areas in the deepest shade. At Giant City State Park, it grows in company with Puttyroot Orchid (*Aplectrum hyemale*), Celandine Poppy (*Stylophorum diphyllum*), and *Valeriana pauciflora*. SP. CIT.: *Mc 463; V 1056; M 55; VM 1804; VM 4793.*

Adiantum L.

Adiantum pedatum L. (1:30). Maidenhair, one of the well-known ferns of Southern Illinois, grows in rich woodlands, commonly associated with *Athyrium pycnocarpon, Polemonium reptans*, species of *Viola* and *Hydrophyllum*, and *Asarum canadense* var. *reflexum*. The Venus'-hair Fern, *Adiantum capillus-veneris* L., grows on limestone bluffs in southeastern Missouri, but is not known in Illinois. Its particular habitat does not seem to be present in Jackson County, but is approached in Randolph County. It may be distinguished easily by its unforked stipe. SP. CIT.: *Steagall 265; Hatcher 787; M 2419.*

Onoclea L.

Onoclea sensibilis L. (1:36). Sensitive Fern, common in rich woodlands in Southern Illinois, is particularly abundant at Lake Murphysboro, where it is associated with Ginseng (*Panax quinquefolia*) and Virginia Snakeroot (*Aristolochia serpentaria*). At one station on Lake Murphysboro it grows in shallow water near the bank. SP. CIT.: *V 1220; Steagall 92.*

Asplenium L.

1. Rachis green throughout . . . . *Asplenium pinnatifidum*
1. Rachis dark, at least at the base 2
    2. Rachis green only near the apex; leaves pinnate-pinnatifid .
     . . . . *Asplenium bradleyi*
    2. Rachis dark throughout 3
        3. Pinnae with a small lobe on the upper margin at the base 4
           4. Pinnae mostly opposite . . *Asplenium resiliens*
           4. Pinnae mostly alternate . *Asplenium platyneuron*
        3. Pinnae without a small lobe on the upper margin at the base . . . . . . *Asplenium trichomanes*

Asplenium pinnatifidum Nutt. (1:39). Pinnatifid Spleenwort is not common in Illinois, but does grow abundantly in Giant City State Park and other isolated areas, generally along the Shawneetown Ridge. Its habitat is crevices of sandstone bluffs. The blades are variable, but usually merely pinnatifid. A new form in which the pinnae are rudimentary and the sori are borne along the rachis has been found among typical plants at Giant City State Park, in the Union County part of the park (Mohlenbrock, 1956). This species also grows in Pope, Johnson, Union, Randolph, and Gallatin counties. SP. CIT.: *Mc 1194; M 4791; BS 714; VM 1806; M 2492.*

Asplenium bradleyi D. C. Eaton (1:41). This fern, one of the rarest of northeastern North America, was first discovered in Illinois in Randolph County in the spring of 1954. It is most abundant along Piney Creek, just west of West Point. It has now been found along the Jackson County extension of the creek and at Panther's Den in Williamson County (Fig. 5). SP. CIT.: along Piney Creek, near West Point, March 6, 1955, *M 4988.*

Asplenium resiliens Kunze (1:41). This, one of the rarest spleenworts of Jackson County, is extremely difficult to find; it grows on limestone bluffs in the Grassy Knob area. It is fairly abundant in the Pine Hills. SP. CIT.: near Grassy Knob, December 26, 1954, *M 4948.*

Asplenium platyneuron (L.) Oakes (1:41). Ebony Spleenwort is a common fern of open woods. A hybrid of it and *Camptosorus rhizophyllus,* called *Asplenium ebenoides* Scott, is known from Alexander County and reported from Jackson County by Steagall who, in 1927, wrote that a specimen collected by Mr. John Marten near Ava was destroyed by fire in 1883. SP. CIT.: *Mc 239; Hatcher 231:19; M 1804; M 1894.*

Asplenium trichomanes L. (1:39). Maidenhair Spleenwort, one of our most delicate ferns, is not common, although more abundant than *A. pinnatifidum, A. bradleyi,* and *A. resiliens.* It grows on moist, shaded sandstone in crevices of rocks. SP. CIT.: *Steagall 215; Mc 613; V 692; VM 4788.*

FIG. 5. *A deep mesophytic defile at Panther's Den.*

Dryopteris ADANS.

1. Veins extending to margins 2
  2. Lowest pair of pinnae much smaller than others . . .
      . . . . *Dryopteris noveboracensis*
  2. Lowest pair of pinnae larger than the others, or at least not
      decidedly reduced 3
    3. All leaf segments (except sometimes lowest pair of pin-
        nae) connected to each other 4
      4. All leaf segments connected to each other by a
          winged rachis . . . *Dryopteris hexagonoptera*
      4. All but lowest pair of leaf segments connected to
          each other by a winged rachis . . . . .
            . . . . . . *Dryopteris phegopteris*
    3. Leaf segments free from each other . . . . .
        . . . . *Dryopteris thelypteris*
1. Veins ending short of margins 5
    5. Blades twice-pinnate . . . *Dryopteris marginalis*
    5. Blades more than twice-pinnate 6
      6. Blades without glands; lowest pinnules longer than the
          ones immediately above them . . . . . .
            . . . . *Dryopteris austriaca* var. *spinulosa*
      6. Blades usually with at least a few glands; lowest pinnules
          shorter than the ones immediately above them . .
            . . . . *Dryopteris austriaca* var. *intermedia*

Dryopteris noveboracensis (L.) A. Gray (1:49) [*Thelypteris noveboracensis* (L.) Nieuwl.]. New York Fern has not been found in Southern Illinois since 1880. SP. CIT.: woods, Makanda (Giant City State Park?), August 19, 1880, *Seymour*.

Dryopteris hexagonoptera (Michx.) C. Chr. (1:49) [*Thelypteris hexagonoptera* (Michx.) Weatherby]. This handsome fern, Broad Beech-fern, is usually confined to moist woodlands, but it sometimes occurs in rather dry situations. It grows in abundance at Lake Murphysboro and Little Grand Canyon, in addition to localities listed below. SP. CIT.: Fountain Bluff, August 22, 1952, *BS 2839;* Giant City State Park, August 30, 1954, *M 4687.*

Dryopteris thelypteris (L.) Gray (1:51) [*Thelypteris palustris* Schott]. Marsh Fern (Fig. 6) is known from a single station in Jackson County where its existence is being threatened since the marsh in which it grows is being drained. It grows there with *Viburnum recognitum, Carex lanuginosa, Carex comosa, Apios americana* f. *pilosa, Solidago patula,* and *Chelone glabra* var. *linifolia.* SP. CIT.:

FIG. 6. *The Marsh Fern* (Dryopteris thelypteris) *grows in abundance at a marshy area one mile north of Murphysboro.*

the "marsh," one mile north of Murphysboro, April 4, 1950, *Hatcher 1157.*

Dryopteris marginalis (L.) Gray (1:55). Marginal Fern, one of our most attractive evergreen ferns, is one of the most variable. The typical form is by far the most abundant, but large forms, known as f. *elegans* (J. Robins.) F. W. Gray, in which the basal pinnules are pinnatifid, may be found (as at Little Grand Canyon, *M 1905*). At Saltpeter Cave, this species grows with *Mitchella repens* and *Luzula bulbosa.* SP. CIT.: *Mc 155; BS 91; Hatcher 789; VM 1801; VM 1807.*

Dryopteris austriaca (Jacq.) Woynar var. *intermedia* (Muhl.) Morton (1:53) [*Dryopteris intermedia* (Muhl.) A. Gray]. This plant is rare in Southern Illinois where it is known from Pope, Union, and Jackson counties, in the last only from Giant City State Park where it grows on moist sandstone bluffs. SP. CIT.: Makanda, August 19, 1880, *Seymour.* The closely allied var. *spinulosa* (Muhl.) Fiori [*Dryopteris spinulosa* (Mull.) Watt.] is known in Illinois mostly from the north, but has been found twice in the south—along a gravel road in Union County (Hatcher *et al.*) and on moist sandstone at Ferne Clyffe State Park in Johnson County (*VM 1906*).

The most rare Long Beech-fern, *Dryopteris phegopteris* (L.) C. Chr. (1:49) has been found once in our area when Mr. G. H. Boewe collected it near Karnak in Pulaski County on August 6, 1946.

Polypodium L.

Polypodium polypodioides (L.) Watt var. *michauxianum* Weatherby (1:35). The densely scaly fronds of this species curl up during drought, like our species of *Cheilanthes*, and revive after a rain; hence the common name Resurrection Fern. SP. CIT.: Leo Rock, September 7, 1921, *Steagall 284;* Fountain Bluff, June 18, 1940, *Welch & Fuller 165;* Giant City State Park, February 23, 1953, *M 47.*

Polypodium virginianum L. (1:35) [P. *vulgare* var. *virginianum* (L.) Eaton]. Polypody, much more common than the preceding species, usually grows on sandstone rocks or bluffs. It lacks scaly appearance on the blades. Like some other ferns of Southern Illinois, it is evergreen. SP. CIT.: *Steagall 276; Hatcher; M 493.*

Polystichum ROTH

Polystichum acrostichoides (Michx.) Schott (1:57). Christmas Fern, one of the commonest and most admired ferns of the area, grows on slopes in shaded woods and on stream banks. Fronds are evergreen, and have been collected for decorations at Christmas time. SP. CIT.: *Steagall 148; Mc 206; Mc 3603; M 492; M 4890.*

Athyrium ROTH

1. Leaves once-pinnate . . . . . *Athyrium pycnocarpon*
1. Leaves more than once-pinnate 2
   2. Leaves pinnate-pinnatifid . . *Athyrium thelypterioides*
   2. Leaves twice-pinnate, the lower segments often free from each other . . . *Athyrium filix-femina* var. *michauxii*

Athyrium pycnocarpon (Spreng.) Tidestr. (1:42). Narrow-leaved Spleenwort, our only *Athyrium,* with simply pinnate leaves, is found in very mesic woodlands, where it is associated with *Athyrium filix-femina* var. *michauxii* and *Aralia racemosa.* SP. CIT.: Fountain Bluff, *Mc 433;* Little Grand Canyon, *BS 818.*

Athyrium thelypterioides (Michx.) Desv. (1:42). Silvery Spleenwort, found in moist woods and along sandstone bluffs, is apparently less common than the preceding. SP. CIT.: Giant City State Park, *M 476.*

Athyrium filix-femina (L.) Roth var. *michauxii* (Spreng.) Farw. (1:42). Lady Fern grows in very moist woods. Specimens with larger pinnules may be segregated as forma *rubellum* (Gilbert) Farw. SP.

CIT.: typical: *Steagall 176; BS 819; M 474;* forma *rubellum:* Little Grand Canyon, *M 4031.*

Cheilanthes SW.

Cheilanthes feei Moore (1:35). Slender Lip-fern is a very woolly plant found only on calcareous bluffs in the southeastern section of Illinois. The fronds are much smaller and more woolly than those of the following species. Like *Cheilanthes lanosa,* however, the fronds curl during dry periods, and revive quickly after a rain. At the Devil's Bake Oven, this species grows with *Pellaea atropurpurea.* SP. CIT.: *Bailey 583.*

Cheilanthes lanosa (Michx.) D. C. Eaton (1:35). This species, apparently more abundant in Southern Illinois than in either Missouri or southern Indiana, grows on dry, exposed, sandstone bluffs along the Shawneetown Ridge, associated with the Prickly Pear (*Opuntia humifusa*), Goat's-rue (*Tephrosia virginica*), and American Aloe (*Agave virginica*). It is called *Cheilanthes vestita* (Spreng.) Swartz in Fernald (1950).

Pellaea LINK

Pellaea atropurpurea (L.) Link (1:32). Purple Cliff-brake is limited in distribution in Southern Illinois because of its preference for calcareous rocks. At Grassy Knob in Jackson County, it is found often with *Solidago drummondii* and *Phlox bifida.* Smooth Cliff-brake, *P. glabella* Mett. [*P. atropurpurea* var. *bushii* Mack.], which differs in the nearly glabrous stipes and rachises, has been found in Southern Illinois only in Pope County. SP. CIT.: *BS 912; M 2428; M 4781.*

Pteridium GLED.

Pteridium aquilinum (L.) Kuhn var. *latiusculum* (Desv.) Underw. (1:29) [*P. latiusculum* (Desv.) Hieron.]. Bracken, by no means common in Southern Illinois, seems to prefer dry woodlands. SP. CIT.: *BS 815; V 986.*

Cystopteris BERNH.

Cystopteris bulbifera (L.) Bernh. (1:47). Bulblet Bladder-fern, so-called because of its ability to produce bulb-like structures along its rachises, grows in calcareous soils and hence is limited in its distribution in Southern Illinois. SP. CIT.: *Steagall 21; Hatcher 820; M 208.*

Cystopteris fragilis (L.) Bernh. var. *protrusa* Weatherby (1:47). Fragile Fern is undoubtedly the most common fern in Southern Illinois for it may be found on almost any moist bluff or in any mesic woods. It resembles two other ferns of this area—*C. bulbifera* and *Woodsia obtusa*—differing from the former in not producing bulblets along the rachis and in having its veins running to the teeth rather than to the sinuses, and from the latter in the absence of scales along the rachis. SP. CIT.: *BS 66; Steagall 74; Hatcher 231:7; Mc 3502.*

Woodsia R. BR.

Woodsia obtusa (Spreng.) Torrey (1:45). Blunt-lobed Woodsia may be distinguished from other ferns by its scaly rachises and stipes. It is usually confined to moist shady bluffs, generally of sandstone. It may be found the year round, since it produces a winter rosette of sterile leaves which are replaced in the spring by fertile ones. SP. CIT.: *Steagall 61; M 27.*

Dennstaedtia BERNH.

Hay-scented Fern, *Dennstaedtia punctilobula* (Michx.) Moore (1:29) is known from Wabash and Pope counties. It occurs in deep sandstone ravines.

9. *Salviniaceae* – Water Fern Family

Azolla LAM.

Azolla mexicana Presl (1:21). This tiny, moss-like fern is found floating on quiet waters. It is not uncommon in LaRue Swamp, Union County, growing with species of *Lemna, Wolffia,* and *Spirodela.* Fernald (1950) considers our species to be *A. caroliniana* Willd., which is restricted to eastern-most North America by Svenson (1944). SP. CIT.: Campbell's Lake, October 11, 1940, *Mc 563.*

10. *Marsileaceae* – Pepperwort Family

Marsilea L.

Marsilea quadrifolia L. (1:24). This peculiar aquatic fern, with leaves resembling a four-leaf clover, is a native of Europe. A collection from a pond in Carbondale by *G. D. Fuller* and *W. B. Welch* (No. *1000*) is our only record for this species from Southern Illinois.

Division Spermatophyta

Class Gymnospermae

11. *Pinaceae* – Pine Family

Pinus L.

Pinus strobus L. (1:63). White Pine is not native to Southern Illinois, although a few stands occur in some of the northern counties of the state. A seedling about one foot tall was found growing in a dry, sandy woods at Giant City State Park. While the White Pine is planted extensively by the Forest Service in Southern Illinois, no plantation is close to this solitary plant. It is distinguished by its needles in groups of five. SP. CIT.: woods, Giant City State Park, *M 317.*

Pinus echinata Mill. (1:65). Shortleaf Pine is not recorded as native to Jackson County, although it has its only Illinois stations in two of the counties adjacent to Jackson. The largest stand of the species occurs in the Pine Hills (Fig. 7) of Union County on limestone bluffs growing with *Rhododendron roseum*. About twenty or thirty large trees are present along Piney Creek in Randolph County, about one-half mile from the boundary between Jackson and Randolph counties. The habitats at both stations are very dry, sandy soil, and plants which grow with this pine are, for the most part, rare in other sections of Illinois. For instance, with the Shortleaf Pine along Piney Creek are found *Ranunculus harveyi* and *Asplenium bradleyi*.

12. *Taxodiaceae* – Bald Cypress Family

Taxodium RICH.

Taxodium distichum (L.) Rich. (1:66). Bald Cypress is one of the Gulf Coastal Plain species which is found in the southern tip of Illi-

FIG. 7. *Limestone bluffs of the Pine Hills in Union County. Red Cedar* (Juniperus virginiana) *is more common at the edge of the bluffs while Yellow Pine* (Pinus echinata) *is found sparsely back from the edge of the bluffs in a few places.*

FIG. 8. *Bald Cypress* (Taxodium distichum) *in a Pulaski County swamp. Photo taken by Irvin Peithman.*

nois (Fig. 8). It is thought (Palmer, 1921) that these species have not come into Southern Illinois by extending their ranges to the north but, on the contrary, are survivors of a great association of plants that flourished here in the remote past. At that time climatic conditions were much more favorable to these plants. This area is confined in Illinois mostly to Alexander, Pulaski, and Massac counties, although some of the species extend northward to Jackson County. Plants which may be classed in this group of Coastal Plain species include Giant Cane (*Arundinaria gigantea*), Overcup Oak (*Quercus lyrata*), Willow Oak (*Quercus phellos*), Swamp Red Maple (*Acer drummondii*), Water Locust (*Gleditsia aquatica*), and Water Elm (*Planera aquatica*), as well as several herbs, such as *Rhexia mariana, Hypericum denticulatum,* and *Triadenum tubulosum.* Bald Cypress attains a large size in swamps of Alexander and Pulaski counties, but in Jackson County, it grows only to a moderate height (25–28 m.) and is less frequently found. It is planted extensively by the Illinois State Highway Department. SP. CIT.: along a stream, Giant City State Park, *M 1924.*

13. *Cupressaceae* – Cypress Family

Juniperus L.

Juniperus virginiana L. var. *crebra* Fern. (1:68). The Red Cedar is a common inhabitant of the dry sandstone blufftops along the Shawneetown Ridge (Fig. 7). It forms a distinct community here with Post Oak (*Quercus stellata*), Black Jack Oak (*Quercus marilandica*), and Winged Elm (*Ulmus alata*) dominating over an understory of Poverty Oat Grass (*Danthonia spicata*), Lespedeza (*Lespedeza stuvei*), Whorled Milkweed (*Asclepias verticillata*), Hairy Ruellia (*Ruellia humilis*), and others. The species is also abundant on the limestone bluffs along the Mississippi and Big Muddy Rivers, often forming one of the borders of the hilltop prairies. It is extensively used as an ornamental in Southern Illinois. SP. CIT.: Midland Hills, *McCree;* Giant City State Park, *M 1925;* Grassy Knob, *M 1971.*

Class Angiospermae

SUBCLASS MONOCOTYLEDONEAE

14. *Typhaceae* – Cat-tail Family

Typha L.

Typha latifolia L. (1:70). Common Cat-tail is a familiar plant in moist ground throughout the area. Leaves of this species exceed 1 cm. in width. Plants from the "marsh" north of Murphysboro sometimes attain a height of twelve feet. SP. CIT.: M *1911*, M *1937*. Narrow-leaved Cat-tail (*Typha angustifolia* L.), (1:70) with leaves 4–8 mm. wide, is known from Williamson County around Crab Orchard Lake.

15. *Sparganiaceae* – Bur-reed Family

Sparganium L.

While no species of Bur-reed have been collected in Jackson County, two are known from Union County where they grow in the LaRue Swamp. These are *Sparganium eurycarpum* Engelm. (Fig. 9) (1:71) with two stigmas and *Sparganium androcladum* (Engelm.) Morong (1:71) with one stigma.

FIG. 9. *Bur-reed* (Sparganium eurycarpum) *growing in the waters of LaRue Swamp in Union County. The floating plants are mostly* Azolla, *a water fern. Photo taken June, 1952.*

16. *Potamogetonaceae* – Pondweed Family

Potamogeton L.

1. Some or all leaves broader than 5 mm. 2
   2. Stems and petioles black-dotted . . *Potamogeton pulcher*
   2. Stems and petioles not black-dotted 3
      3. Widest leaves three times as long as broad or longer .
      . . . . *Potamogeton americanus*
      3. Widest leaves at most only twice as long as broad . .
      . . . . . . *Potamogeton diversifolius*
1. Leaves at most only 3 mm. broad 4
   4. Leaves 1-nerved . . . . *Potamogeton pectinatus*
   4. Leaves 3- to 7-nerved 5
      5. Spikes 2 to 6 mm. long . . . .
      . . . . *Potamogeton foliosus* var. *genuinus*
      5. Spikes over 8 mm. long . . *Potamogeton friesii*

Potamogeton pulcher Tuckerm. (1:76). This species is rare in Illinois. Our specimen from Campbell's Lake near Elkville (Jackson County) is the first recorded from Illinois since the collections of Elihu Hall in Mason and Menard counties in 1861. SP. CIT.: Campbell's Lake, October 2, 1952, *BS 3109.*

Potamogeton americanus Cham. & Schlect. (1:77) [*P. nodosus* Poir.]. This species is known in our area from Jackson, Williamson, and Pope counties. SP. CIT.: Thompson's Lake, Carbondale, *BS 920.*

Potamogeton diversifolius Raf. (1:83). This Pondweed is recorded from more Southern Illinois counties than any other in our area. It is usually found in shallow water. SP. CIT.: Crab Orchard Lake, Williamson County, *Hankla 1.*

Potamogeton pectinatus L. (1:83). This species is a recent addition to the Southern Illinois flora. It was found sparingly in Lake Murphysboro during the summer of 1955. SP. CIT.: *Weber.*

Potamogeton foliosus Raf. var. *genuinus* Fern. (1:80). This rather uncommon Pondweed is recorded from Williamson (Crab Orchard Lake) and Jackson (Lake Murphysboro and Campbell's Lake) counties. SP. CIT.: *BS 1093; BS 3059.*

Potamogeton friesii Rupr. (1:80). Among our rare species in Jackson County is Fries' Pondweed. Besides our station, this species is known in Illinois only in the far northeast (Lake and Cook counties) (See Jones, 1955).

17. *Naiadaceae* – Naias Family

Naias L.

Naias flexilis (Willd.) Rostk. & Schmidt (1:85). This is the common Bushy Pondweed in our area, but it is nowhere abundant. It is distinguished only with difficulty by its long tapering leaves and shiny seeds. We have it from Williamson and Jackson counties. SP. CIT.: *Weber*.

At Crab Orchard Lake in Williamson County is found *Naias guadalupensis* (Spreng.) Magnus. (1:85). This species has acute or obtuse leaves and dull seeds. While it is attributed to Jackson County (Jones *et al.*, 1955) on a specimen collected by Hankla (No. 6), the station actually is in Williamson County.

18. *Alismaceae* – Water Plantain Family

1. Some of the flowers with both stamens and pistils   .   .   .   2
1. Flowers either staminate or pistillate, never with both stamens and pistils .   .   .   . *Sagittaria*
    2. Upper flowers entirely staminate .   . *Lophotocarpus*
    2. Upper flowers with stamens and pistils   3
        3. Flowers not in heads   .   .   .   .   . *Alisma*
        3. Flowers in heads .   .   .   .   . *Echinodorus*

Sagittaria L.

1. None of the leaves sagittate or hastate (some young specimens of S. *rigida* may be looked for here)   .   . *Sagittaria graminea*
1. Some or all of the leaves sagittate or hastate   2
    2. Bracts over 1 cm. long   .   . *Sagittaria brevirostrata*
    2. Bracts less than 1 cm. long   3
        3. Stamens more than 22; leaves hastate or sagittate .   .
            .   .   .   . *Sagittaria latifolia*
        3. Stamens less than 22; leaves entire, rarely hastate or sagittate .   .   .   .   .   . *Sagittaria rigida* °

Sagittaria graminea Michx. (1:90). Narrow-leaved Arrowhead is uncommon in Southern Illinois although it is abundant at its only station in Jackson County. Here it grows in a roadside ditch which is filled with water for a good portion of the year. Associated with it are *Ranunculus pusillus, Callitriche heterophylla,* and *Spermacoce glabra.* SP. CIT.: along Illinois highway 3, *BS 235.*

Sagittaria brevirostrata Mack. & Bush (1:91) [*S. engelmanniana* J. G. Smith]. This arrowhead is relatively abundant in our area. It occurs profusely at Crab Orchard Lake. SP. CIT.: *BS 833.*

Sagittaria latifolia Willd. (1:93). By far the most common of Southern Illinois arrowheads, this species may be found along shores of nearly every pond or lake. SP. CIT.: *M 1922.*

A fourth species of *Sagittaria* occurs in Southern Illinois. It is *Sagittaria rigida* Pursh (1:89). We have collected it from Alexander and Union counties.

Lophotocarpus   T. DURAND

Lophotocarpus calycinus (Engelm.) J. G. Smith (1:90) [*Sagittaria calycina* Engelm.] [*S. montevidensis* Cham. & Schlect.]. This plant takes three forms in our area. The typical form occurs locally at Lake Murphysboro. Specimens with tiny unlobed narrow leaves occur at Lake Glendale in Pope County. This is forma *depauperatus* (Engelm.) Fern. A specimen from a drainage ditch in Murphysboro with leaves over 3 dm. wide is forma *maximus* (Engelm.) Fern. There is striking contrast between the two extreme forms. SP. CIT.: typical: *M 1923.* Forma *maximus: M 4735.*

Alisma   L.

Alisma plantago-aquatica L. var. *parviflorum* (Pursh) Torr. [*Alisma subcordatum* Raf.] (1:89). The common Water Plantain is found in low wet ground throughout the southern counties. SP. CIT.: *M 4622.*

Echinodorus   RICH.

Echinodorus cordifolius (L.) Griseb. (1:93). This species is rare in our area, being recorded only from Jackson and Union counties. There are twelve stamens and the flowering stalk is erect. SP. CIT.: Howardton Cutoff, September 11, 1954, *M 4782.*

Echinodorus radicans (Nutt.) Engelm. (1:93). The only Southern Illinois station for this species is shallow water of the lagoon in Riverside Park, Murphysboro. This species has twenty or more stamens and the flower stalk is repent. SP. CIT.: Murphysboro lagoon, August 13, 1954, *M 4618.*

19. *Hydrocharitaceae* – Frog's-bit Family

1. Leaves opposite. . . . .*Anacharis*
1. Leaves basal or alternate, heart-shaped . . .*Limnobium* *

Anacharis RICH.

Anacharis occidentalis (Pursh) Vict. (1:95) [*Elodea nuttallii* (Planch.) St. John]. This species formerly was abundant in the foundry pond in southwestern Murphysboro. However, in the past few years, the pond has been overrun by *Lemna minor* until the *Anacharis* is no longer found. SP. CIT.: *M 5727*.

FIG. 10. *LaRue Swamp in Union County is fed by a large spring which comes from beneath a huge bluff of Bailey limestone. Photo taken June, 1952.*

LaRue Swamp (Fig. 10) in Union County is our only Southern Illinois station for *Anacharis canadensis* (Michx.) Planch. (1:95) [*Elodea canadensis* Michx.]. Leaves of *A. canadensis* are about two millimeters broad while those of *A. occidentalis* average somewhat narrower.

Limnobium RICH.

One of our most curious and rarest plants is the Sponge Plant, *Limnobium spongia* (Bosc) Steud. (1:95). It occurs locally in the LaRue Swamp and at Wolf Lake (Union County), and in Alexander County. It is another southern element in our flora.

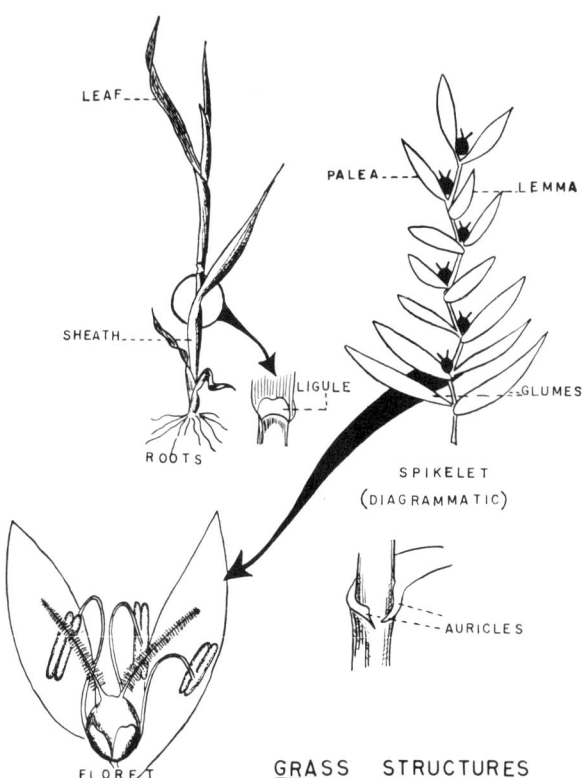

LEAF

PALEA

LEMMA

SHEATH

LIGULE

ROOTS

GLUMES

SPIKELET
(DIAGRAMMATIC)

AURICLES

FLORET

PLATE 7.

GRASS STRUCTURES

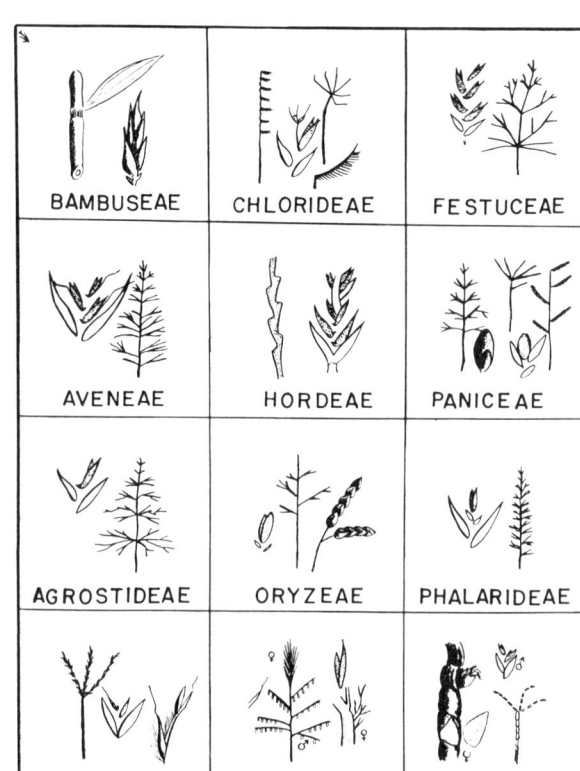

BAMBUSEAE	CHLORIDEAE	FESTUCEAE
AVENEAE	HORDEAE	PANICEAE
AGROSTIDEAE	ORYZEAE	PHALARIDEAE
ANDROPOGONEAE	ZIZANIEAE	TRIPSACEAE

PLATE 8. *Characteristics of selected grass tribes.*

20. *Gramineae* – Grass Family

Members of the grass family in Southern Illinois are distributed among twelve of the fourteen tribes found in North America. Only tribes *Zoysiae* and *Melinideae* are missing. In our area, one hundred sixty-five species belonging to fifty-seven genera occur. In Jackson County, we have recorded one hundred forty-eight species and fifty-five genera. Of these, thirty-nine species have been introduced into or are adventive in Southern Illinois, while the remaining one hundred seven are native species. Grasses may be found in all habitats where vegetation exists. Roadsides and railroad right-of-ways are particularly fine collecting sites for adventive weeds. The more uncommon ones found in such areas in Jackson County include *Holcus lanatus, Agropyron smithii, Aegilops cylindrica, Eriochloa contracta,* and *Chloris verticillata.* Characteristic grasses of moist woods are *Cinna arundinacea, Leersia virginica, Diarrhena americana, Poa sylvestris, Glyceria striata,* and *Panicum polyanthes.* In dry oak-hickory woods, the following grasses are common: *Bromus purgans, Bromus ciliatus, Eragrostis capillaris, Festuca obtusa, Muhlenbergia sobolifera,* and *Panicum xalapense.* Along the dry, exposed sandstone bluffs of the Shawneetown Ridge, an interesting flora is encountered. Grasses adapted to this dry ridge are *Agrostis elliottiana, Danthonia spicata, Muhlenbergia capillaris, Aristida oligantha,* and *Panicum leibergii.* Many of our grasses occur along shores, in marshes, and in swamps. Some of these are *Glyceria septentrionalis, Leersia oryzoides, Leersia lenticularis, Paspalum geminum, Eragrostis hypnoides, Agrostis palustris, Leptochloa filiformis, Leptochloa attenuata,* and *Diplachne fascicularis.* A large number of prairie grasses occur on our hilltop prairies and in the prairie remnants along the railroads in the northern part of Jackson County. Included in this group are *Bouteloua curtipendula, Andropogon scoparius, Andropogon gerardi, Sporobolus asper,* and *Sorghastrum nutans.* Many of the cultivated cereals, although seldom persisting, sometimes escape into waste ground and are therefore included in the keys.

TRIBES OF GRAMINEAE

1. Plants perennial and culms woody, usually 1 to several meters tall
   .   .   .   . 1. *Bambuseae*
1. Plants herbaceous, annuals or perennials   2

2. Flowers bisexual 3
   3. Spikelets 2- to many-flowered 4
      4. Inflorescence an open or contracted panicle or spike-like raceme or, if spikes, the inflorescence digitate 5
         5. Spikelets on one side of the rachis, pedicelled or sessile (those that are sessile resemble *Paniceae*) . . . . 2. *Chlorideae*
         5. Spikelets not as above 6
            6. Glumes unequal, shorter than the lowest floret . . . . 3. *Festuceae*
            6. Glumes equal to subequal, longer than the lowest floret (except *Sphenopholis* and *Koeleria*). . . . . . . 4. *Aveneae*
      4. Inflorescence a true spike. . . . 5. *Hordeae*
   3. Spikelets with 1 perfect floret (when two florets are present, one is staminate or reduced) 7
      7. Spikelets all alike, 1-flowered and perfect . . 8
        8. Flower enclosed by 3 to 4 scales . . . 9
           9. Spikelets turgid, falling entire; glumes reduced . . . . . . 6. *Paniceae*
           9. Spikelets laterally compressed, falling free from the glumes 10
              10. Glumes unequal in length . . . . 7. *Agrostideae*
              10. Glumes absent . . . 8. *Oryzeae*
        8. Flower enclosed by a fifth scale; scales 3 and 4 are usually reduced . . . 9. *Phalarideae*
      7. Spikelets of two kinds in pairs, one pedicelled (staminate or neuter) and one sessile and perfect . . . . 10. *Andropogoneae*
2. Flowers unisexual 11
   11. Stamens 6 . . . . 11. *Zizanieae*
   11. Stamens 3 . . . . 12. *Tripsaceae*

TRIBE 1. BAMBUSEAE

Arundinaria MICHX.

Arundinaria gigantea (Walt.) Chapm. (1:01) and (Fig. 11). The Giant Cane which belongs to the bamboo tribe is our only woody grass. It grows usually in swampy areas, although in the Kinkaids it often is found about one-third of the way up some of the south-facing slopes. Most of our specimens were originally determined as *Arundinaria tecta,* but this southeastern plant does not reach our area. The

FIG. 11. *Giant Cane* (Arundinaria gigantea) *forms heavy canebreaks in some lowland areas.*

cane is the tallest of Southern Illinois grasses, sometimes attaining a height of twelve feet in the Cave Valley area. The species becomes abundant in the most southern counties of Illinois. It begins to flower in our area as early as April 24 (Winterringer, 1953) and continues to bloom until mid-June although flowering is infrequent. SP. CIT.: Fountain Bluff, *Welch & Fuller 173, 188;* Pomona (Cave Valley), *Hatcher and Stewart, V 731.*

TRIBE 2. CHLORIDEAE

1. Spikelets with two or more perfect florets 2
   2. Spikes slender, numerous (more than 6) . .*Leptochloa*
   2. Spikes thicker, few (4 to 6) . . . . .*Eleusine*
1. Spikelets with one perfect flower 3
   3. Spikes digitately arranged 4
     4. Spikes 4 to 6 . . . .*Cynodon*
     4. Spikes more than 6. . . . . . .*Chloris*
   3. Spikes not digitately arranged 5
     5. Leaves 5 to 10 mm. broad . . . .*Spartina*
     5. Leaves 2 to 5 mm. broad . . . .*Bouteloua*

Leptochloa BEAUV.

Leptochloa filiformis (Lam.) Beauv. (1:189). This grass is not uncommon along the Mississippi River near Grand Tower. It grows in moist, sandy soil. Typical plants have stiff spikes and acute glumes which do not exceed the florets. Specimens with lax spikes and mucronate glumes which surpass the florets may be segregated as

Leptochloa attenuata (Nutt.) Steud. (1:189). These occur in similar habitats. SP. CIT.: *Leptochloa filiformis:* Big Muddy River levee, *BS 3056. L. attenuata:* Mississippi River bank, near Grand Tower, *Collins.*

Eleusine GAERTN.

Eleusine indica (L.) Gaertn. (1:191). Goose-grass is an aggressive weed of moist ground. It is naturalized from Eurasia. SP. CIT.: *BS 176.*

Cynodon RICH.

Cynodon dactylon (L.) Pers. (1:191). This extensively creeping perennial is often very difficult to eliminate from lawns. Although called Bermuda Grass, it has been naturalized from Europe. SP. CIT.: *V 693.*

Spartina SCHREB.

Spartina pectinata Link. (1:193). Prairie Cord Grass is rare in Southern Illinois where it has been collected only in moist roadside ditches in Jackson and Alexander counties. SP. CIT.: one-fourth mile north of Carbondale, *BS 774.*

Chloris SW.

Chloris verticillata Nutt. (1:195). Windmill Grass is adventive from the West. It was first discovered in Jackson County in 1953 in Murphysboro where it has become rather common. Elsewhere in Illinois it is known from Peoria and Tazewell counties. SP. CIT.: *M 1636.*

FIG. 12. *Side-oats Grama* (Bouteloua curtipendula) *from a hill prairie at Government Rock in Pine Hills of Union County. Photo taken July, 1952.*

Bouteloua LAG.

Bouteloua curtipendula (Michx.) Torr. (1:197). Side-oats Grama occurs on the limestone bluff-top prairies of southwestern Illinois (Fig. 12) where it grows with *Allium stellatum, Isanthus brachiatus, Petalostemum purpureum,* and others. We have it from Randolph, Jackson, Union, and Hardin counties. SP. CIT.: *BS* 892.

TRIBE 3. FESTUCEAE

1 Rachilla prolonged, bearing 2 to 3 empty scales, each enclosing the one above . . . . *Melica*

1. Rachilla not prolonged, bearing 2 to 3 empty and overlapping scales. 2

  2. Inflorescence an open or narrow panicle, not strict or spike-like 3

    3. Lemmas 5-nerved or more (the nerves may be obscure) 4

      4. Spikelets not in one-sided clusters; leaves boat-tipped 5

      4. Spikelets on one side of panicle branches and crowded . . . . *Dactylis*

        5. Callus with a cobwebby mass of hairs; lemmas keeled near the apex . . . . *Poa*

        5. Callus without a cobwebby mass of hairs (also *Poa annua*) 6

          6. Lemmas over 6 mm. long 7

            7. Spikelets subterete; lemmas awned from a bifid apex or, if not awned from a bifid apex, then leaves with a mark like an inverted "W" on the blade . *Bromus*

            7. Spikelets strongly compressed laterally, thin . . . . . *Uniola*

          6. Lemmas less than 6 mm. long . . 8

            8. Lemmas pointed or awned from the apex . . . . . *Festuca*

            8. Lemmas not pointed or owned, broad, obtuse, scarious at apex . *Glyceria*

    3. Lemmas 1- to 3-nerved 9

      9. Lemmas with short-pointed teeth; spikelets purplish-tinged . . . . . . *Tridens flavus*

      9. Lemmas entire or not pointed or awned . . . 10

        10. Sterile florets present; caryopsis beaked . . . . *Diarrhena*

        10. All florets perfect 11

          11. Lateral nerves of lemma glabrous . . . . *Eragrostis*

> *11.* Lateral nerves of lemma hairy .   .   .
>   .   .   .   . *Diplachne*
> 2. Inflorescence a strict or spike-like panicle; spikelets purplish;
>   lemmas with short-pointed teeth .   .   .*Tridens strictus*

Melica L.

Melica mutica Walt. (1:135). This Melic-grass is rare, having been collected only along sandstone bluffs at Giant City State Park. It has hairy leaf sheaths, leaves two to six millimeters wide, and nearly equal glumes. SP CIT.: Giant City State Park, *M 210*.

Melica nitens Nutt (1:137). This species, which also inhabits bluffs, has glabrous leaf sheaths, leaves five to eleven millimeters wide, and unequal glumes. It may be found across Southern Illinois. SP. CIT.: *M 102*.

Dactylis L.

Dactylis glomerata L. (1:135). Orchard-grass is very common in waste ground in this area. SP. CIT.: *Mc 29*. The uncommon variety *ciliata* Peterm. has been collected at Giant City State Park SP. CIT.: *M 149*.

Poa L.

Although only six species of the Bluegrasses are found here, identification is sometimes difficult. Some species are commonly cultivated for lawns and pastures while others are usually confined to rich woodlands. *Poa pratensis* L. is one of the most abundant grasses in the state and is found in almost all pastures of Southern Illinois. It is also common along railroads. *Poa compressa* L. is occasionally found in pastures, particularly those with poorer soils. It is recognized by its blue-green color and its compressed stem. Two of the Jackson County *Poas* are annuals—*P. annua* L. and *P. chapmaniana* Scrib. The former is very common and is often a troublesome lawn grass. It is not infrequently found in woods. *Poa chapmaniana* has been found in two habitats in Jackson County. One location is in waste ground around a pond, the other is in moist depressions atop a sandstone bluff. At the latter station it is associated with *Isoetes butleri*. The remaining *Poas* of the area are found in moist woodlands. *Poa sylvestris* Gray has been collected at three or four stations in Jackson County, being most abundant at Giant City State Park. *Poa alsodes*

Gray is one of the rarest grasses in Illinois. It has been reported from St. Clair County (Eggert in 1891) and from the Giant City station where it grows with the rare *Synandra hispidula* and *Carex albursina*. Our specimen of *P. alsodes* has many short upcurved hairs on the pedicels.

1. Lemmas with no long cottony hairs at base; annual . . .
    . . . . *Poa annua* (1:119)
1. Lemmas with long cottony hairs at base; annual or perennial   2
    2. Nerves of lemmas obscure; plants blue-green, perennial, with long rhizomes; culms very much flattened . . . .
        . . . . *Poa compressa* (1:119)
    2. Nerves of lemmas 3 to 5, distinct; plants green, annual or perennial, without rhizomes; culms not flattened . . .   3
        3. Lemmas with pubescence only on the keel . .
            . . . . *Poa alsodes* (1:120)
        3. Lemmas with pubescence on keel and some of the nerves   4
            4. Distinct nerves of lemma 3; annual . . . .
                . . . . . *Poa chapmaniana* (1:119)
            4. Distinct nerves of lemma 5; perennial . . .   5
                5. Stolons present; lemmas pubescent on some of the nerves; ligule about 2 mm. long; plants usually of waste places . . . . . .
                    . . . . . *Poa pratensis* (1:119)
                5. Stolons absent; lemmas pubescent on and between the nerves and on the keel; ligule at most 1 mm. long; plants of rich woods . . .
                    . . . . . *Poa sylvestris* (1:121)

SP. CIT.: *Poa annua:* Peter's Cave, *M 1909;* Giant City State Park, *M 4750. Poa compressa:* one mile north of Murphysboro, *Landolt;* Giant City State Park, *BS 725. Poa alsodes:* Giant City State Park, rich mesic woods, May 12, 1953, *M 1974. Poa chapmaniana:* Giant City State Park, *M 329;* campus of Southern Illinois University, *V 1325. Poa pratensis:* one-half mile south of Carbondale, *Mc 6. Poa sylvestris:* Saltpeter Cave, *M 2436.*

Bromus L.

Two of the nine species of *Bromus* which occur in Jackson County are native woodland grasses, the other seven are introductions from Europe. *Bromus ciliatus* L., which ranges throughout northern and eastern sections of the United States, is uncommon in Southern Illi-

Comparative characteristics of Southern Illinois species of Poa

	Blade width (mm.)	Panicle length (cm.)	Spikelet length (mm.)	Fl. per spikelet	Nerves of lemma	Lemmas with cobwebs	Pubescence of lemma	Habit
pratensis	2–5	5–20	3–5	3–5	5	present	on some nerves	tufted, stolons
annua	1–3	1–8	3–7	3–6	5	none	on nerves and keel	tufted, annual
chapmaniana	1–2	2–8	2.5–4.5	3–6	3	present	on nerves and keel	tufted, annual
compressa	1–4	2–10	4–8	3–6	obscure	sparse	on nerves	rhizomes culms fla
sylvestris	3–6	10–20	3–4	2–5	5	present	on and between nerves and on keel	tufted
alsodes	2–4	about 15	3–5	2–3	5	present	on keel	tufted

nois, but does occur in a rich woods at Giant City State Park associated with *Poa alsodes, Poa sylvestris,* and the common spring wild flowers. The other native species, *Bromus purgans* L., also grows in mesic woods but sometimes is found in moist roadside ditches. It is common throughout the southern third of Illinois. The most common introduced *Bromus* is *B. tectorum* L. which is found along every roadway. The drooping panicle branches, along with the soft pubescence of the blades, give the species a very attractive appearance. It is the first Brome to bloom, sometimes being found in early April. *Bromus inermis* Leyss. is cultivated occasionally in our area and sometimes escapes into waste ground. The other weedy Bromes are *B. secalinus* L., *B. racemosus* L., *B. commutatus* Schrad., and *B. mollis* L. *Bromus commutatus* is sometimes included under *B. racemosus* (Gleason, 1952) but some of the Jackson County specimens with very long lemmas have been referred to *B commutatus.* Railway embankments, roadsides, and abandoned fields provide the common habitats for most of these species of *Bromus.*

1. First glume 1-nerved, second glume 3-nerved 2
  2. Leaves 2 to 4 mm. wide, soft pubescent; awn 12 to 17 mm. long; introduced species. . .*Bromus tectorum* (1:103)
  2. Leaves 5 to 15 mm. wide, glabrous or pubescent; awn none or up to 8 mm. long 3
    3. Spikelets on erect pedicels; lemmas glabrous; awn none or up to 2 mm. long; introduced species . . . .
      . . . . *Bromus inermis* (1:103)
    3. Spikelets on drooping pedicels; lemmas pubescent; awn 2 to 8 mm. long; native species 4
      4. Lemmas usually 5-nerved; spikelets 1 to 2 cm. long
        . . . . . . *Bromus ciliatus* (1:104)
      4. Lemmas usually 7-nerved; spikelets 2 to 3 cm. long
        . . . . . . *Bromus purgans* (1:104)
1. First glume 3-nerved (sometimes 5 in *B. secalinus*), second glume 5-nerved (sometimes 7 in *B. secalinus*); introduced species . 5
  5. Lemmas pubescent; spikelets on erect pedicels . . .
    . . . . *Bromus mollis* (1:107)
  5. Lemmas usually glabrous; spikelets drooping, at least at maturity (erect at first in *B. racemosus*) 6
    6. Sheaths glabrous; spikelets 1 to 2 cm. long . . .
      . . . . . . *Bromus secalinus* (1:107)
    6. Sheaths pubescent; spikelets 2.0 to 3.5 cm. long . . 7
      7. Some of the spikelets on erect pedicels (at least at first), the pedicels shorter than the spikelets . .
        . . . . . *Bromus racemosus* (1:105)
      7. All of the spikelets on drooping pedicels, the pedicels mostly longer than the spikelets . . . .
        . . . . *Bromus commutatus*

SP. CIT.: *Bromus tectorum:* along Illinois Central Railroad, Carbondale, *Mc 5, 3490;* one mile north of Murphysboro, *Landolt;* four miles north of Carbondale, Big Muddy River bottoms, *Mc 33;* two miles west of Carbondale, *M 1666;* near Etherton, *M 2196. Bromus inermis:* Pomona, *V 700. Bromus ciliatus:* Giant City State Park, *BS 771, M 226;* two miles north of Murphysboro, *M 1386;* three miles east of Murphysboro, *M 1690;* Lake Murphysboro, *M 2654* (in rich woods). *Bromus mollis:* near Etherton, along railway, *M 5300. Bromus secalinus:* four miles north of Carbondale, *BS 700;* Giant City State Park, *M 283;* near Etherton, along railway, *M 2517. Bromus racemosus:* Giant City State Park, *M 283. Bromus commutatus:* Little Grand Canyon, *V 697;* two miles south of Elkville, *M 1676.*

Comparative characteristics of Southern Illinois species of Bromus

	Sheaths	Blade width (mm.)	Spikelet length (mm.) & position	Nerves of 1st glume	Nerves of 2nd glume	Nerves of Lemma	Lemma	Length of awn (mm.)	Native of
tectorum	softly pubescent	2–4	3, drooping	1	3	5 or 7	pubescent	12–17	*Europe*
inermis	glabrous	8–15	3, erect	1	3	3 or 5	glabrous	0–2	*Europe*
ciliatus	glabrous or pubescent	5–10	1–2, drooping	1	3	about 5	pubescent	3–5	*N. Am.*
purgans	glabrous or pilose	8–15	2–3, drooping	1	3	7	pubescent	2–8	*N. Am.*
racemosus	pubescent	3–6	2–3, suberect or drooping	3	5	7 or 9	glabrous	3–9	*Europe*
commutatus	pubescent	3–6	2.5–3.5, suberect, drooping	3	5	7 or 9	glabrous	5–10	*Europe*
secalinus	glabrous	3–8	1–2, drooping	3 or 5	5 or 7	7 or 9	Usually glabrous	0–6	*Europe*
mollis	softly pubescent	3–6	1.5–2.0, erect	3	5	7	pubescent	6–10	*Europe*

Glyceria   R. BR.

The genus *Glyceria* is composed principally of hydrophytic species. There are four species reported from Southern Illinois, with *Glyceria pallida* (Torr.) Trin., and *Glyceria arkansana* Fern. not known from Jackson County. *G. striata* (Lam.) Hitchc. is common along streams in woods and is particularly abundant in the marshy area north of Murphysboro. Unlike the other three species, this one seldom is found in standing water. *Glyceria septentrionalis* Hitchc. (Fig. 13) grows in swampy woods while *G. pallida* and *G. arkansana* are known in Southern Illinois only from LaRue Swamp in Union County. *G. pallida* was added to the Illinois flora in 1951, *BS 1500* (Swayne, 1953).

1. Spikelets 1 to 2 cm. long, with 8 or more flowers 2
    2. Lemmas obscurely nerved . . . .
        . . . . . . *Glyceria septentrionalis* (1:114)
    2. Lemmas strongly nerved . . .*Glyceria arkansana* *
1. Spikelets 1 cm. long or less, usually 6-flowered 3
    3. Glumes very tiny, less than 1 mm. long; plants usually grow-
        ing in moist soil . . . .*Glyceria striata* (1:117)
    3. Glumes 1 to 2 mm. long; plants usually growing in standing
        water . . . . . .*Glyceria pallida* (1:117) *

SP. CIT.: *Glyceria striata:* one mile north of Murphysboro, *Stewart and Hatcher 113;* one mile south of Murphysboro, *Hardy 11;* Giant City State Park, *BS 1417;* near Etherton, wet ditch along Gulf, Mobile and Ohio Railroad, *M 2506. G. septentrionalis:* swampy woods, three miles northeast of Gorham, *M 5376.*

Tridens ROEM. & SCHULTES

Tridens flavus (L.) Hitchc. (1:137). Purple-top is very common in waste ground and along roads throughout Southern Illinois. The inflorescence is a spreading panicle. SP. CIT.: *Mc 32.*

Tridens strictus (Nutt.) Nash (1:137) with a narrow spike-like panicle, has been collected in Williamson and Franklin counties.

Diarrhena BEAUV.

Diarrhena americana Beauv. (1:137). This grass is localized in Southern Illinois and is known only from Pope, Hardin, and Jackson counties. SP. CIT.: Little Grand Canyon, *BS 824.*

FIG. 13. *Swampy woods in LaRue Scenic Area where a tall hydrophytic grass* (Glyceria septentrionalis) *grows.*

Eragrostis HOST.

The members of this genus are popularly known as Love-grasses. Eight species are found in Southern Illinois, with seven of these known from Jackson County. *Eragrostis frankii* C. A. Mey. has been collected in moist sandy soil from several southern counties. *Eragrostis reptans* (Michx.) Nees, also a plant of moist soil, is infrequent to rare in Southern Illinois.

Eragrostis spectabilis (Pursh) Steud. with purplish panicles which sometimes reach a length of 34 centimeters, is one of our most attractive grasses. The panicle, which is easily detached from the plant, is often seen being blown across fields much in the same manner as the western tumbleweeds. The name "Tumbleweed" is sometimes applied to it in Southern Illinois. The species is common along roads, and flowers from July to October. Our only Love-grass of woodlands is *Eragrostis capillaris* (L.) Nees. This species, with very thread-like panicle branches, is known in Jackson County only from sandstone bluffs at Giant City State Park. Other grasses characteristic of the dry sandstone bluffs of the Shawneetown Ridge include *Agrostis elliottiana, Andropogon virginicus, Aristida dichotoma, Danthonia spicata,* and *Festuca octoflora. Eragrostis cilianensis* (All.) Link, along with *E. poaeoides* (L.) Beauv. and *E. pectinacea* (Michx.) Nees, forms an important part of the roadside grass flora of Southern Illinois. *Eragrostis hypnoides* (Lam.) BSP. is a prostrate grass which creeps along the ground, rooting at the nodes. It is found in very moist soil and is abundant in the sandy beaches along the Mississippi River. Near Bluff Lake in Union County, it completely covers a two-acre field. This land during some periods of the year is under water.

1. Culms creeping, rooting at the nodes   2
    2. Lemmas glabrous, 1.5 to 2.0 mm. long .   .   .   .   .
       .   .   .   .   .   . *Eragrostis hypnoides* (1:126)
    2. Lemmas pubescent on the nerves, 2 to 4 mm. long .   .
       .   .   .   .   .   . *Eragrostis reptans*° (1:126)
1. Culms cespitose, erect or ascending   3
    3. Spikelets usually 2- to 5-flowered, 2 to 4 mm. long .   .   4
       4. Panicle very broad, nearly as broad as long; panicle
          branches very slender . *Eragrostis capillaris* (1:129)
       4. Panicle longer than broad; panicle branches not thread-
          like .   .   .   .   . *Eragrostis frankii* (1:128)

3. Spikelets usually more than 5-flowered, sometimes up to 40-flowered, 3 to 17 mm. long 5
   5. Lemmas 2 to 2.5 mm. long . . . . . .
       . . . . . *Eragrostis cilianensis* (1:129)
   5. Lemmas 1.5 to 2 mm. long 6
       6. Spikelets purplish; plants perennial . . . .
           . . . . . *Eragrostis spectabilis* (1:127)
       6. Spikelets usually not purplish; plants annual . . 7
           7. Panicle usually 3 to 10 cm. long; spikelets with 10 to 20 flowers; keel of lemmas without glands
               . . . . *Eragrostis pectinacea* (1:128)
           7. Panicle usually 10 to 30 cm. long; spikelets with 5 to 11 flowers; keel of lemmas glandular . .
               . . . . *Eragrostis poaeoides* (1:129)

SP. CIT.: *Eragrostis hypnoides:* near Grand Tower, along the Mississippi River, sandbar, *Collins;* Riverside Park, Murphysboro, *Hatcher 114;* Carbondale Reservoir, *Bell;* Giant City State Park, waste ground, *M 556;* Grand Tower Slough, *M 4638. Eragrostis capillaris:* Giant City State Park, sandstone bluffs, *M 580. Eragrostis frankii:* seven miles west of Murphysboro, *M 5731. Eragrostis cilianensis:* Carbondale, *BS 177;* along railroad, Murphysboro, *Hatcher 782;* Giant City State Park, roadsides, *M 386. Eragrostis spectabilis:* five miles south of Carbondale, *Welch et al., 491;* Giant City State Park, roadside, *M 551. Eragrostis poaeoides:* Carbondale, *V 836;* Murphysboro, *M 4516.*

Uniola L.

Uniola latifolia Michx. (1:133). Sea-oats is a common and attractive grass throughout the southern counties. SP. CIT.: *French 3519.*

Festuca L.

The Fescue-grasses are represented in Jackson County by three native species and two introductions, *Festuca ovina* and *Festuca elatior.*

Festuca ovina L. has been collected along a roadside at Giant City State Park. *Festuca elatior* L., often cultivated for forage, may be found along roads and levees. *Festuca octoflora* Walt. is a plant of dry sandstone bluffs. It is found all along the Shawneetown Ridge and is commonly associated with *Oxalis violacea, Oenothera linifolia,* and *Agrostis elliottiana.* This species, unlike the other *Festucas* of Southern Illinois, is an annual and often is placed in the genus *Vulpia.*

Plants with awns less than 3 mm. long and which are smaller in every respect than the typical species occur with it throughout Southern Illinois. It is segregated as var. *tenella*.

The remaining species, *Festuca obtusa* Spreng. and *F. paradoxa* Desv., are usually found in rich woods. The former is common at Giant City State Park, Fountain Bluff, and similar areas, while the latter is local and has been collected in the county only at Giant City State Park and in low woods near Campbell's Lake (Elkville), where it is associated with *Carex muskingumensis, Ptilimnium costatum,* and *Sium suave.*

1. Leaves flat, 3 mm. wide or wider; lemmas without an awn (at
    least in our specimens) 2
    2. Lemmas 6 to 9 mm. long . . .*Festuca elatior* (1:107)
    2. Lemmas 3 to 5 mm. long 3
        3. Spikelets remote, few of them overlapping; panicles open
            . . . . *Festuca obtusa* (1:108)
        3. Spikelets overlapping; panicles suberect . . . .
            . . . .*Festuca paradoxa* (1:108)
1. Leaves involute, about 1 mm. wide; lemmas with an awn 1 to 7
    mm. long 4
    4. Culms tufted; perennial; stamens 3 . *Festuca ovina* (1:110)
    4. Culm single; annual; stamen 1. 5
        5. Awns 3 to 7 mm. long . .*Festuca octoflora* (1:109)
        5. Awns 1 to 3 mm. long; plants very slender . . .
            . . . . . . *Festuca octoflora* var. *tenella*

SP. CIT.: *Festuca obtusa:* one-half mile south of Carbondale, *Mc 32.* Giant City State Park, *V 1260;* ten miles northwest of Murphysboro, *BS 1513;* along Gulf, Mobile and Ohio Railroad, near Etherton, *M 2205. Festuca paradoxa:* Giant City State Park, *V 694, V 695, Mc 19, Mc 3504;* Giant City State Park, *M 1221. Festuca octoflora* var. *tenella:* Giant City State Park, *M 1060. Festuca elatior:* Finney, *M 54.*

Diplachne BEAUV.

Diplachne fascicularis (Lam.) Beauv. (1:189) [*Leptochloa fascicularis* (Lam.) Gray]. This grass is known in Southern Illinois only from moist sand along the Mississippi River near Grand Tower. SP. CIT.: near Grand Tower, August 2, 1952, *Collins.*

TRIBE 4. AVENEAE

1. Glumes as long as or longer than the first lemma 2
   2. One floret staminate, the other perfect . . .*Holcus*
   2. All florets perfect 3
      3. Hairs at base of floret; spikelets pendulous . .*Avena*
      3. Hairs absent from base of floret; spikelets not pendulous
      . . . .*Danthonia*
1. Glumes shorter than the first lemma 4
   4. Second glume narrow and acute . . .*Koeleria*
   4. Second glume broad, obtuse at apex . .*Sphenopholis*

Holcus L.

Holcus lanatus L. (1:158). Velvet-grass is not common in Southern Illinois but it does occur occasionally along roads. It is naturalized from Europe. SP. CIT.: Carbondale, *V 1360.*

Avena L.

Avena sativa L. (1:156). The cultivated Oat occurs commonly in waste places but seldom, if ever, persists. SP. CIT.: *M 5390.*

Danthonia LAM.

Danthonia spicata (Lam.) Beauv. (1:158). Poverty-grass is one of the characteristic grasses of the Shawneetown Ridge bluff tops. Often it is the only species growing under the scrub oaks and hickories. SP. CIT.: Giant City State Park, *M 140, M 72.*

Koeleria PERS.

Koeleria cristata (L.) Pers. (1:150). This is one of the grasses found on hilltop prairies. It is not common. SP. CIT.: *BS 1872.*

Sphenopholis SCRIBN.

Two of the three Southern Illinois species of *Sphenopholis* have been recorded from Jackson County. The other, *S. intermedia* Rydb., is expected to be recorded. *S. obtusata* (Michx.) Scribn. grows along borders of woods or edges of streams. *S. nitida* (Spreng.) Scribn. is common in rich and often rocky woods.

1. Panicle strict . . . . .*Sphenopholis obtusata* (1:150)
1. Panicle lax. 2

2. Glumes equal; sheaths pubescent . . . . . .
    . . . . *Sphenopholis nitida* (1:150)
2. Glumes unequal; sheaths glabrous . . . . . .
    . . . . . . *Sphenopholis intermedia*° (1:150)

SP. CIT.: *S. obtusata:* V *1258;* Mc *178.* S. *nitida:* M *780.*

TRIBE 5. HORDEAE

1. One spikelet at each rachis node 2
    2. Spikelets placed edgewise to the rachis; first glume absent
        except in the terminal spikelet . . . . . *Lolium*
    2. Spikelets placed flatwise to the rachis 3
        3. Spikelets cylindric. . . . . . . *Aegilops*
        3. Spikelets not cylindric. 4
            4. Plants perennial; long rhizomes present . . .
                . . . . *Agropyron*
            4. Plants annual. 5
                5. Glumes broad, obtuse . . . . *Triticum*
                5. Glumes narrow, pointed . . . *Secale*
1. More than 1 spikelet at each rachis node 6
    6. Spikelets 3 at a node, the lateral pair reduced, sterile, and
        pedicelled . . . . *Hordeum*
    6. Spikelets 1 to 4 at a node, all alike 7
        7. Glumes absent or reduced; spikelets spreading . *Hystrix*
        7. Glumes present; spikelets ascending . . *Elymus*

Lolium L.

Three species of *Lolium* occur in Jackson County. These species, *L. temulentum* L., *L. perenne* L., and *L. multiflorum* Lam. are all introduced from Europe.

1. Plants annual; lemmas awned 2
    2. Glumes not as long as the spikelet . . . . . .
        . . . . *Lolium multiflorum* (1:149)
    2. Glumes as long as the spikelet . *Lolium temulentum* (1:149)
1. Plants perennial; lemmas not awned . *Lolium perenne* (1:149)

SP. CIT.: *Lolium multiflorum:* BS *1416.* L. *temulentum:* Walker Hill, M *5381.* L. *perenne:* V *671.*

Aegilops L.

Aegilops cylindrica Host. (1:142). Goat-grass, rare in Illinois, has been found along railroads in Jackson County. Our specimens have very hirsute spikelets. SP. CIT.: *Collins 1225; Stewart 74.*

Agropyron J. GAERTN.

Agropyron repens L. (1:138). This species is adventive along the Gulf, Mobile and Ohio Railroad in Murphysboro. The leaf blades are six to ten millimeters wide. SP. CIT.: *V 680.*

Agropyron smithii Rydb. (1:138). One colony of this species is known to occur two miles south of Murphysboro along the Gulf, Mobile and Ohio Railroad. The leaves are two to four millimeters wide. SP. CIT.: *M 5498.*

Triticum L.

Triticum aestivum L. (1:141). Wheat is often found along roads.

Secale L.

Secale cereale L. (1:142). Cultivated Rye, occasionally found escaped from cultivation, is often planted by the Illinois State Highway Department for prevention of soil erosion. SP. CIT.: *M 53.*

Hordeum L.

Three species of *Hordeum* are found in Jackson County. *H. jubatum* L. (Fig. 14), which blooms in early summer, is very showy with its white awns two to six centimeters long. Small Wild Barley, *H. pusillum* Nutt. (1:146), is very common throughout this area. *H. vulgare* L. (1:148), the cultivated Barley, may be found occasionally escaped from cultivation. SP. CIT.: *Hordeum jubatum* L.: *V 685. H. pusillum: BS 729. H. vulgare: V 1452.*

FIG. 14. *Waving in unison, a community of Big Wild Barley* (Hordeum jubatum) *creates a spectacular ripple in a sea of grass.*

Hystrix MOENCH

Hystrix patula Moench (1:146). Bottle-brush Grass grows in or along the edges of woods and at the bases of bluffs throughout Southern Illinois. A form with hairy lemmas, var. *bigeloviana* (Fern.) Deam, is common on the limestone bluffs of Jackson County just north of the Pine Hills. SP. CIT.: *H. patula:* V *702. H. patula* var. *bigeloviana: M 5466.* A specimen resembling a hybrid between *Elymus canadensis* and *Hystrix patula* has been determined as *Elymus interruptus* Buckl. The spikelets are more crowded than in *Hystrix* and are nearly sessile. SP. CIT.: Little Grand Canyon, V *877,* and Pine Hills (Union County), V *1311.*

Elymus L.

Five species and several varieties of Rye occur in Southern Illinois. The rather rare *E. riparius* Wiegand grows in low rich woods. *E. villosus* Muhl. may be found in either moist or dry woods. The extremely variable *E. virginicus* L. is common along the edges of woods, in waste ground, and along railroads. *E. canadensis* L. occurs in open ground and in woods (Fig. 15). *Elymus interruptus* Buckl. has been found in Union and Jackson counties.

1. Glumes over 1 mm. wide   2
    2. Awns straight at maturity   3
        3. Glumes hirsute, green, about 1 mm. wide .   .   .
            .   .   .   . *Elymus villosus* (1:144)
        3. Glumes usually glabrous (hirsute in *E. virginicus* f. *hirsutiglumis*), yellowish, about 2 mm. wide.   .   .   .   4
            4. Glumes glabrous   5

FIG. 15. *Heavy and arching spikes of Nodding Wild Rye* (Elymus canadensis) *proclaim the arrival of summer.*

    5. Spikes included in base of sheath . . .
            . . . . *Elymus virginicus* (1:145)
    5. Spikes exserted . . . .
            . . . . *Elymus virginicus* var. *jejunus*
  4. Glumes hirsute . . . .
            . . . *Elymus virginicus* var. *hirsutiglumis*
2. Awns recurved upon drying; heads arching . . . .
  . . . . *Elymus canadensis* (1:144)
1. Glumes less than 1 mm. wide   6
  6. Awns arching; palea 8.5 to 9 mm. long . *Elymus interruptus*
  6. Awns straight; palea 7.5 to 8 mm. long . . . . .
      . . . . . . *Elymus riparius* (1:145)

SP. CIT.: *E. villosus:* V 402. *E. virginicus:* V 768. *E. virginicus* var. *jejunus:* M 287. *E. virginicus* f. *hirsutiglumis:* BS 510. *E. canadensis:* M 229. *E. riparius:* V 636.

TRIBE 6. PANICEAE

1. Spikelets with basal bristles or enclosed by a spiny bur . . .   2
  2. Spikelets with basal bristles . . . . . . *Setaria*
  2. Spikelets inclosed by a spiny bur . . . *Cenchrus*
1. Spikelets without basal bristles   3
  3. Spikelets borne in one-sided spikes or racemes . . .   4
    4. Spikelets longer than wide; rachis not widened . .
        . . . . *Echinochloa*
    4. Spikelets nearly as wide as long, plano-convex; rachis widened . . . . *Paspalum*
  3. Spikelets not borne in one-sided spikes or racemes . . .   5
    5. Lower rachilla joint forming a ring-like or cup-like callus below the spikelet . . . . . *Eriochloa*
    5. Not as above   6
      6. Inflorescence a true panicle   7
        7. First glume absent . . . *Leptoloma* °
        7. First glume usually present . . *Panicum*
      6. Inflorescence of racemes which are paniculate or digitate   8
        8. Spikelets copiously long villous, and tawny .
            . . . . *Trichachne* °
        8. Spikelets with short pubescence . *Digitaria*

Setaria BEAUV.

Five species of Foxtail Grass occur in Southern Illinois. All are rather common along roads and in waste ground except *Setaria geniculata* (Lam.) Beauv. which is our only native species. *S. italica* (L.) Beauv., the Italian Millet, is often cultivated in our area. Yellow Foxtail (*S. lutescens* (Weigel) Hubb) [*S. glauca* (L.) Beauv. of Fernald (1950)], Green Foxtail (*S. viridis* Beauv.) and Giant Foxtail (*S. faberii* Herrm.) are all attractive grasses but in many cases their aggressiveness makes them noxious weeds.

1. Panicle conspicuously nodding   2
   2. Panicle not conspiciously branched .   .   .*Setaria faberii*
   2. Panicle conspicuously short-branched .   .   .   .   .
   .   .   .   .   *Setaria italica* (1:235)
1. Panicle erect or only slightly nodding   3
   3. Bristles 1 to 3 per spikelet .   .   .*Setaria viridis* (1:235)
   3. Bristles 5 to 20 per spikelet   4
      4. Perennial; plants not glaucous .   .   .   .   .   .
      .   .   .   .   .   .   *Setaria geniculata* (1:235)
      4. Annual; plants usually glaucous.   .   .   .   .   .
      .   .   .   .   .   .   *Setaria lutescens* (1:235)

SP. CIT.: *Setaria geniculata: French. S. lutescens: M 437; Bell. S. viridis: M 378. S. faberii: M 387; V 885.*

Cenchrus L.

Cenchrus pauciflorus Benth. (1:237) [*Cenchrus longispinus* (Hack.) Fern.]. Sandbur is occasional in sandy soil particularly along the larger rivers. SP. CIT.: *Bailey and Hankla 610; Hatcher.*

Echinochloa BEAUV.

The five species of *Echinochloa* which occur in Southern Illinois all are found in wet waste ground. The extreme scabrousness of the leaf blade margins makes a patch of these species very uncomfortable to walk through. *E. walteri* (Pursh) Nash and the Japanese Millet (*E. frumentacea* (Roxb.) Beauv.) are known only in Union County. The following three species are common: *E. pungens* (Poir.) Rydb., *E. crusgalli* (L.) Beauv., *E. occidentalis* (Wieg.) Rydb.

1. Lowest sheaths hirsute .   .   .*Echinochloa walteri°* (1:233)
1. Lowest sheaths glabrous   2
   2. Second glume and lemmas papillose-hispid .   .   .   .
   .   .   .   .   .   .   *Echinochloa pungens* (1:232)
   2. Second glume and lemmas hispid but not papillose .   .   3

    3. Nerves of first glume parallel 4
        4. Panicle slender, green or straw-colored . . .
        . . . . . *Echinochloa crusgalli* (1:234)
        4. Panicle thicker, purplish-brown . . . . .
        . . . . . *Echinochloa frumentacea*°
    3. Nerves of first glume curved inward . . . .
    . . . . . *Echinochloa occidentalis* (1:234)

SP. CIT.: *Echinochloa pungens: BS 186; V 984; Bell; M 556. E. crusgalli: V 406.*

Paspalum L.

Species of *Paspalum* are common in moist ground along streams, edges of woods, or waste places. *Paspalum dissectum* L. is known in our area only from Pulaski County where it has not been collected since 1853. *P. geminum* Nash and *P. fluitans* (Ell.) Kunth occur locally throughout Southern Illinois. *P. laeve* Michx. [*P. circulare* Nash] and its variety *pilosum* Scribn. are abundant along with *P. pubescens* Muhl. Our records show *P. stramineum* Nash to be known only from Jackson, Union, Pope, and Alexander counties.

1. Rachis dilated, thin, foliaceous 2
    2. Racemes 20 or more . . . *Paspalum fluitans* (1:207)
    2. Racemes usually 3 to 5 . . *Paspalum dissectum* ° (1:207)
1. Rachis narrow to expanded, not thin and foliaceous . . . 3
    3. Spikelets borne in pairs 4
        4. Spikelets in two rows. 5
            5. Sheaths and leaves sparsely pilose to ciliate at the
            margins . . . *Paspalum stramineum* (1:208)
            5. Sheaths pubescent, leaves softly pubescent on both
            surfaces. . . . *Paspalum pubescens* (1:208)
        4. Spikelets in more than two rows . . . . . .
        . . . . . *Paspalum geminum* (1:208)
    3. Spikelets borne singly 6
        6. Sheaths and leaves glabrous . . . .
        . . . . . . *Paspalum laeve* (1:207)
        6. Sheaths and leaves pilose . . . .
        . . . . . *Paspalum laeve* var. *pilosum*

SP. CIT.: *Paspalum fluitans: BS 2886. P. stramineum: V 746. P. pubescens: M 701; BS 869. P. laeve: V 758; M 455; V 385. P. geminum: BS 868.*

Eriochloa HBK.

Eriochloa contracta Hitchc. (1:205). Prairie Cup-grass was first collected in Illinois along a grassy roadside south of Grand Tower, July 25, 1954, by H. E. Ahles (No. *8111*). On October 13, 1954, the species was again found in Union County along the levee road leading to Pine Hills (Voigt, 1955).

Leptoloma CHASE

Fall Witch Grass, *Leptoloma cognatum* (Schultes) Chase (1:205) has been recorded in Southern Illinois from only Johnson County. It grows in sandy soil.

Panicum L.

Of the twenty-nine species of *Panicum* occurring in Southern Illinois, twenty-five are known from Jackson County. *Panicum praecocius* has been found in Southern Illinois only on dry hillsides in Hardin County. A rich woods at Belle Smith Springs (Pope County) provides the only station in our area for *P. barbulatum*. *Panicum philadelphicum* has been collected only in Union County while *P. commutatum* is recorded from Union, Johnson, Pope, Hardin, and Massac counties. The rare *P. oligosanthes* has been collected in a moist woods at Giant City State Park. The usually more northern *P. implicatum* has been found atop dry sandstone bluffs in Randolph, Jackson, and Johnson counties where it is associated with *Cyperus filiculmis*. The species of *Panicum* in our area are listed below according to their usual habitats.

MOIST OPEN GROUND	MOIST OR DRY WOODS	DRY OPEN GROUND
P. agrostoides Spreng.	P. boscii Poir.	P. praecocius Hitchc. & Chase
P. anceps Michx.	P. dichotomum L.	
P. gattingeri Nash	P. barbulatum Michx.	WASTE GROUND
P. microcarpon Muhl.	P. oligosanthes Schult.	P. capillare L.
P. polyanthes Schult.	P. scribnerianum Nash	P. dichotomiflorum Michx.
P. tennesseense Ashe	P. sphaerocarpon Ell.	P. philadelphicum Bernh.
P. huachucae Ashe		P. virgatum L.

MOIST WOODS	SANDY SOIL IN WOODS	
P. clandestinum L.	P. depauperatum Muhl.	P. leibergii (Vasey) Scribn.
P. commutatum Schult.	P. flexile Gatt.	P. lindheimeri Nash
P. yadkinense Ashe	P. implicatum Scribn.	P. linearifolium Scribn.
	P. latifolium L.	P. xalapense HBK.

1. Spikelets glabrous 2
   2. Spikelets under 3 mm. long 3
      3. Sheaths glabrous 4
         4. Panicle small (5 to 10 cm. long) 5
            5. Nodes bearded . . . .
                . . . *Panicum barbulatum* ° (1:229)
            5. Nodes glabrous . . . *Panicum dichotomum*
         4. Panicle large (10 to 30 cm. long) 6
            6. Sheaths pale, glandular-dotted . . . .
              . . . . *Panicum yadkinense* (1:230)
            6. Sheaths not as above . . . .
              . . . . *Panicum agrostoides* (1:215)
      3. Sheaths pubescent 7
         7. Nodes retrorsely bearded . . . . . .
            . . . . . *Panicum microcarpon* (1:229)
         7. Nodes not retrorsely bearded 8
            8. Blades less than 1 cm. wide. 9
               9. Main panicle branches without hairs; fruit
                dark . *Panicum philadelphicum* ° (1:213)
               9. Main panicle branches with hairs; fruit
                straw-colored . . . .
              . . . *Panicum gattingeri* (1:214)
            8. Blades over 1 cm. wide; leaves pubescent . .
              . . . . *Panicum capillare* (1:214)
   2. Spikelets over 3 mm. long 10
     10. Sheaths glabrous 11
        11. Panicle branches each bearing a gland at its base;
        plants annual . . . .
          . . . *Panicum dichotomiflorum* (1:214)
        11. Panicle branches without glands at base; plants
        perennial . . . *Panicum virgatum* (1:215)
     10. Sheaths pubescent 12
       12. Leaves pubescent 13
         13. Panicle narrow and small (up to 10 cm. long) 14
           14. Leaves under 1 cm. wide . . .
            . *Panicum depauperatum* (1:216)
           14. Leaves mostly over 1 cm. wide . .
            . . . *Panicum scribnerianum*
         13. Panicle large (over 10 cm. long), loose,
          flexuous . . . *Panicum flexile* (1:213)
       12. Leaves glabrous; stems flattened at base . .
         . . . . *Panicum anceps* (1:214)

1. Spikelets pubescent (glabrous or nearly so in *Panicum oligosan-*
   thes) 15
   15. Spikelets 3 mm. or more long. 16

16. Sheaths glabrous . *Panicum commutatum* ° (1:223)
16. Sheaths pubescent 17
    17. Leaves 7 to 12 mm. wide, upper ones sometimes crowded . . . . *Panicum scribnerianum*
    17. Leaves usually well over 10 mm. wide, not crowded 18
        18. Nodes retrorsely bearded . . . .
        . . . . *Panicum boscii* (1:224)
        18. Nodes not retrorsely bearded . . . 19
            19. Spikelets not papillose-hispid, 3.5 mm. or more long. . . . . . .
            . . . *Panicum leibergii* (1:231)
            19. Spikelets papillose-hispid . . . 20
                20. Leaves pubescent or scabrous; sheaths harshly scabrous . .
                Panicum clandestinum (1:229)
                20. Leaves glabrous . . . .
                . *Panicum latifolium* (1:224)
15. Spikelets under 3 mm. long 21
21. Sheaths pubescent 22
    22. Upper leaves crowded toward the upper one-third of plant 23
        23. Blades glabrous above or nearly so, stiff . . . *Panicum tennesseense* (1:221)
        23. Blades pubescent (both surfaces) . . 24
            24. Upper surface of leaves with erect hairs 3.0 to 3.5 mm. long . . .
            . . *Panicum implicatum* (1:221)
            24. Upper surface of blades with short appressed hairs; panicles well-exserted above crowded leaves . . .
            . . *Panicum huachucae* (1:221)
    22. Upper leaves not crowded at upper one-third of plant 25
        25. Blades over 5 mm. wide 26
            26. Nodes retrorsely bearded . . . 27
                27. Plants densely tufted, soft; sheaths pilose . . . .
                . *Panicum xalapense* (1:217)
                27. Plants not densely tufted and soft, 30 to 100 cm. tall; spikelets long-pedicelled . . . .
                . *Panicum lindheimeri* (1:221)
            26. Nodes not bearded . . . .
            . . *Panicum oligosanthes* (1:218)
        25. Blades under 5 mm. wide 28

28. Panicle branches not pilose   .   .
.   . *Panicum linearifolium* (1:217)
28. Panicle branches pilose   .   .   .
.   . *Panicum praecocius* ° (1:222)
21. Sheaths glabrous; nodes glabrous 29
29. Panicle about 3 times as long as broad   .   .
.   .   .   *Panicum polyanthes* (1:222)
29. Panicle about twice as long as broad .   .   .
.   .   .   *Panicum sphaerocarpon* (1:222)

SP. CIT.: *Panicum dichotomum: V 1256; French 3587. P. yad-kinense: M 289. P. agrostoides: French 3577; BS 2883; Bell. P. micro-carpon: Stewart and Hatcher 114; VM 675; Hatcher 139. P. gattingeri: BS 871. P. capillare: M 477. P. dichotomiflorum: M 595; Bell. P. vir-gatum: BS 2678. P. depauperatum: V 773; V 769; M 241. P. scribner-ianum: Mc 275. P. flexile: BS 3120. P. anceps: Stewart 835; V 983; M 411; BS 578; Bell. P. boscii: BS 767; BS 1413; M 2448; M 1046; V 1259. P. leibergii: M 211; M 2648. P. clandestinum: M 267; BS 1435; V 701; V 781. P. latifolium: M 268; V 545. P. implicatum: BS 2463. P. huachucae: M 2778. P. xalapense: Stewart and Hatcher 139; M 218; Voigt. P. lindheimeri: Stewart 101. P. oligosanthes: M 4297. P linearifolium: M 126; V 487. P. tennesseense: Stewart and Hatcher 113; M 164. P. implicatum: French. P. polyanthes: M 285; BS 505; Bell. P. sphaerocarpon: M 284; BS 768.*

Digitaria HEISTER

Digitaria ischaemum (Schreb.) Muhl. (1:205). Smooth Crab Grass occurs sparingly in waste ground throughout Illinois. It is distin-guished from the following species of Crab Grass by its glabrous sheaths. SP. CIT.: *M 530*.

Digitaria sanguinalis (L.) Scop. (1:205). One of our most abun-dant waste ground grasses is the Common Crab Grass. It is often difficult to eradicate from lawns. SP. CIT.: *BS 178*.

Trichacne NEES

A recently discovered grass for our area is the southern *Trichacne insularis* (L.) Nees. Two clumps of this handsome grass were col-lected from a roadside half a mile south of Cambria (Williamson County), October 13, 1954, by J. W. Voigt. This is the only record of this adventive grass in Southern Illinois.

TRIBE 7. AGROSTIDEAE

1. Lemmas 3-nerved or 1-nerved 2
  2. Inflorescence an open or narrow panicle 3
    3. Culms with a bulbous base; leaves about 10 mm. wide
      . . . . *Cinna*
    3. Culms without bulbous bases; leaves usually less than
      8 mm. wide 4
      4. Glumes scabrous on keel, longer than lemmas . .
        . . . . *Agrostis*
      4. Glumes pubescent or smooth, shorter than lemmas 5
        5. Lemmas awned, falling with grain . . . 6
          6. Awn not twisted or coiled at base or
            branched . . . . *Muhlenbergia*
          6. Awn twisted or bent, coiled at base, 3-
            branched, lateral ones often short . *Aristida*
        5. Lemmas not awned, grain falling free of lemmas
          . . . . *Sporobolus*
  2. Inflorescence a strict, cylindric, spike-like panicle . *Phleum*
1. Lemmas 5-nerved 7
  7. Plants perennial; awn of lemmas 1 to 3 cm. long . . .
    . . . . *Brachyelytrum*
  7. Plants annual; awn of lemmas less than 5 mm. long . .
    . . . . *Alopecurus*

Cinna L.

Cinna arundinacea L. (1:169). This tall grass is not uncommon throughout Southern Illinois where it grows in rich woods and on the floors of some of the deeper ravines. SP. CIT.: Little Grand Canyon, *BS 823;* Giant City State Park, *M 385.*

Agrostis L.

The genus *Agrostis* is represented in Southern Illinois by five species. Redtop (*Agrostis alba*) is commonly found in waste ground while *A. palustris* Huds. grows along streams. They flower throughout most of the summer. Tickle-grass (*Agrostis hyemalis* (Walt.) BSP.) grows in dry soil of thin woods or open areas. It produces its flowers during May and June. *Agrostis perennans* (Walt.) Tuckerm. which produces spikelets in August may be found in moist or dry woods or occasionally in waste places. One of the most interesting grasses of the area is *Agrostis elliottiana* Schult. This species grows only on the very dry exposed sandstone bluffs of the Shawneetown Ridge. It is not common.

1. Lemmas conspicuously awned, awn 5 to 10 mm. long . . .
   . . . . *Agrostis elliottiana* (1:166)
1. Lemmas not conspicuously awned 2
    2. Plants with creeping rhizomes and rooting at the nodes . . 3
        3. Panicles open; stems usually erect . . . . .
           . . . . *Agrostis alba* (1:166)
        3. Panicle contracted; stems decumbent . . . .
           . . . . *Agrostis palustris* (1:166)
    2. Plants annual 4
        4. Stems erect; leaf blades about 1 mm. wide or less . .
           . . . . *Agrostis hyemalis* (Walt.) BSP. (1:167)
        4. Stems weak, often decumbent; leaf blades 1 to 6 mm.
           wide . *Agrostis perennans* (Walt.) Tuckerm. (1:167)

SP. CIT.: *Agrostis alba: V 107; V 655; BS 1527; M 362. A. hyemalis: M 207; M 1658. A. perennans: V 399; V 886; BS 2838; BS 861. A. elliottiana: Mc 453; M 321.*

Muhlenbergia SCHREB.

Eight species and one variety are listed from Southern Illinois, but some of these are rare. *Muhlenbergia sylvatica* Torr. has been found only in Williamson and Pope counties. *Muhlenbergia sobolifera* (Muhl.) Trin. has been collected in Jackson County from Little Grand Canyon in an oak-hickory woods. The robust *M. racemosa* (Michx.) BSP. has been picked up only from a sandstone bluff along the north end of Fountain Bluff. *M. mexicana* (L.) Trin. and *M. frondosa* (Poir.) Fern. usually grow in moist soil along streams. The awned form of *M. frondosa*, named forma *commutata* (Scribn.) Fern. has been collected near the Ava Cave. A specimen found near Grand Tower by Agnes Chase is *M. glabriflora* Scribn. The delicate *M. capillaris* (Lam.) Trin. is known in Jackson County only from sandstone bluffs of Giant City State Park (Fig. 16). Our most common *Muhlenbergia* is *M. schreberi* J. F. Gmel. It not only is an aggressive weed of lawns, but is common on sandstone throughout Southern Illinois.

1. Awns as long as or longer than the lemmas 2
    2. Panicle purple, one-third to one-half the length of the plant
       . . . . *Muhlenbergia capillaris* (1:173)
    2. Not as above 3
        3. Panicle 3 to 10 mm. thick . . . .
           . . . . . *Muhlenbergia racemosa* (1:174)
        3. Panicle less than 3 mm. thick 4

FIG. 16. *Pennsylvania sandstone forms a great bluff at Giant City State Park.*

SP. CIT.: *Muhlenbergia schreberi:* M 553. *M. sobolifera:* BS 849, BS 822. *M. frondosa:* V 991, M 558. *M. frondosa* f. *commutata:* Ava Cave, M. 4653. *M. mexicana: Nat. Hist. Survey. M. glabriflora:* near Grand Tower, *Chase. M. racemosa:* BS 913. *M. capillaris:* V 8371.

Aristida L.

Three-Awns are confined to dry, sandy, usually sterile soil, although one species, *Aristida ramosissima* Engelm., has been found growing in a moist roadside ditch. *Aristida dichotoma* Michx. is common in fields and on sandstone bluffs. *A. oligantha* Michx. usually may be found in dry pastures and waste ground. *A. longespica* Poir. has been collected in a field at Giant City State Park. More common is var. *geniculata* (Raf.) Fern. *A. purpurascens* Poir. has been found in a dry sandy upland woods at Giant City State Park.

1. Plants perennial; spike usually about one-half the length of the plant . . . . . . *Aristida purpurascens* (1:187)
1. Plants annual; spike usually less than one-half the length of the plant 2
   2. Central awn coiled 3
      3. Lateral awns exserted about 1 mm. beyond the glumes; central awn less than 15 mm. long . . . . *Aristida dichotoma* (1:185)
      3. Lateral awns reduced (to 6 mm. long); central awn 15 to 20 mm. long . . *Aristida ramosissima* (1:185)
   2. Central awn not coiled 4
      4. Awns nearly equal in length, 2 to 5 cm. long . . . . *Aristida oligantha* (1:187)
      4. Awns unequal in length, to 2 cm. long 5
         5. Lateral awns 1 to 4 mm. long . . . . *Aristida longespica* (1:186)
         5. Lateral awns 4 to 15 mm. long . . . . *Aristida longespica* var. *geniculata*

SP. CIT.: *Aristida purpurascens:* Giant City State Park, *Voigt.* *A. dichotoma: M 315.* *A. ramosissima:* at the junction of Illinois highways 3 and 144, *M 4334.* *A. oligantha: Bell.* *A. longespica: M 695.* *A longespica* var. *geniculata: V 994.*

Phleum L.

Phleum pratense L. Timothy, like many other grasses, is a naturalized plant in Illinois. SP. CIT.: *M 80, M 1689.*

Sporobolus R. BR.

The Dropseeds grow in dry soil. Two of our four species, *Sporobolus neglectus* Nash and *S. vaginiflorus* (Torr.) Wood, have been found along railroads in Jackson County. *Sporobolus cryptandrus*

(Torr.) Gray grows on the limestone bluffs of Devil's Bake Oven and Devil's Backbone (Fig. 17 and Fig. 18). *Sporobolus asper* (Michx.) Kunth belongs to such a prairie community as that found south of Elkville and near Finney. Panicles of nearly all members of this genus are either wholly or partially included within the sheath.

1. Panicle contracted, partly enclosed by a sheathing leaf; sheath of
    leaves not copiously bearded at nodes   2
    2.   Plants perennial; leaves long attenuate .   .   .   .   .
          .   .   .   .   *Sporobolus asper* (1:177)
    2.   Plants annual; leaves not long attenuate   3
        3.   Lemmas glabrous .   .   *Sporobolus neglectus* (1:178)
        3.   Lemmas sparsely pubescent   .   .   .   .   .   .
          .   .   .   .   .   *Sporobolus vaginiflorus* (1:177)
1. Panicle open, usually not sheathed; sheath of leaves copiously
    bearded at nodes   .   .   .   *Sporobolus cryptandrus* (1:179)

SP. CIT.: *Sporobolus asper: BS 878. S. cryptandrus: M 5430. S. vaginiflorus: NHS. S. neglectus: NHS.*

Brachyelytrum BEAUV.

Brachyelytrum erectum (Schreb.) Beauv. (1:181). This grass is not uncommon in oak-hickory woods in Southern Illinois. It flowers throughout the summer. SP. CIT.: *V 699; BS 1116; BS 825; M 68; M 1476; M 4077.*

Alopecurus L.

Alopecurus carolinianus Walt. (1:169). The common Foxtail grows along roads, in fields, or along the edges of woods. It is an annual with awns extending two to three millimeters beyond the glumes. SP. CIT.: *V 1072, M 959, M 1471, M 1668. Alopecurus aequalis* Sobol. (1:169) has been collected in the LaRue Swamp of Union County. This species is a perennial with awns not extended or scarcely extended beyond the glumes.

TRIBE 8. ORYZEAE

Leersia SW.

Leersia lenticularis Michx. (1:201). This species of Cut Grass grows in swampy ground in most of our southern counties. It may be

FIG. 17. *A tilted calcareous promontory near Grand Tower known locally as Devil's Bake Oven.*

FIG. 18. *A sharp limestone ridge near Grand Tower called Devil's Backbone.*

segregated from other *Leersia* species in our area by its spikelets which are more than half as long as wide. SP. CIT.: *BS 874.*

Leersia oryzoides (L.) Sw. (1:201). Rice Cut Grass, a common grass of swamps, has very scabrous sheaths and leaf blades. The principal blades are at least fifteen centimeters long. The spikelets are never half as wide as long. SP. CIT.: *BS 858; V 1001; Bell.*

Leersia virginica Willd. (1:201). White Grass is found occasionally in low moist woods throughout our area. Blades of this species usually do not exceed ten centimeters in length. The spikelets are never half as wide as long. Specimens with long ciliated lemmas are var. *ovata* (Poir.) Fern. SP. CIT.: typical: *BS 857.* Var. *ovata:* Little Grand Canyon, *M 5576.*

TRIBE 9. PHALARIDEAE

Phalaris L.

Phalaris canariensis L. (1:200). Canary Grass, an introduction from Europe, has been found in moist ground at two or three stations in Jackson County. SP. CIT.: Cave Valley, *Voigt;* two miles north of Oraville, *M 5283.*

FIG. 19. *Plume-grass* (Erianthus alopecuroides) *is a southern grass. It is found only sparingly in Southern Illinois. Photo taken in the fall of 1953.*

TRIBE 10. ANDROPOGONEAE

1. Spikelets all perfect; panicles densely woolly, 2 to 3 dm. long
   .   .   .   . *Erianthus*
1. Sessile spikelets perfect, the pedicelled ones staminate or neuter   2
   2. Inflorescence an open or closed panicle 3
       3. Leaves usually less than 1 cm. broad .   . *Sorghastrum*
       3. Leaves often over 1 cm. broad .   .   .   . *Sorghum*
   2. Inflorescence a spike-like raceme   .   .   . *Andropogon*

Erianthus MICHX.

Erianthus alopecuroides (L.) Ell. (1:238) (Fig. 19). The huge attractive Plume Grass is uncommon in Southern Illinois, being known from only Jackson, Union, Saline, and Alexander counties. Its principal habitat is open woods. SP. CIT.: Fountain Bluff, *BS 636;* Little Grand Canyon, *Evers 21963;* five miles south of Murphysboro, *Hardy* 116.

Sorghastrum NASH

Sorghastrum nutans (L.) Nash (1:244). Indian Grass is one of the species so characteristic of hilltop prairies and prairie strips along railroads in the area. Few such areas are without this species. It begins to flower the first of August. SP. CIT.: *M 529; Bell.*

Sorghum MOENCH

Sorghum halepense (L.) Pers. (1:244). This perennial species, popularly known as Johnson Grass, is very common in fields and waste places in Southern Illinois. It is naturalized from Europe. SP. CIT.: *BS 175; M 384; V 407; Mc 515.*

Sorghum vulgare Pers. Sorghum, an annual plant with leaves closely resembling *Zea mays*, is an occasional escape from cultivation. SP. CIT.: *M 382.*

Andropogon L.

The genus *Andropogon* is composed primarily of large prairie grasses of which four species occur in Southern Illinois. *Andropogon elliottii* Chapm. is locally abundant in open woods or in old fields. Broom-sedge, *Andropogon virginicus* L., is very common in similar habitats (Fig. 20). The other two species of *Andropogon* are more typically prairie plants. Big Bluestem (*A. gerardi* Vitman) [*A. furcatus* Muhl.] is known from Jackson, Williamson, Saline, Union, Hardin, and Alexander counties. Little Bluestem (*A. scoparius* Michx.) is much more common than the preceding one. All species bloom in late summer and fall.

1.   Racemes 3 to 6, borne on a long exserted terminal peduncle; lower leaf sheath hairy .   .   . *Andropogon gerardi* (1:242)
1.   Racemes borne on slender, short (2 to 6 cm.) peduncles .   .   2
    2.   Upper sheaths inflated, crowded, spathe-like, coppery-colored (an early inflorescence of paired racemes is long-exserted from the sheaths) .   .   . *Andropogon elliottii* (1:242)

FIG. 20. *A very conspicuous grass is Broomsedge* (Andropogon virginicus).

2. Upper sheaths not crowded nor inflated, tawny or brownish   3
    3. Racemes 2 to 4, partly included in the sheaths; awns
        straight, 1 to 2 cm. long . . . .
        .   .   .   .   .   *Andropogon virginicus* (1:242)
    3. Racemes single; awn bent, 1 to 2 cm. long   .   .   .
        .   .   .   .   .   *Andropogon scoparius* (1:241)

SP. CIT.: *Andropogon gerardi: M 316; BS 809. A. elliottii: V 500.
A. virginicus: V 992; M 663; BS 573. A. scoparius: M 802; Bell;
V 1020; BS 837.*

TRIBE 11. ZIZANIAE

Zizania L.

Wild Rice, *Zizania aquatica* L. (1:202), was reported from Union
County by A. B. Seymour, Bluff Lake, in water, August 17, 1880
(ILL).

TRIBE 12. TRIPSACEAE

1. Pistillate and staminate flowers terminal   .   .   . *Tripsacum*
1. Pistillate flowers axillary, staminate flowers terminal   .   . *Zea*

Tripsacum L.

Tripsacum dactyloides L. (1:245). Gama Grass is uncommon in
Southern Illinois, being recorded from only Jackson and Union counties. SP. CIT.: *BS 738.*

Zea L.

Zea mays L. Corn is extensively cultivated in our area and sometimes escapes into waste ground.

21. *Cyperaceae* – Sedge Family

The family of sedges (*Cyperaceae*) is the third largest plant family in Southern Illinois with ten genera and one hundred fifteen species recorded from this area. The genus *Carex* has sixty-nine species
in Southern Illinois. These are found in a wide variety of habitats,
but moist ground provides the site for the greatest number of species.
Some species of *Carex* are confined to swamps (*C. decomposita* and
C. comosa), others to low swampy woods (*C. lupuliformis* and
C. grayii), some to mesic woods (*C. rosea* and *C. albursina*), a few to

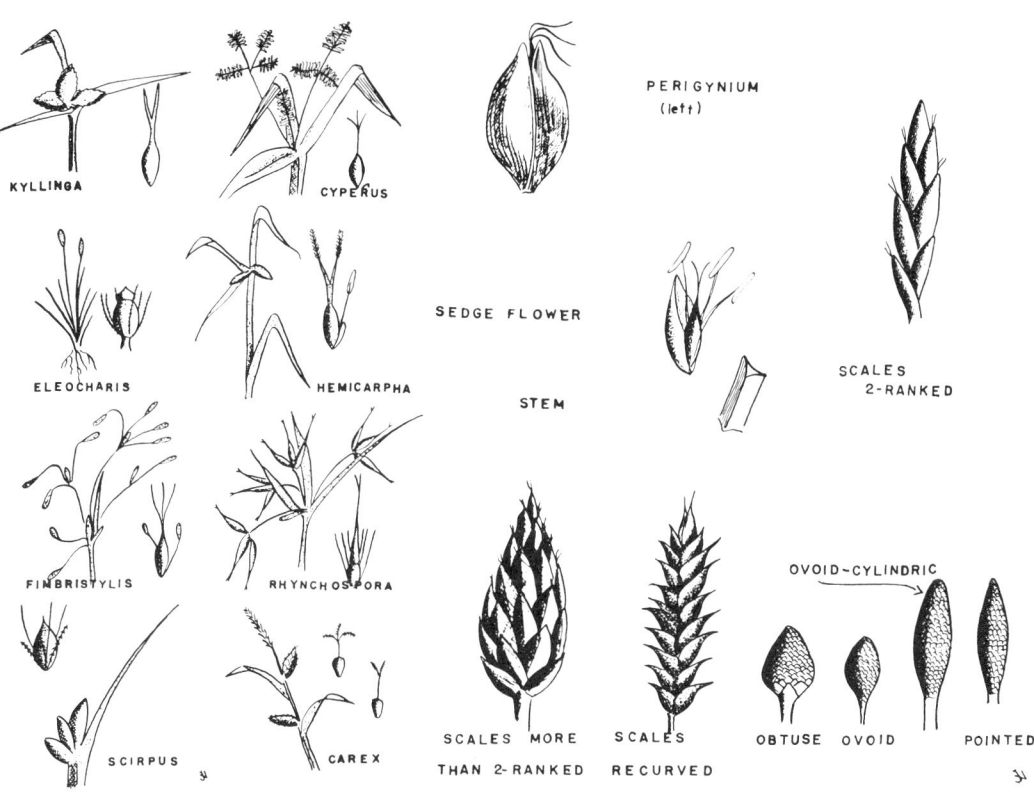

KYLLINGA

CYPERUS

PERIGYNIUM
(left)

ELEOCHARIS

HEMICARPHA

SEDGE FLOWER

STEM

SCALES
2-RANKED

FIMBRISTYLIS

RHYNCHOSPORA

SCIRPUS

CAREX

SCALES MORE
THAN 2-RANKED

SCALES
RECURVED

OVOID-CYLINDRIC

OBTUSE   OVOID

POINTED

PLATE 9. *General aspect of inflorescences of the more important genera of* Cyperaceae *in Southern Illinois.*

PLATE 10. *Structure of sedge flowers, fruits, and spikelets. In the lower right-hand corner four figures show shapes of spikes of Spike-rush* (Eleocharis).

dry woods (*C. retroflexa* and *C. swanii*), and some even to open ground (*C. cephalophora* and *C. leavenworthii*). Several species are known only from a single station in our area—*C. debilis, C. oxylepis, C. substricta, C. brevior, C. bromoides, C. sparganioides, C. caroliniana, C. gravida, C. hirtifolia, C. interior, C. lacustris,* and *C. retrorsa.* The genus *Cyperus* has seventeen species in Southern Illinois. Of these, five are extremely rare. Of the eleven species of *Eleocharis* here, five are known from a single station only. Other sedges in our area are *Scirpus* with ten species, *Rhynchospora, Fimbristylis;* genera represented by only one species include *Bulbostylis, Hemicarpha, Scleria,* and *Dulichium.* Twenty-eight per cent of our species of *Cyperaceae* are known from only a single station.

GENUS	No. of Species in Our Area	No. of Species from One Station	Percent of Those from One Station
Bulbostylis	1	0	0
Carex	69	17	25

GENUS	No. of Species in Our Area	No. of Species from One Station	Percent of Those from One Station
Cyperus	17	5	29
Dulichium	1	0	0
Eleocharis	11	5	45
Fimbristylis	2	1	50
Hemicarpha	1	0	0
Rhynchospora	2	1	50
Scirpus	10	3	30
Scleria	1	0	0
TOTALS	115	32	28

H. N. Patterson (1876) attributes a George Vasey collection of *Fuirena squarrosa* Michx. to Union County (Jonesboro), but there are no specimens of this plant in Illinois herbaria.

A specimen of *Dichromena latifolia* Baldw. is in the Southern Illinois University herbarium with the following data: Makanda, Illinois, 1872. This is from the George Hazen French collection. Since this species occurs only in southeastern United States, we are assuming that a mix-up of labels has occurred since Mr. French did carry on an extensive exchange with southern collectors such as C. Mohr and William Harvey.

1. Achene enclosed in a perigynium (Plate 10) . . . . *Carex*
1. Achene not enclosed in a perigynium 2
  2. Achenes white and shiny . . . . . . *Scleria* °
  2. Achenes brown, green, or straw-colored, not white and shiny 3
    3. Spikelet solitary and terminal; no involucral bracts present . . . . *Eleocharis*
    3. Spikelets several (if solitary, then one or more involucral bracts present) 4
      4. Leaves thread-like, never more than 1 mm. broad . 5
        5. Spikelets sessile or subsessile; plants glabrous . 6
          6. Involucral bract one . *Scirpus koilolepis* °
          6. Involucral bracts 2 to 3 . . *Hemicarpha*
        5. Some or all of the spikelets on peduncles; sheaths pubescent or glabrous 7
          7. Sheaths pubescent . . . *Bulbostylis*
          7. Sheaths glabrous or merely ciliate . .
           . . . . *Fimbristylis*

4. Leaves over 1 mm. broad, never thread-like . . 8
  8. Spikelets with scales borne in two ranks (Plate 10) 9
    9. Inflorescence terminal 10
      10. Spikes 1 to 3, sessile; plants at most 12 cm. tall; sweet smelling . . . . . . *Cyperus densicaespitosus*
      10. Spikes more than 3, if only 1 to 3, then either peduncled or plant over 12 cm. tall and not sweet smelling . *Cyperus*
    9. Inflorescence axillary . . *Dulichium* °
  8. Spikelets with scales 3- to several-ranked (spiral) 11
    11. Plants 5 to 40 cm. tall (usually much smaller); achene not beaked; achenes 0.5 to 1.5 mm. long . . . *Fimbristylis*
    11. Plants over 40 cm. tall (if less, then either the achene beaked or the spikelet lateral and subtended by a single terete bract which appears as a continuation of the culm); achenes over 1.5 mm. long . . 12
      12. Spikes 1- to 2-flowered; achenes beaked . . . *Rhynchospora*
      12. Spikes 3- to many-flowered; achenes not beaked . . . . *Scirpus*

Cyperus L.

Species of *Cyperus* are chiefly plants of moist ground. *Cyperus aristatus* Rottb., while commonly growing in roadside ditches, does occur in our area in moist depressions atop very dry sandstone bluffs. Also on the dry bluffs in a few of our counties is *C. filiculmis* Vahl var. *macilentus* Fern. *Cyperus rivularis* Torr. has been collected in Southern Illinois only from the shores of Lake Glendale in Pope County. The northern *C. schweinitzii* Torr. was found as a railroad waif in Murphysboro. A plant of the east coast, *C. filicinus* Vahl, was collected in a roadside ditch north of De Soto where it probably was brought in by bird. The most common species of *Cyperus* are *C. strigosus* L., *C. esculentus* L., *C. erythrorhizos* Muhl., and *C. ferruginescens* Boeckl. which occur in practically any low field, and *Cyperus ovularis* (Michx.) Torr. which may be found in a wide number of habitats. *Cyperus acuminatus* Torr. and Hook., *C. engelmannii* Steud., *C. flavescens* L. var. *poaeformis*, (Pursh) Fern., *C. pseudovegetus* Steud., and *C. densicaespitosus* Mattf. and Kükenth. occur locally throughout our area. Massac and Pulaski counties are the only Illinois stations for *C. lancastriensis* Porter.

1. Spikelets 1-flowered; stamen 1   .   .   .   .
    .   .   .   . *Cyperus densicaespitosus* (1:250)
1. Spikelets 2- to several-flowered   2
    2. Spikelets not pectinately arranged (mostly radiating from a single point) or, if somewhat pectinate, then the spikelets 2 mm. broad or broader, *or* spikelets purple   3
        3. Scales of the spikelets spreading or reflexed (Plate 10)   4
            4. Scales of the spikelets strongly recurved at the tips; plants less than 10 cm. tall   .   .   .   .   .
                .   .   .   .   .   . *Cyperus aristatus* (1:252)
            4. Scales of the spikelets spreading but not strongly recurved at the tip   5
                5. Main branches of inflorescence 2 to 5 cm. long, the spikes therefore conspicuously pedunculate
                    .   .   .   *Cyperus pseudovegetus* (1:257)
                5. Main branches of the inflorescence less than 2 cm. long, the spike therefore appearing sessile or subsessile   6
                    6. Spikelets longer than broad .   .   .   .
                        .   .   .   *Cyperus schweinitzii* (1:253)
                    6. Spikelets as broad as long or broader .   .
                        .   .   .   *Cyperus acuminatus* (1:257)
        3. Scales of the spikelets appressed-ascending or, if somewhat loosely arranged, then the larger spikelets over 12 mm. long   7
            7. Spikes globose or subglobose, the spikelets compactly arranged   8
                8. Spikelets 4- to 11-flowered .   .   .   .   .
                    .   *Cyperus filiculmis* var. *macilentus* (1:252)
                8. Spikelets 2- to 3-flowered   .   .   .   .   .
                    .   .   .   .   . *Cyperus ovularis* (1:255)
            7. Spikes not of compact globose heads, the spikelets usually loose and spreading   9
                9. Spikelets distinctly purplish-brown   .   .   .
                    .   .   .   .   . *Cyperus rivularis* (1:251)
                9. Spikelets straw-colored or tinged with golden-brown   10
                    10. Scales tinged with golden-brown; stamens 2   .   .   . *Cyperus filicinus* (1:251)
                    10. Scales straw-colored; stamens 3 .   .   .
                        Cyperus flavescens var. *poaeformis* (1:250)
    2. Spikelets pectinately arranged (radiating from an elongated

axis); spikelets never purplish-brown (although sometimes reddish-brown) 11

   11.   Scales 5-nerved or less . *Cyperus erythrorhizos* (1:255)

   11.   Scales 7- to many-nerved 12

      12.   Scales 4 mm. long or longer . . . . .

           . . . . *Cyperus lancastriensis* (1:255)

      12.   Scales less than 4 mm. long 13

         13.   Scales not overlapping . . . . .

             . . . *Cyperus engelmannii* (1:252)

        13.   Scales overlapping 14

           14.   Scales at most 3 mm. long . . . 15

              15.   Perennial with tubers; upper scales more or less obtuse . . 16

                  16.   Spikelets to 15 mm. long, 1.5–3.0 mm. broad; achenes obovoid . . . .

                       Cyperus esculentus (1:256).

                  16.   Spikelets 20–35 mm. long, less than 1.5 mm. broad; achenes narrowly oblongoid . . . . *Cyperus esculentus* var. *leptostachyus*

              15.   Annual with fibrous roots; upper scales more or less acute . .

                 Cyperus ferruginescens (1:255)

           14.   Some or all of the scales over 3 mm. long . . *Cyperus strigosus* (1:255)

SP. CIT.: *Cyperus densicaespitosus* [*Kyllinga pumila* Michx.]: *M 552; M 4689. C. aristatus* [*C. inflexus* Muhl.]: *M 4641; M 4650. C. pseudo-vegetus* [*C. virens* Michx.]: *BS 743; M 4827. C. schweinitzii: M 4514. C. acuminatus: Hardy 128; BS 761; M 4332. C. filiculmis* var. *macilentus: BS 1512. C. ovularis: Hardy 33; M 373. C. filicinus:* north of De Soto, September 23, 1950, *V 391. C. flavescens* var. *poaeformis: V 1300; Hatcher; M 4741. C. erythrorhizos: Collins; Bell; M 4646; M 4690; M 4734. C. strigosus: M 1508; M 4691; Bell; Mc 396; M 222. C. esculentus: Hardy 40; BS 870. C. esculentus* var. *leptostachyus* Boeckl.: Fountain Bluff, *Cranwill. C. ferruginescens* [*C. odoratus* L.]: *Bell; Bailey and Hankla 641.*

Dulichium RICH.

Dulichium arundinaceum (L.) Britt. (1:258) is found in low swampy woods of Johnson and Massac counties. This species should be sought in the Oakwood Bottoms of Jackson County near Grand Tower.

Eleocharis R. BR.

The genus *Eleocharis* is easily recognized by the single terminal spikelet. Three species occur in our area which have very capillary stems and all are locally abundant around lakes and ponds. *Eleocharis acicularis* (L.) Roem. & Schultes attains a height of four to six inches while *E. tenuis* (Willd.) Schultes var. *verrucosa* Svenson and *E. intermedia* Schultes attain a somewhat greater stature. Species usually found in roadside ditches or in wet meadows include the following: *E. obtusa* (Willd.) Schultes, *E. engelmannii* Steud., *E. quadrangulata* (Michx.) Roem., *E. smallii* Britt., *E. calva* Torr., *E. palustris* (L.) R. & S., *E. macrostachya* Britt., and *E. compressa* Sulliv.

1. Spikelets scarcely thicker than the culm   .   .   .   .   .
    .   .   .   .   .   .   *Eleocharis quadrangulata* ° (1:260)
1. Spikelets thicker than the culm   2
    2. Spikelets to 2 mm. long but no longer .   .   .   .   .
        .   .   .   .   .   .   *Eleocharis acicularis* (1:263)
    2. Spikelets 2 mm. long or longer   3
        3. Plants without rhizomes   4
            4. Culms capillary .   *Eleocharis intermedia* ° (1:264)
            4. Culms thicker; spikelets obtuse, ovoid, or cylindric   5
                5. Bristles longer than the achene .   .   .   .
                    .   .   .   .   *Eleocharis obtusa* (1:261)
                5. Bristles shorter than or equalling the achene   .
                    .   .   .   *Eleocharis engelmannii* (1:261)
        3. Plants with rhizomes; spikelets acute or obtuse, ovoid or
            lanceolate   6
            6. Spikelets obtuse, ovoid to ovoid-cylindric (Plate 10).   7
                7. Culms 4- to several-angled .   .   .   .   .
                    .   . *Eleocharis tenuis* var. *verrucosa* (1:265)
                7. Culms flattened . *Eleocharis compressa* (1:265)
            6. Spikelets pointed, lanceolate to narrowly ovoid
                (Plate 10).   8
                8. Basal scale solitary, encircling the base of spike-
                    let .   .   .   .   *Eleocharis calva* (1:262)
                8. Basal scales two or three   9
                    9. Culms flattened, soft   .   .   .   .   .
                        .   .   *Eleocharis macrostachya* (1:261)

9. Culms subterete, firm or wiry   .   .   .   10
  10. Tubercle of achene longer than broad
      .   .   *Eleocharis palustris* (1:261)
  10. Tubercle of achene as broad as long
      .   .   .   *Eleocharis smallii* (1:261)

SP. CIT.: *Eleocharis acicularis: Voigt; Bell; M 1506. E. obtusa: M 212; Hatcher 146; V 450; Bell. E. engelmannii: BS 728. E. tenuis var. verrucosa: Mc 741; V 1330; Hatcher 147; M 749. E. compressa: V 1330-A. E. calva: BS 733. E. macrostachya: Hatcher 148. E. palustris:* Cave Valley, *Mohlenbrock. E. smallii: BS 762.*

Bulbostylis KUNTH

Bulbostylis capillaris (L.) C. B. Clarke (1:266). This slender sedge occurs in sandy soil locally in Southern Illinois. At Giant City State Park (Union County) it grows on sandstone blufftops with *Polygonum tenue* and *Isanthus brachiatus*. We also have it from Jackson, Pope, and Johnson counties. SP. CIT.: Carbon Lake, June 15, 1950, *Hatcher 840.*

Hemicarpha NEES & ARN.

Hemicarpha micrantha (Vahl) Pax (1:283). This small sedge grows in sandy soil along the Mississippi River. The two Jackson County stations are many miles from their nearest known stations in Illinois (Lawrence and Macoupin counties). SP. CIT.: Cora Cutoff, August 24, 1954, *M 4648;* Grand Tower Slough, August 20, 1954, *M 4640.*

Fimbristylis VAHL

Fimbristylis autumnalis (L.) Roem. and Schultes var. *mucronulata* (Michx.) Fern. (1:268). This sedge is relatively common in Southern Illinois but is no doubt often overlooked. While usually occurring in roadside ditches, it also has been found growing in moist depressions on dry sandstone bluffs (Belle Smith Springs, Pope County) with *Cyperus inflexus* and *Oenothera linifolia*. The style is triple cleft and the achenes are about 0.5 millimeter long. SP. CIT.: *Bell; M 4649; M 4733.* A specimen of *F. baldwiniana* (Schultes) Torr. (1:268) was collected in a lumber yard in Massac County, August 10, 1950, by Harry E. Ahles. This species has a double cleft style and achenes one millimeter long or longer.

Scirpus L.

Ten species of *Scirpus* occur in Southern Illinois but only three of these have been recorded from Jackson County. The small S. *koilolepis* (Steud.) Gleason was added to the Illinois flora in 1951 by H. E. Ahles (Alexander County, *Ahles 3846*). *Scirpus fluviatilis* (Torr.) Gray a species with spikelets over ten millimeters long, has been collected in Randolph County. *Scirpus purshianus* Fern. was found along Lake Glendale (Pope County). It also is known in Illinois from Hancock, Lawrence, and Mason counties. Low swampy woods in Massac and Pope counties provide the only stations in our area for S. *polyphyllus* Vahl. *Scirpus americanus* Pers. occurs on marshy ground and along lake shores in Randolph, Williamson, and Hardin counties. Three species of Bulrush are widely distributed in Southern Illinois. These are S. *lineatus* Michx., S. *atrovirens* Willd. var. *georgianus* (Fern.) Harper, and S. *cyperinus* (L.) Kunth. The giant S. *validus* Vahl is found in standing water locally throughout our area. There is only one known station of S. *acutus* Muhl. in Illinois south of Champaign County. This is around Crab Orchard Lake in Williamson County. While the species is attributed to Jackson County (Jones *et al.*, 1955), the specimen upon which this is based (*D. J. Hankla 10*) actually is from Williamson County.

1. Involucral bract 1 (rarely 2 or 3); spikelets appearing lateral   2
    2. Spikelets 1 to 12   3
        3. Leaves thread-like; plants 5 to 15 cm. tall   .   .   .
            .   .   .   .   .   *Scirpus koilolepis* ° (1:273)
        3. Leaves broader; plants over 15 cm. tall   4
            4. Culms sharply 3-angled   .   .   .   .   .   .
                .   .   .   .   . *Scirpus americanus* ° (1:270)
            4. Culms terete or nearly so .   .   .   .   .   .
                .   .   .   .   . *Scirpus purshianus* ° (1:272)
    2. Spikelets more than 12; inflorescence umbellate   .   .   .   5
        5. Spikelets pendulous at maturity, 3 to 10 mm. long .   .
            .   .   .   . *Scirpus validus* (1:273)
        5. Spikelets erect or somewhat spreading at maturity, usually over 10 mm. long .   . *Scirpus acutus* ° (1:273)
1. Spikelets subtended by 2 or more well-developed involucral bracts, thus appearing terminal   6
    6. Spikelets over 10 mm. long   . *Scirpus fluviatilis* ° (1:279)
    6. Spikelets less than 10 mm. long   7
        7. Spikelets in umbellate heads; bristles retrorsely barbed   8
            8. Leaves more than 10; leaves less than 4 mm. wide
                .   .   .   .   . *Scirpus polyphyllus* ° (1:277)

8. Leaves less than 10; leaves more than 4 mm. wide
   .   .   *Scirpus atrovirens* var. *georgianus* (1:276)
7. Spikelets in decompound umbels; bristles not retrorsely
   barbed   9
   9. Involucral bracts delicate, seldom exceeding the
      spikelets   .   .   .   .   *Scirpus lineatus* (1:276)
   9. Involucral bracts more robust, usually overtopping
      the spikelets   .   .   .   *Scirpus cyperinus* (1:275)

SP. CIT.: *Scirpus atrovirens* var. *georgianus: BS 727; M 265; M 1687; Hardy 30. S. lineatus: M 195; Hardy 3; Hatcher. S. cyperinus* (including *S. rubricosus* Fern.): *BS 784; BS 1115; Hatcher 88.*

Rhynchospora   VAHL.

Rhynchospora corniculata (Lam.) Gray var. *interior* Fern. (1:284). This large, interesting Beaked-rush occurs in very wet roadside ditches in most of our southern counties. It grows with species of *Scirpus* and *Carex, Ludwigia glandulosa,* etc. It attains a height of one and one-half meters in our area. SP. CIT.: along Illinois highway 3, near Grand Tower, September 20, 1949, *BS 888.* A smaller *Rhynchospora* is present at Belle Smith Springs (Pope County). It grows in moist sandy soil along Bay Creek. While both *R. capitellata* (Michx.) Vahl (1:285) and *R. glomerata* (L.) Vahl are recorded from Pope County (Jones *et al.,* 1955), they are based on plants from the same colony. This colony appears to consist wholly of *R. capitellata. Rhynchospora glomerata* is not known from Illinois. Specimens from Pope County of *R. capitellata* which we have seen are *Ahles 4686, BS 1769,* and *Brewer and Hardy.*

Scleria   BERG.

Nut-rush, *Scleria pauciflora* Muhl., (1:290) has been collected in Johnson and Hardin counties. It is easily distinguished from all other *Cyperaceae* in our area by its glossy white achenes.

Carex   L.

1. Some of the leaves (at least the basal) over 10 mm. broad .   .   2
   2. Some or all of the pistillate spikes over 1 cm. broad   .   .
      .   .   .   . Section A
   2. None of the pistillate spikes over 1 cm. broad .   . Section B
1. Leaves less than 10 mm. broad   3
   3. A distinct male spike present .   .   .   . Section C

3. Staminate flowers not on a separate spike but are either above or below pistillate flowers on some or all of the spikes   4
    4. Plants completely glabrous *and* with leaf-sheaths not cross-puckered *and* with the pistillate spikes less than 1 cm. broad .   .   .   .   .   . *Section D*
    4. Plants either pubescent *or* with cross-puckered leaf-sheaths *or* with the pistillate spikes over 1 cm. broad *or* a combination of these .   .   .   . *Section E*

SECTION A   Leaves over 10 mm. broad, some or all of the pistillate spikes over 1 cm. broad.

1. Perigynia and sometimes the foliage short-hairy .   .   .   .
     .   .   .   . *Carex grayii* var. *hispidula*
1. Perigynia and foliage glabrous   2
    2. Leaf-sheaths conspicuously spotted with purple; perigynia strongly inflated at base .   .   .   . *Carex crus-corvi*
    2. Leaf-sheaths not conspicuously spotted with purple; perigynia sometimes enlarged but not strongly inflated at base .   .   3
       3. Pistillate spikes globose .   .   .   . *Carex grayii*
       3. Pistillate spikes longer than broad   4
          4. Pistillate spikes on drooping peduncles .   .   .   5
             5. Perigynia with two sharp teeth; staminate spike one .   .   .   .   . *Carex comosa*
             5. Perigynia without two sharp teeth; staminate spikes usually 2 to 4 .   .   . *Carex crinita*
          4. Pistillate spikes erect or spreading   6
             6. Pistillate spikes over 15 mm. broad .   .   .   7
                7. Staminate spike sessile or subsessile .   .
                   .   .   .   .   .   . *Carex lupulina*
                7. Staminate spike pedunculate .   .   .
                   .   .   .   .   . *Carex lupuliformis*
             6. Pistillate spikes less than 15 mm. broad .   .   8
                8. Bases of sheaths purplish-tinged .   .
                   .   .   .   .   .   . *Carex lacustris*
                8. Bases of sheaths green . *Carex hyalinolepis*

SECTION B   Leaves over 10 mm. broad, none of the pistillate spikes over 1 cm. broad.

1. Leaf-sheaths cross-puckered   2
    2. No distinct staminate spike present . *Carex sparganioides*
    2. A distinct staminate spike present   3

3. Peduncle of lowest pistillate spikes 4 to 5 cm. long . .
. . . . *Carex careyana*
3. Peduncle of lowest pistillate spikes less than 3 cm. long . 4
    4. Some of the leaves over 15 mm. broad . . .
    . . . . *Carex albursina*
    4. Leaves at most 12 mm. broad . .*Carex blanda*
1. Leaf-sheaths not cross-puckered 5
    5. Pistillate spikes on drooping peduncles . .*Carex crinita*
    5. Pistillate spikes erect or slightly spreading 6
        6. Leaves very glaucous; staminate spike overtopping pistil-
        late spikes . . . . . .*Carex glaucodea*
        6. Leaves only slightly glaucous; staminate spike shorter
        than pistillate spikes . . . *Carex flaccosperma*

SECTION C Leaves less than 10 mm. broad and a distinct
staminate spike present.

1. Some pistillate spikes 1 cm. or more broad 2
    2. Perigynia and sometimes the foliage short-hairy . . .
    . . . . . . *Carex grayii* var. *hispidula*
    2. Perigynia and foliage glabrous 3
        3. Pistillate spikes globose . . . . .*Carex grayii*
        3. Pistillate spikes longer than broad 4
            4. Pistillate spikes on drooping peduncles . . . 5
                5. Perigynia with two sharp teeth . *Carex comosa*
                5. Perigynia without two sharp teeth (although the
                perigynia may be long-beaked) . *Carex crinita*
            4. Pistillate spikes erect or ascending 6
                6. Some pistillate spikes over 15 mm. broad . . 7
                    7. Peduncle of staminate spike surpassing the
                    pistillate spikes 8
                        8. Perigynia up to 12 mm. long . .
                        . . . . . *Carex louisianica* *
                        8. Perigynia over 12 mm. long . . .
                        . . . . . *Carex lupuliformis*
                    7. Peduncle of staminate spike not surpassing
                    the pistillate spikes 9
                        9. Perigynia less than 10 mm. long . .
                        . . . . . . *Carex lurida*
                        9. Perigynia over 10 mm. long . . .
                        . . . . . .*Carex lupulina*
                6. No pistillate spikes 15 mm. broad . . . 10
                  10. Perigynia up to 3.5 mm. long . . .
                  . . . . . . *Carex frankii*

drooping; staminate spikes
1 to 3 . . . . .
Carex aquatilis var. *altior* °

22. Pistillate scales not green and
purplish 24

    24. Plants with slender rhi-
zomes . .*Carex woodii*

    24. Plants cespitose . . .
. . .*Carex digitalis* °

21. None of the leaves overtopping the
spikes . . . .*Carex meadii*

20. Staminate spikes sessile to subsessile (at
least never on peduncles 2 cm. long) so
that the male spikes are scarcely if at all
elevated above the pistillate spikes . . 25

    25. Some or all of pistillate spikes on
slender drooping peduncles . . 26

        26. Spikes 8 mm. thick or thicker .
. . . . .*Carex crinita*

        26. Spikes less than 8 mm. thick .
. . . . *Carex debilis* °

    25. Pistillate spikes short pedunclate or,
if on longer peduncles, then not
drooping 27

        27. None of the pistillate spikes
overlapping .*Carex oligocarpa*

        27. At least the uppermost pistillate
spikes approximate and over-
lapping 28

            28. Pistillate spikes 5 to 6 mm.
broad; plants usually over
60 cm. tall; common . .
. . .*Carex granularis*

            28. Pistillate spikes 3 to 5 mm.
broad; plants less than 60
cm. tall; rare . . .
. . . *Carex haleana*

SECTION D Leaves less than 8 mm. broad; staminate flowers not
on a separate spike; plants completely glabrous; leaf-
sheaths not cross-puckered; pistillate spikes less than
1 cm. broad.

1. Spikes slender and elongate, perigynia loosely arranged . .
. . . . *Carex bromoides*

1. Spikes with perigynia compactly arranged in a dense head .   .   2
   2. Spikes over 15 mm. long, pointed at both ends .   .   .
      .   .   .   .   *Carex muskingumensis*
   2. Spikes less than 15 mm. long   3
      3. Inflorescence capitate, spikes scarcely distinguishable .   4
         4. Staminate flowers at apex of some of the spikes   .   5
            5. None of the leaves over 3 mm. wide (averaging
               2 mm. or less); perigynia widest near the base .
               .   .   .   .   .   .   *Carex leavenworthii*
            5. Some of the leaves 3 mm. wide; perigynia widest
               just below the middle .   .*Carex cephalophora*
         4. Staminate flowers not at apex of any of the spikes .
            .   .   .   .   *Carex brevior*
      3. Inflorescence spicate, some spikes plainly distinguishable   6
         6. Staminate flowers at apex of some of the spikes   .   7
            7. Spikelets less than 4 mm. wide; leaves at most
               1.5 mm. broad; beak of perigynia smooth .   .
               .   .   .   .   *Carex texensis*
            7. Spikelets usually more than 4 mm. wide; some
               of leaves over 1.5 mm. broad (except sometimes
               in *Carex rosea* which has a roughened beak of
               the perigynia)   8
               8. Only upper 2 or 3 spikes overlapping; beak
                  of perigynia roughened   9
                  9. Broadest leaves 1 to 2 mm. wide .   .
                     .   .   .   .   .   .   *Carex rosea*
                  9. Broadest leaves 2 to 3 mm. wide .   .
                     .   .   .   .   .   *Carex convoluta*
               8. Usually all but lowest spikes overlapping;
                  beak of perigynia smooth or, if roughened,
                  then almost every spike subtended by a con-
                  spicuous setaceous bract   10
                  10. Beaks of perigynia smooth; none or
                      only 1 or 2 spikes subtended by con-
                      spicuous setaceous bracts .   .   .
                      .   .   .   .   .   *Carex retroflexa*
                  10. Beaks of perigynia roughened; almost
                      all of the spikes subtended by con-
                      spicuous setaceous bracts .   .   .
                      .   .   .   .   *Carex muhlenbergii*
         6. Staminate flowers never at apex of any of the spikes   11
            11. Leaves 1 to 2 mm. wide .   .*Carex interior* °
            11. Most leaves over 2 mm. wide or, if less, peri-
                gynia ascending and not spreading .   .   .   12
               12. Perigynia lanceolate (about one-third as
                   wide as long)   13

13. Leaves about 2 mm. wide; inflorescence somewhat moniliform   .   .
.   .   .   .   . *Carex scoparia*

13. Leaves mostly over 2 mm. wide;
inflorescence a compact head .   .
.   .   .   . *Carex tribuloides*

12. Perigynia ovate, oblong, or orbicular, at
least not lanceolate 14

14. Perigynia 3 to 4 mm. broad .   .
.   .   .   .   . *Carex brevior*

14. Perigynia at most 2.5 mm. broad .   15

15. Pistillate scales as long as
perigynia 16

16. Perigynia broadest at
middle   .   .   .   .
.   .*Carex albolutescens*

16. Perigynia broadest below
middle, near the base .
.   . *Carex normalis*

15. Pistillate scales much shorter
than perigynia 17

17. Perigynia over 4 mm.
long 18

18. Perigynia widest
near the middle .
.*Carex festucacea* °

18. Perigynia widest
near the base .   .
. *Carex normalis*

17. Perigynia less than 4 mm.
long 19

19. Tips of perigynia
spreading or re-
curved   .   .   .
. *Carex cristatella*

19. Tips of perigynia
erect or ascending .   20

20. Spikes brown
at maturity;
inflorescence
not more than
2 cm. long   .
. *Carex bebbii*

20. Spikes green-
ish at matu-
rity; inflores-

cence often
longer than
2 cm. . . 21
21. Perigynia
widest
near the
middle .
. *Carex
tribu-
loides*
21. Perigynia
widest
near the
base .
. *Carex
normalis*

SECTION E Leaves less than 8 mm. broad, staminate flowers not on a separate spike, plants either pubescent *or* with cross-puckered leaf-sheaths *or* with the pistillate spikes over 1 cm. broad *or* a combination of these.

1. Foliage (sheaths especially) or perigynia or both pubsecent   . 2
  2. Some of the lowest spikes on long slender drooping pedicels; perigynia pubsecent . . . . . .*Carex swanii*
  2. Lowest spikes not on long drooping pedicels; perigynia glabrous 3
    3. Pistillate inflorescence over 2 cm. long . .*Carex davisii*
    3. Pistillate inflorescence less than 2 cm. long . . . 4
      4. Leaf blades soft pubescent 5
        5. Perigynia 2.5 to 3.5 mm. long, shorter than scales . . . . . .*Carex bushii*
        5. Perigynia 2.0 to 2.5 mm. long, equalling or exceeding the scales . . .*Carex hirsutella*
      4. Leaf blades glabrous to glabrate .*Carex caroliniana*
1. Foliage and perigynia completely glabrous 6
  6. Leaf sheaths neither cross-puckered nor red-spotted . . 7
    7. Inflorescence in globose or oblong heads 8
      8. Spikes over twice as long as broad .*Carex typhina*
      8. Spikes less than twice as long as broad, nearly globose . . . .*Carex squarrosa*
    7. Inflorescence in elongated spikes, usually more than three times as long as broad . . .*Carex laevivaginata*

6. Leaf sheaths cross-puckered or red-spotted 9
   9. Inflorescence over 1 cm. across 10
      10. Inflorescence 4 cm. long 11
         11. Perigynia not dilated to a spongy disk at base
           . . . . *Carex stipata*
         11. Perigynia dilated to a spongy disk at base .
           . . . . . . *Carex crus-corvi*
      10. Inflorescence less than 4 cm. long (usually 1 to 3
        cm.) . . . .*Carex gravida*
   9. Inflorescence less than 1 cm. across 12
      12. Staminate flowers at base of the spikes and con-
        spicuous . . . . . .*Carex shortiana*
      12. Staminate flowers at summit of some or all of the
        spikes, often unnoticeable 13
         13. Leaf sheaths red-dotted; perigynia black at
           maturity . . . .*Carex decomposita* [*]
         13. Leaf-sheaths not red-dotted; perigynia pale
           to deep brown at maturity 14
           14. Bracts subtending spikes none, or mi-
             nute, or only the lowest up to 1 cm. long
             . . . . . *Carex cephaloidea*
           14. Bracts subtending spikes numerous and
             obvious, several over 1 cm. long . . 15
             15. Beak of perigynium about as long
               as the body . *Carex vulpinoidea*
             15. Beak of perigynium about one-
               third as long as body . . . 16
               16. Perigynia yellowish-brown,
                 1.7 to 2.5 mm. long, con-
                 spicuously nerved on the con-
                 vex face . *Carex annectens*
               16. Perigynia deep brown, 1.4 to
                 1.7 mm. long, obscurely
                 nerved on the convex face .
                 . . *Carex brachyglossa*

JACKSON COUNTY CARICES

SECTION A.

Carex grayii Carey (1:364). Low moist woods, rare. V *1354*.

Carex grayii Carey var. *hispidula* Gray. Low moist woods, occasional, more common than the preceding. *Hatcher.*

Carex crus-corvi Shuttlw. (1:312). Low swampy woods, often with *Carex muskingumensis*, occasional. *Hardy 35, M 4309*.

Carex comosa Boott (1:358). The "marsh," two miles north of Murphysboro, now possibly extinct at this station. *M 1678.*

Carex crinita Lam. (1:356). The "marsh," rare. *V 688.* Also in Pope County.

Carex lupulina Muhl. (1:365). Wet ground, occasional. *French.*

Carex lupuliformis Sartw. (1:365). Lester Swamp east of Cora, rare. *M 4703.*

Carex lacustris Willd. (1:359). Wet roadside ditch north of Carbondale, locally abundant. *V 1327.*

Carex hyalinolepis Steud. (1:259). Low woods, not common. Near Howardton, *M 3909.*

SECTION B.

Carex sparganioides Muhl. (1:310). Rich mesic woods, growing with *Carex grayii* var. *hispidula* and *Spigelia marilandica* at Lake Murphysboro. This is our only Southern Illinois station for the species. *M 2679.*

Carex careyana Torr. (1:335). Rich mesic woods, occasional. *Hatcher.*

Carex albursina Sheldon (1:334). Rich mesic woods, not rare. *M 1215, M 4042.*

Carex blanda Dewey (1:334). Low rich woods, common. *BS 694, M 2192, V 1324.*

Carex crinita Lam. (Section A.)

Carex glaucodea Tuckerm. (1:340). Rich rocky woods, occasional. *M 2435, M 4050.*

Carex flaccosperma Dewey (1:340). Low wet woods seven miles southwest of Murphysboro. *BS.*

SECTION C.

Carex lurida Kunth (1:363). Wet ground, common. *Voigt.*

Carex frankii Wahl. (1:361). Wet soil in ditches and woods, common. *Hardy 81; M 271; Bell.*

Carex gracilescens Steud. (1:337). [*C. laxiflora* Lam. var. *angustifolia* Dewey.] Moist alluvial soil, rare. Giant City State Park, *M 231.*

Carex grisea Wahl. (1:339). Dry woods, occasional. *M 209; Voigt.*

Carex umbellata Schk. (1:329). Dry sandstone bluffs, occasional. Saltpeter Cave, *M 5030;* near Elkville, *M 5101.* Also in Randolph and Union counties.

Carex pennsylvanica Lam. (1:328). Dry woods, not common. Giant City State Park, M 691.

Carex artitecta Mack. (1:327) [*C. varia* Muhl.; *C. nigro-marginata* Schw. var. *muhlenbergii* (Gray) Gl.]. Dry sandstone bluffs, common. M 1480; V 1321.

Carex hirtifolia Mack. (1:331). Rich woods, rare. Lake Murphysboro, M 2451.

Carex lanuginosa Michx. (1:349) [*C. lasiocarpa* Ehrh. var. *latifolia* (Boeck.) Gl.]. The "marsh," two miles north of Murphysboro, is the only Southern Illinois station for this species. M 1697.

Carex jamesii Schw. (1:326). Rich mesic woods, common. M 453, V 1387.

Carex torta Boott (1:354). This species grows in running streams; not common (Fig. 21). Midland Hills, V 506. Also in Randolph, Pope, and Hardin counties.

Carex meadii Dewey (1:333). Along Illinois Central Railroad, a half mile south of Elkville; very rare. This is the only Southern Illinois station. BS 971.

Carex oligocarpa Schk. (1:338). Rich woods in calcareous areas, rare. Near Grassy Knob, M 5467.

Carex granularis Muhl. (1:338). Wet ground, common. M 5359.

Carex haleana Olney (1:338). This rare species, previously known in Illinois only from the northeastern section, has one station in a woods at Giant City State Park. M 629.

FIG. 21. *Patches of* Carex torta *in a tributary of Poplar Creek at Midland Hills in Jackson County.*

SECTION D.

Carex bromoides Schk. (1:318). Wet ground, rare. Near Carbondale, *Mc 18.*

Carex muskingumensis Schw. (1:319). Low pin-oak woods, occasional. *M 3789.*

Carex leavenworthii Dewey (1:307). Moist or dry soil, not common. *M 330; M 5369.*

Carex cephalophora Muhl. (1:307). Dry woods, common. *M 1023; M 2338.*

Carex brevior (Dewey) Mack. (1:325). Pasture at Midland Hills, *Voigt.*

Carex texensis (Torr.) Bailey (1:308). Open ground, rare. Two miles north of Carbondale, *V 1326.*

Carex rosea Schk. (1:309). Rich mesic woods, common. *M 1335.*

Carex convoluta Mack. (1:309). Dry open woods, not common. *M 775; M 2363.*

Carex retroflexa Muhl. (1:308). Dry woods, often atop sandstone bluffs, common. *M 3105.*

Carex muhlenbergii Schk. (1:309). Dry woods, not common. *M 1510.*

Carex scoparia Schk. (1:320). Woods, not uncommon. *M 111; V 1328.*

Carex tribuloides Wahl. (1:321). Low woods and marshes, common. *M 152; M 2364; M 3790; M 5505.*

Carex albolutescens Schw. (1:322). Rich mesic woods, rare. *M 154.*

Carex normalis Mack. (1:324). Rich woods, common. *M 240, M 2340.*

Carex cristatella Britt. (1:321). Low woods, rare. *M 156; Bell.*

Carex bebbii Olney (1:323). Our specimen was growing along a stream in Giant City State Park. This species previously had been known in Illinois only in the northern section. *M 797.*

SECTION E.

Carex swanii (Fern.) Mack. (1:347). This species grows in dry soil near the "marsh." Two other specimens, originally determined as *Carex virescens* Muhl., have tentatively been placed with *C. swanii. M 3327.*

Carex davisii Schw. & Torr. (1:341). Rich calcareous woods, rare.

Along "Granny" Railroad, seven miles south of Murphysboro, *M 4696.*

Carex bushii Mack. (1:348). Fields, dry open woods, occasional. *M 1025.*

Carex hirsutella Mack. (1:348) [*C. complanata* Torr. and Hook. var. *hirsuta* (Bailey) Gl.]. Woods, common. *Mc 191; Hardy 47; M 1178.*

Carex caroliniana Schw. (1:348). The only station for this species in Illinois is along Illinois highway 3 about one mile south of the junction of Illinois highways 3 and 144. *BS 1025.*

Carex typhina Michx. (1:361). Wet ground, rare. *Mc 776.*

Carex squarrosa L. (1:361). Wet ground, common. *V 846; BS 2679; BS 502; M 174.*

Carex laevivaginata (Kükenth.) Mack. (1:312). Low swampy woods, not common. *M 1385; M 1672.*

Carex stipata Muhl. (1:312). Low wet woods, rare. *Stewart & Hatcher 40.*

Carex gravida Bailey (1:210). Dry woods, occasional. *M 700.*

Carex shortiana Dewey (1:250). Among streams and in wet ground, common. *Mc 788; M 77; M 2191.*

Carex cephaloidea Dewey (1:310) [*C. sparganioides* Muhl. var. *cephaloidea* (Dewey) Carey]. Wooded slopes, known in our area only from Little Grand Canyon. *M 4835.*

Carex vulpinoidea Michx. (1:311). Wet ground, common. *M 266; M 1682; M 3105; M 4834; Mc 796; Hardy 41.*

Carex annectens Bickn. (1:311). The only station in Southern Illinois for this species is in a wet ditch along Illinois highway 3 near the junction of Illinois highways 3 and 144. *M 5364.*

Carex brachyglossa Mack. (1:311). This species has its only occurrence in Southern Illinois around Walker Hill pond near Grand Tower. *M 4753.*

CARICES OF SOUTHERN ILLINOIS AS YET UNKNOWN FROM JACKSON COUNTY

Carex louisianica Bailey (1:265). Swampy woods, Pulaski County, *BS 2848.*

Carex aquatilis Wahl. var. *altior* (Mack.) Fern. (1.233) [*C. substricta* (Kukenth.) Mack.]. Along Piney Creek, Randolph County, *M 5632.*

Carex digitalis Willd. (1:335). Beech-maple valley, Belle Smith Springs, Pope County, *BS 1650*.

Carex debilis Michx. (1:343). Rich woods. Near Cave-in-Rock, Hardin County, *M 4257*. The only station for this species in Illinois.

Carex interior Bailey (1:317). On sandstone boulder, Lusk Creek, Pope County. *BS 2546*.

Carex decomposita Muhl. (1:311). Deep swamps, often growing on stumps of *Taxodium*. Union County, *BS 2317;* Pulaski County, *V 1809*.

Carex oxylepis Torr. & Hook. (1:342). Low woods, rare, near Dongola, Union County, the only station in Illinois. *VM 5553*.

22. *Araceae* – Arum Family

1. Leaves compound with 3 to several leaflets .   .   .*Arisaema*
1. Leaves simple and entire   2
      2. Leaves over 10 cm. broad .   .   .   .   .*Peltandra* °
      2. Leaves 1 to 5 cm. broad, many times longer than wide .   .
          .   .   .   .   *Acorus*

Arisaema MARTIUS

Arisaema dracontium (L.) Schott (1:367). Green Dragon, which flowers from May to early June, is common in moist woods. It is distinguished by its leaves which are divided into five to several leaflets. SP. CIT.: *BS 388; Welch 181; V 592; M 64; M 2162*.

Arisaema triphyllum (L.) Schott (1:367). Throughout most moist woods may be found the interesting Jack-in-the-Pulpit or Indian Turnip. Flowers are produced from April through the middle of May while the brilliant scarlet fruits mature the last of August. The lateral leaflets of the triple foliolate leaf are very unequal at the base. SP. CIT.: *BS 10, V 508, M 666, M 1917, M 4085*. The small *Arisaema pusillum* (Peck) Nash [*A. triphyllum* (L.) Schott var. *pusillum* Peck] has been collected in Alexander County in a low woodland near Tamms (*Ahles 3848*). Its lateral leaflets are nearly symmetrical at the base.

Peltandra RAF.

Arrow-Arum (*Peltandra virginica* (L.) Kunth [1:368]) grows in shallow water in four Southern Illinois counties (Saline, Union, Johnson, and Alexander). It is particularly abundant in Wolf Lake (Union County). It flowers at the end of May.

Acorus L.

Acorus calamus L. (1:369). Sweet Flag is very aromatic in all its parts, and the thick rhizomes were, in the past, often used for medicinal purposes (for colic, etc.). It is locally abundant in low marshy areas. SP. CIT.: *M 2165.*

23. *Lemnaceae* – Duckweed Family

1.   Thallus without roots   2
   2.   Thallus at most only twice as long as broad   .   .*Wolffia*
   2.   Thallus 2 to several times longer than broad .   .*Wolffiella* °
1.   Thallus with 1 to several rootlets   3
   3.   Thallus with one rootlet   .   .   .   .   .   .*Lemna*
   3.   Thallus with 2 to many rootlets   .   .   .   .*Spirodela*

Wolffia HORKEL

Wolffia columbiana Karst. (1:370) is commonly known as Columbian Water-meal. It and the next species are the smallest flowering plants in Illinois. They float as little specks (one-half to one and one-half millimeters) on the water. The thallus of the Columbian Water-meal is not punctate. We have it only from Jackson County. SP. CIT.: Campbell's Lake, near Elkville, July 22, 1952, *BS 270.*

Wolffia papulifera C. H. Thompson (1:370). Spotted Water-meal, so named because of its punctate leaves, is known in Illinois from Union, Jackson, and Wabash counties. SP. CIT.: Campbell's Lake, near Elkville, July 22, 1952, *BS 2671.*

The strap-shaped *Wolffiella floridana* (J.C. Sm.) C. H. Thompson (1:370) has been collected in Illinois only at Wolf Lake, Union County, November 4, 1950, *Swayne 1163.*

Lemna L.

1.   Thallus over 6 mm. long   .   .   .   .   .*Lemna trisulca* °
1.   Thallus at most only 5 mm. long, usually much smaller   .   .   2
   2.   Thallus 1-nerved, linear to elliptic   .   .*Lemna valdiviana*
   2.   Thallus 3-nerved, ovate, or suborbicular   3
      3.   Thallus symmetrical; root-tip rounded   . *Lemna minor*
      3.   Thallus asymmetrical; root-tip pointed .   .   .   .
       .   .   .   . *Lemna perpusilla* °

Lemna valdiviana Phil. (1:370). Only two stations are known in Illinois for this rare Duckweed. One is at Wolf Lake (Union County) and the other in the Lester Swamp (Jackson County). SP. CIT.: Lester Swamp, September 2, 1954, *Mohlenbrock*.

Lemna minor L. (1:370). This Duckweed is the most common one in our area. It often completely covers the surfaces of small ponds. SP. CIT.: Lester Swamp, September 2, 1954, *Mohlenbrock*.

Two other species of Duckweed occur in Southern Illinois. *Lemna perpusilla* Torr. (1:370) has been found in Union and Alexander counties while *Lemna trisulca* L. (1:370) grows in Union County.

Spirodela SCHLEID.

Spirodela polyrhiza (L.) Schleid. (1:370). This species is quite common in slow streams. SP. CIT.: *BS 2672*.

24. *Commelinaceae* – Spiderwort Family

1. Flowers regular, not subtended by a spathe .   .*Tradescantia*
1. Flowers irregular, subtended by a spathe .   .   .*Commelina*

Tradescantia L.

Tradescantia ohiensis Raf. (1:379). This species of Spiderwort occurs in open ground, often along railroad right-of-ways. We have it from Jackson, Union, Alexander, and Pulaski counties. Leaves are less than two centimeters broad and the pedicels are glabrous. It begins to flower at the end of May. SP. CIT.: *M 2482*.

Tradescantia virginiana L. (1:378). Common Spiderwort is abundant in meadows and open woods. The leaves are less than two centimeters broad and the pedicels are pubescent. Its flowers, which often are rose-colored, are produced in May and early June. SP. CIT.: *BS 49; M 655*.

Tradescantia subaspera Ker (1:378). Wide-leaved Spiderwort with some of its leaves over two centimeters broad is found in moist woods in most of our southern counties. It blooms from mid-June until mid-August. The plant often attains a height of one meter. SP. CIT.: *V 1249; M 4739*.

Commelina L.

Commelina communis L. (1:380). Dayflower is relatively common in moist shady ground in our area. The spathe which is free at the base subtends a corolla of one white and two blue petals. Our collection dates range from July 1 to September 4. SP. CIT.: *M 170; M 4701; Mc 293.*

Commelina diffusa Burm. f. (1:380). This smaller creeping Dayflower is common in moist waste ground. All three petals are blue. SP. CIT.: *Mc 1176; BS 1811.*

Commelina virginica L. (1:380). This species is rare in Southern Illinois where it grows along streams in low woods. It occurs with *Gratiola neglecta, Mimulus alatus, Lycopus rubellus,* etc. SP. CIT.: along Cedar Creek, nine miles south of Murphysboro, August 6, 1954, *M 4717.*

25. *Pontederiaceae* – Pickerel-weed Family *

1. Flowers regular; stamens 3 . . . .
   . . . .*Heteranthera reniformis* Ruiz & Pavon ° (1:385)
1. Flowers irregular; stamens 6 .*Pontederia cordata* L.° (1:384)

Neither of these species is listed from Jackson County although each is known to occur in Wolf Lake (Union County) where they have their only station in Southern Illinois (Fig. 22). The handsome Pickerel-weed (*Pontederia cordata*) flowers during June and early July (Fig. 23).

FIG. 22. *Wolf Lake, an old oxbow of the Mississippi River, in Union County. Photo taken May, 1955.*

FIG. 23. *Pickerel-weed* (Pontederia cordata) *growing on the east shore of Wolf Lake. Photo taken May, 1955.*

26. *Juncaceae* – Rush Family

1. Plants completely glabrous . . . . . . *Juncus*
1. Plants pubescent or at least ciliate . . . . . *Luzula*

Juncus L.

1. Inflorescence lateral; leaves reduced to sheaths . . . .
   . . . *Juncus effusus* L. var. *solutus* Fern. & Wieg. (1:389)
1. Inflorescence terminal; leaves present 2
   2. Leaves flat, not septate 3
      3. Plants annual; inflorescence one-half to one-third the length of the plant . . *Juncus bufonius* L. (1:390)
      3. Plants perennial; inflorescence shorter, proportionately . 4
         4. Flowers subtended by a pair of small bracteoles . 5
            5. Flowers distinctly secund; leaves only about one-third the length of the stem . . . .
               . . . . *Juncus secundus* Beauv. (1:392)
            5. Flowers scarcely secund (if slightly secund, then the leaf sheath with a distinct auricle); leaves well over one-third the length of the stem . . 6
               6. Auricles of leaf-sheath 1 to 3 mm. long, delicate in texture . . . . .
                  . . . *Juncus tenuis* Willd. (1:392)
               6. Auricles of leaf-sheath less than 1 mm. long or absent, firmer in texture and often colored 7
                  7. Bracteoles obtuse at apex . . .
                     . . *Juncus dudleyi* Wieg. (1:392)
                  7. Bracteoles pointed at apex . . .
                     . . *Juncus interior* Wieg. (1:393)
         4. Flowers without a pair of bracteoles at the base . 8
            8. Plants with over 20 heads; principal leaves 4 to 6 mm. wide . *Juncus biflorus* Ell. (1:394)
            8. Plants with fewer than 20 heads; principal leaves 1 to 3 mm. wide . . . .
               . *Juncus marginatus* Rostk. var. *setosus* Coville
   2. Leaves terete, septate 9
      9. Flowers in dense globose heads 10
         10. Involucral leaf overtopping the heads which are 10 to 15 mm. in diameter . . . . . .
            . . . . *Juncus torreyi* Coville (1:397)
         10. Involucral leaf shorter than the heads which are 7 to 11 mm. in diameter . . . . . .
            . . *Juncus brachycarpus* Engelm. (1:398)
      9. Flowers in narrower heads or, if subglobose, then not densely crowded 11

11. Inflorescence at maturity with more than 50 heads   12
    12. Inflorescence with more than 200 heads.   .
        .   .   .*Juncus nodatus* Coville ° (1:398)
    12. Inflorescence with less than 200 heads .   .
        .   .   .*Juncus diffusissimus* Buckl. (1:399)
11. Inflorescence at maturity with 5 to 50 heads .   .   13
    13. Seeds with a definite appendage .   .   .
        .   .   . *Juncus canadensis* Gay (1:395)
    13. Seeds without appendages.   .   .   .   .
        .   .   .*Juncus acuminatus* Michx. (1:398)

Fourteen species of *Juncus* occur in Southern Illinois with thirteen
of these being reported from Jackson County. While most of these
grow in moist ground (often roadside ditches), some seem to do well
in drier soils. *Juncus interior* is not uncommon in sandy soil while
Juncus secundus grows on very dry limestone on Devil's Bake Oven
near Grand Tower. *Juncus diffusissimus* is known in Illinois from two
stations in Jackson County where it was first discovered in 1953. The
very common *Juncus tenuis* is often found in waste ground, particularly along paths in woods and pastures. The widely branching *Juncus nodatus* is known in our area only from Williamson and Pope
counties while *Juncus canadensis* is found only in Jackson County.
A common plant of standing water or very marshy situations is *Juncus
effusus* var. *solutus*. Other species of *Juncus* in Southern Illinois and
the counties from which they are known are the following: *Juncus
bufonius* (Randolph, Jackson), *Juncus acuminatus* (Jackson, Williamson, Pope), *Juncus dudleyi* (Jackson, Randolph, Union, Alexander,
Pulaski), *Juncus marginatus* var. *setosus* (Jackson, Pope), and *Juncus
torreyi* (Jackson, Randolph, Hardin, Alexander). The species are
quite difficult to determine and mature seeds should be available for
identification. SP. CIT.: *Juncus effusus* var. *solutus: Voigt; M 76;
M 2202; BS 457; Hatcher. J. bufonius:* Murphysboro, Riverside Park
lagoon, *M 4809. J. secundus:* Devil's Bake Oven, *Mohlenbrock;*
Round Bluff (Johnson County), *BS 2467. J. tenuis: Voigt; M 2355;
M 89; Mc 3509. J. dudleyi: M 129; M 1696; M 2201. J. interior: Hardy
57; BS 506; M 4734. J. brachycarpus: M 303; Hatcher. J. canadensis:*
(See Jones, 1955). *J. acuminatus: Voigt; M 763; BS 456; Hatcher. J.
diffusissimus:* Giant City State Park, September 19, 1953, *VM 1507;*
roadside ditch five miles north of Carbondale, *J. Garrison and
W. Weber.*

Luzula DC.

Luzula multiflora (Retz.) Le Jeune (1:402). This is our most common species of Wood Rush in Illinois. It occurs in rocky woods in most southern counties. It is distinguished by its erect or ascending inflorescence rays and its noncormose base. SP. CIT.: *Hatcher; BS 311; V 664; M 4788.*

Luzula bulbosa (Wood) Rydb. This species is similar to the preceding but possesses small white corms at the base. It is less frequent than either of the other wood rushes. SP. CIT.: Saltpeter Cave, *M 1896, BS 1995.*

Luzula echinata (Small) F. J. Herm. (1:402). This species with strongly divergent inflorescence rays is known from Jackson, Gallatin, Union, Johnson, Pope, Hardin, and Alexander counties. Like other species of *Luzula*, it flowers in April and May. SP. CIT.: *Mc 682.*

27. *Liliaceae* – Lily Family

The Lily family is represented in Southern Illinois by eighteen genera and thirty-five species, of which seventeen genera and thirty-one species may be found in Jackson County. *Chamaelirium luteum, Muscari racemosum, Allium mutabile,* and *Trillium viride* are the Southern Illinois species which have not been found in Jackson County. In Illinois, twenty-four genera and forty species of *Liliaceae* are known.

1. Leaves absent at flowering time   .   .   . *Allium tricoccum*
1. Leaves present at flowering time   2
    2. Leaves all basal or near base of the plant .   .   .   3
        3. Plants, including flower stalk, over two feet tall .   .   4
            4. Flowers orange .   .   .   .   . *Hemerocallis*
            4. Flowers white, cream, yellow, or pink .   .   .   5
                5. Flowers yellowish, with two glands near the claw of the petals .   .   . *Melanthium*
                5. Flowers white, cream, or pink, without glandular-clawed petals   6
                    6. Leaves thick and leathery, over 3 cm. broad .   .   .   . *Yucca*
                    6. Leaves not thick and leathery, less than 2.5 cm. broad.   7
                        7. Plants with odor and taste of onion .   .   .   .   . *Allium*

7. Plants without odor and taste of onion
. . . . . . *Stenanthium*
3. Plants less than 2 feet tall 8
8. Plants with odor and taste of onion. . . *Allium*
8. Plants without odor and taste of onion . . . 9
9. Flowers deep blue . . . . . *Muscari*
9. Flowers white, greenish, yellow, or lilac . . 10
10. Flowers lilac; plants of prairies and low woods . . . . . . *Camassia*
10. Flowers white, greenish, or yellowish. . 11
11. Flowers solitary; leaves 2 . . . . *Erythronium*
11. Flowers more than one; leaves more than 2 12
12. Perianth segments each with a green stripe down the back . . . . . *Ornithogalum*
12. Perianth segments not variegated . . . *Nothoscordum*
2. At least most of the leaves cauline and not at base of the plant 13
13. Leaves in whorls of 3 to 8 14
14. Leaves in one whorl of three . . . *Trillium*
14. Leaves in more than one whorl of 3, usually 4 to 8 in a whorl . . . . . . *Lilium*
13. Leaves not whorled 15
15. Leaves less than six times as long as broad . . 16
16. Leaves net-veined; plants often climbing . . . . *Smilax*
16. Leaves parallel-veined; plants never climbing 17
17. Flowers large (2.5 to 4.0 cm. long), yellow . . . . . *Uvularia*
17. Flowers smaller, white or greenish . 18
18. Flowers borne in a terminal inflorescence . . . *Smilacina*
18. Flowers borne from the leaf axils . . . . *Polygonatum*
15. Some of the leaves 6 or more times as long as broad. 19
19. Leaves filiform . . . . *Asparagus*
19. Leaves (at least the upper) linear to lanceolate . . . . . *Chamaelirium* *

Stenanthium KUNTH

Stenanthium gramineum (Ker) Morong (1:405). This handsome

species has been collected only once in Southern Illinois. It flowers during the first week in July. SP. CIT.: one and one-half miles north of Makanda, *BS 509*.

Melanthium L.

Melanthium virginicum L. (1:410). Bunchflower is rare in Southern Illinois and is known from only one station. About a dozen plants exist in a thick stand of grasses and sedges between U.S. highway 51 and the Illinois Central Railroad about five miles north of Carbondale. It blooms in early June. SP. CIT.: five miles north of Carbondale, *Mc 820, BS 5730, M 5387*.

Chamaelirium WILLD.

Only Massac and Hardin counties provide the known stations in Illinois for *Chamaelirium luteum* (L.) Gray (1:405).

Uvularia L.

Uvularia grandiflora Sm. (1:429). The Large-flowered Bellwort is one of the characteristic species of rich mesic woods in Southern Illinois. Leaves of this species are minutely pubescent and clasp the stem. The plant is common in all southern counties, blooming from the second week in April to the last week in May. SP. CIT.: Giant City State Park, *M 658, M 1464;* Midland Hills, *V 1039.*

Uvularia sessilifolia L. (1:429). This species is less common than the preceding and differs from it by having glabrous, sessile leaves. This plant has been collected only a few times in Southern Illinois. Besides Jackson County, it is known in our area from Williamson, Pope, and Johnson counties. It usually grows in floodplain woods. SP. CIT.: Chapman's Hollow, Murphysboro, *Mohlenbrock.*

Allium L.

1. Leaves 2, up to 6 cm. wide, usually disappearing before flowering time; plants of rich mesic woods . . *Allium tricoccum*
1. Leaves usually more than 2, less than 0.5 cm. wide, present at flowering time 2
    2. Umbel without bulblets 3
        3. Outer bulb-coats not fibrous; stamens exserted; plants blooming after the middle of August . . . . .
        . . . . *Allium stellatum*
        3. Outer bulb-coats fibrous; stamens included; plants bloom-

ing during the first weeks of June . . . . .
. . . . *Allium mutabile*
2. Umbel with bulblets 4
   4. Leaves terete and hollow; outer bulb-coats not fibrous
      . . . . *Allium vineale*
   4. Leaves flat and not hollow; outer bulb-coats fibrous .
      . . . . *Allium canadense*

Allium tricoccum Ait. (1:414). Wild Leek is known in Southern Illinois only in rich woods at Little Grand Canyon where it grows on the valley floor or on sandstone ledges with *Hydrophyllum macrophyllum* and *Stylophorum diphyllum*. It flowers the last of June. SP. CIT.: Little Grand Canyon, *Mohlenbrock*.

Allium stellatum Fraser (1:415). This beautiful pink-flowered onion is confined in Illinois to limestone bluff prairies of the southwestern section of the state. It grows in crevices in bluffs as well as being firmly entrenched in prairie soil which is characteristic of the bluff-tops. It is recorded from Union, Jackson, and Monroe counties. SP. CIT.: hilltop prairie, two miles north of the Pine Hills, *Mohlenbrock*.

Another species of *Allium, Allium mutabile* Michx. (1:415), was first discovered in Illinois in 1952 in thin soil atop a sandstone bluff in Johnson County (*V 1109*). It has also been found in Alexander County (*Garrison*).

Allium vineale L. (1:414). Field Garlic, an introduction from Europe, is abundant throughout Southern Illinois along roads, in pastures, and in vacant fields. SP. CIT.: two miles west of Murphysboro, *Mc 862;* Giant City State Park, *Mc 198;* Midland Hills, *Voigt*.

Allium canadense L. (1:414). This is the common Wild Garlic of Illinois. It grows in waste ground, particularly in abandoned fields and along roads. SP. CIT.: Giant City State Park, *M 4;* two miles north of Fountain Bluff, *BS 409*.

Nothoscordum KUNTH

Nothoscordum bivalve (L.) Britt. (1:416). False Garlic, closely resembling members of the genus *Allium*, but differing from them chiefly in its lack of an alliaceous odor, is common along the dry sandstone bluffs of the Shawneetown Ridge. It usually blooms in early April with *Hypoxis hirsuta* and *Oxalis violacea* but, like the latter, has been collected in flower during September and October. SP. CIT.: Giant City State Park, *M 2585;* north of Boskydell, *Mc 644.*

Hemerocallis L.

Hemerocallis fulva L. (1:413). Orange Day-Lily is common along highways and railroads. It was introduced from Europe and has become naturalized. SP. CIT.: two miles north of De Soto, *Mc 217;* Giant City State Park, *M 306.*

Lilium L.

Lilium michiganense Farw. (1:418). Turk's-Cap Lily is one of our most attractive summer wild flowers. It commonly grows in low rich woods with *Spigelia marilandica.* It is often found along railroad cinder banks. Besides Jackson County, we have collected it in Union, Williamson, and Franklin counties. SP. CIT.: Giant City State Park, *M 327;* Cave Valley (Pomona), *BS 1515;* Campbell's Lake, *Mc 859;* four miles west of Murphysboro, *Pulliam.*

Erythronium L.

Erythronium americanum Ker (1:420). Yellow Dog's-tooth Violet is found in mesic woods or along moist ledges of sandstone bluffs throughout the entire length of the Shawnee Hiking Trail. Our collection dates for this species range from March 13 to April 18. SP. CIT.: Giant City State Park, *M 452;* one mile south of Carbondale, *BS 321.*

Erythronium albidum Nutt. (1:420). Unlike the preceding species, White Dog's-tooth Violet is less frequently found along sandstone bluffs but is usually confined to floors of rich woods or on wooded hillsides. It blooms slightly later than the yellow one. SP. CIT.: Lake Murphysboro, *M 2401;* Giant City State Park, *M 667;* one mile south of Carbondale, *BS 322.*

Camassia LINDLEY

Camassia scilloides (Raf.) Cory. Wild Hyacinth is found very locally throughout Southern Illinois in moist woods or more commonly in railroad prairies, particularly near Pyatts. In Jackson County, it has been collected only in a low woods where it grows with *Carex grayii.* We also have recorded it from Perry, Randolph, and Union counties.

It begins to bloom in late April. SP. CIT.: along northern part of "Granny" Railroad, *Mohlenbrock*.

Ornithogalum L.

Ornithogalum umbellatum L. (1:423). Star-of-Bethlehem is an introduction from Europe and is found frequently along highways and railroads. SP. CIT.: Giant City State Park, *M 130*; north part of Carbondale, *Mc 44*.

Muscari MILL.

Muscari botryoides (L.) Mill. (1:423). This very attractive Grape Hyacinth has been found in a woods near the Stonefort at Giant City State Park. Since it does persist at this location, it is included in this flora. The species is introduced from Europe. The leaves are 4 to 7 mm. broad. SP. CIT.: Giant City State Park, *M 121*.

Muscari racemosum (L.) Mill. (1:423), an introduction from Europe, was found covering a one acre field on the farm of Phinneus Friese in Union County. Leaves of this species are less than three millimeters broad.

Yucca L.

Yucca filamentosa L. (1:424). This species is common along roads as an escape, and it sometimes makes its way into woodlands. There is scarcely a cemetery in Southern Illinois without it. SP. CIT.: along a stream, Giant City State Park, *M 201*.

Asparagus L.

Asparagus officinalis L. (1:424). Abandoned fields and roadsides are the habitats for this introduction from Europe. SP. CIT.: Giant City State Park, *M 681*; one-half mile south of Carbondale, *Swayne 706*.

Smilacina DESF.

Smilacina racemosa (L.) Desf. (1:427). False Solomon's-seal is a common plant of Southern Illinois woodlands. It begins to flower in our area in mid-April. Fruits mature during early July. SP. CIT.: Fountain Bluff, *BS 41*; Giant City State Park, *M 8*; four miles north of Carbondale, *Mc 3604*; three miles south of Murphysboro, *Mc 721*.

FIG. 24. *Nodding blossoms of Solomon's Seal* (Polygonatum canaliculatum). *Photo taken June, 1955.*

Polygonatum MILL.

Polygonatum biflorum (Walt.) Ell. (1:431). This Solomon's-seal in all respects is smaller than the following one. The leaves are narrow (the lateral ones in our specimens with fifty-five to seventy-five nerves) and the rhizome does not exceed 1.5 centimeters in thickness. We have it only from Hardin, Pope, and Jackson counties where it grows in mixed mesophytic woods. SP. CIT.: Giant City State Park, *M 2092*.

Polygonatum canaliculatum (Muhl.) Pursh (1:431) (Fig. 24). The lateral leaves of this species sometimes attain a width of six centimeters and have over one hundred nerves (best seen through transmitted light). The rhizome is two to three centimeters thick. This species is common in mesophytic woods throughout Southern Illinois. SP. CIT.: near Boskydell, *Hall 1033*; Giant City State Park, *M 356*.

Trillium L.

1. Flowers white or yellow 2
    2. Flowers white, borne on pedicels; leaves nearly sessile . .
        . . . . *Trillium gleasoni*
    2. Flowers yellow, sessile; leaves narrowed at base to a distinct petiole. 3
        3. Stamens and sometimes base of the petals maroon . .
            . . . . . *Trillium recurvatum* forma *viridiflorum*
        3. Stamens and petals entirely bright yellow . . .
            . . . . . *Trillium recurvatum* forma *shayi*
1. Flowers maroon or green. 4
    4. Leaves petioled; sepals reflexed at maturity; flowers maroon
        . . . . *Trillium recurvatum*
    4. Leaves sessile; sepals spreading at maturity; flowers maroon or green 5
        5. Flowers usually maroon, petals not clawed . . .
            . . . . *Trillium sessile*
        5. Flowers usually green, petals distinctly clawed . .
            . . . . *Trillium viride* °

FIG. 25. *Gleason's Trillium* (Trillium *gleasoni*) *in Giant City State Park. Photo taken May, 1956, by Marvin Rensing.*

FIG. 26. *The rare yellow-flowered form of Wake Robin* (Trillium recurvatum *forma* shayi).

Trillium gleasoni Fern. [*T. flexipes* Raf.] (1:435). The beautiful White Trillium is not uncommon in rich mesic woodlands of Southern Illinois (Fig. 25). It begins to bloom in early April and continues through mid-May. Leaves on some of our specimens measure sixteen centimeters across. Although many of the herbarium specimens were determined originally as *Trillium grandiflorum* (Michx.) Salisb., this species does not extend into the southern part of the state. SP. CIT.: Fountain Bluff, BS *332;* Giant City State Park, V *1202.*

Trillium recurvatum Beck (1:432). The typical maroon-flowered Recurved Wakerobin is common in woods of Southern Illinois. Immature specimens are often mistaken for the much rarer *T. sessile.* Two yellow-flowered forms of *T. recurvatum* occur in Jackson County. Forma *viridiflorum* Beyer grows with the species at a few Jackson County stations while the very rare forma *shayi* Palmer & Steyerm. (Fig. 26) has been collected in woods at Lake Murphysboro. SP. CIT.: typical: Fountain Bluff, BS *20;* Giant City State Park, V *1204;* Lake Murphysboro, M *2402;* Midland Hills, V *851;* Saltpeter Cave, M *1939.* Forma *shayi:* Lake Murphysboro, M *2207.* Forma *viridiflorum:* six miles west of Murphysboro, M *5061.*

Trillium sessile (1:432). This species has been collected only a few times in Southern Illinois. We have it from Jackson and Union counties. It grows in oak woods, usually on sloping terrain. SP. CIT.: Fountain Bluff, *BS 320.*

Green Trillium (*Trillium viride* Beck) (1:432) was collected first in Southern Illinois in 1951 in Union County by Mr. Harry Ahles (No. *3636*); it later was found in an open oak woods in the Union County section of Giant City State Park (*M 5120*), scarcely more than five hundred yards from Jackson County.

Smilax L.

```
1.  Stems woody, often bearing prickles .   .     .    .    .    2
    2.  Leaves whitened on the lower surface .    . Smilax glauca
    2.  Leaves green on both sides .  .  .  .  .  .  .    3
        3.  Prickles black or entirely lacking .    .    .
            .    .    .    .    .Smilax tamnoides var. hispida
        3.  Prickles green or yellow, seldom lacking .   .   .   .    4
            4.  Leaves very large (often over 5 cm. broad), orbicu-
                lar, eciliate; prickles green . Smilax rotundifolia
            4.  Leaves up to 5 cm. broad, with some marginal bristly
                cilia; prickles yellowish  .  .  .  .  .  .    5
                5.  Leaves with broad, rounded, basal lobes, margins
                    bristly-ciliate throughout .   .Smilax bona-nox
                5.  Leaves without rounded basal lobes, margins
                    bristly-ciliate only near the base.    .    .    .
                    .  .  . Smilax bona-nox var. hederaefolia
1.  Stems herbaceous, not bearing prickles .   .    .    .    .    6
    6.  Leaves green on both sides .   .    .Smilax pulverulenta
    6.  Leaves pale beneath .   .  .  .  .  .  .  .  .    7
        7.  Plants erect, usually without tendrils  .  .  .  .
            .  .  .  .  .   .   .   . Smilax ecirrhata
        7.  Plants twining, tendril-bearing .    . Smilax lasioneura
```

Smilax glauca Walt. (1:437). This species of Cat-brier is common in dry open woods throughout Southern Illinois. It is very abundant on Devil's Backbone near Grand Tower. SP. CIT.: Giant City State Park, *M 65.*

Smilax bona-nox L. (1:437). The typical plants with lobed leaves are found on the very dry sandstone bluffs along the Shawneetown Ridge. Equally abundant in our area is var. *hederaefolia* (Beyrich) Fern. It grows with the lobed form. SP. CIT.: typical: Fountain Bluff, *BS 2308* (in part); Giant City State Park, *M 772.* Var. *hederaefolia:* Fountain Bluff, *BS 2308* (in part); Giant City State Park, *M 368.*

Smilax tamnoides L. var *hispida* (Muhl.) Fern. [*S. hispida* Muhl.] (1:436). This common Cat-brier often forms dense mats on dry hill-sides in oak-hickory woods. It may be found also on limestone talus, particularly in the Pine Hills of Union County. SP. CIT.: Fountain Bluff, *BS 2309;* one mile south of Carbondale, *Bailey 579.*

Smilax rotundifolia L. (1:437). Round-Leaved Cat-brier is not common in Southern Illinois, but where it grows, it often forms dense thickets. Stems of this species are usually shining green. SP. CIT.: Giant City State Park, *M 461.*

Smilax pulverulenta Michx. (1:436). This species is found in or along edges of woods. It seems to be more plentiful in limestone areas. We have it from Johnson, Pope, Franklin, and Jackson counties. SP. CIT.: one mile south of Carbondale, *Bailey 582;* three miles south of Murphysboro, *Mc 720.*

Smilax ecirrhata (Engelm.) S. Wats. (1:436). This Cat-brier is common farther to the north, but in Southern Illinois, it has been found only in Jackson County. It usually is erect, and though an annual it sometimes becomes slightly woody near the base. SP. CIT.: Giant City State Park, *M 204; M 415.*

Smilax lasioneura Hook. (1:436). This species has been collected only a few times in our area. It grows in rich woods and thickets. SP. CIT.: north of De Soto, *BS 776.*

28. *Dioscoreaceae* – Yam Family

Dioscorea L.

Dioscorea villosa L. (1:441). Wild Yam (Fig. 27) is found in rich

FIG. 27. *Wild Yam* (Dioscorea villosa) *growing in upland woods at Horseshoe Bluff in Jackson County. Photo taken in late July, 1954.*

mesophytic woods across Southern Illinois. It is often difficult to distinguish from the following species on the basis of vegetative characteristics, but most leaves of *D. villosa* are alternate while those of *D. quaternata* are opposite or occasionally whorled. Seeds of *D. villosa* average ten millimeters broad. This species begins to flower in late May and to mature its fruit at the end of July. SP. CIT.: Fountain Bluff, *BS 2313;* Giant City State Park, *M 183.*

Dioscorea quaternata (Walt.) Gmel. (1:441). This species seems to be most abundant in open oak woods. Like the preceding, it is found throughout Southern Illinois. It flowers in mid-May and begins to mature its fruit in mid-June. Seeds are over fifteen millimeters broad. SP. CIT.: one mile southwest of Pomona Natural Bridge, *Swayne 715.*

29. *Amaryllidaceae* – Amaryllis Family

1. Perianth parts and leaves pubescent; flowers yellow . *Hypoxis*
1. Perianth and leaves glabrous; flowers white or greenish, or if yellow, then with a crown   2
    2. Leaves thick, often spiny-toothed .   .   .   . .*Agave*
    2. Leaves thin, or if somewhat thick, margins entire .   .   3
        3. Flowers yellow, usually nodding; leaves 1 to 3 cm. broad
        .   .   .   . *Narcissus*
        3. Flowers white, erect; leaves over 3 cm. broad.   .   .
        .   .   .   . *Hymenocallis*

Hymenocallis SALISB.

Hymenocallis occidentalis (Le Conte) Kunth (1:443). One of the most attractive flowers in our flora is produced by the rare Spider Lily. The plants grow in low, wet woods. In our area we have it from Jackson, Union, Johnson, and Pulaski counties. The delicate flowers begin to bloom the middle of August. SP. CIT.: Carbondale, *Mc Cree and Wilson 1161;* Lake Murphysboro, *M 4724.*

Agave L.

Agave virginica L. (1:444). American Aloe, characteristic of sandstone bluffs of the Shawneetown Ridge, begins to flower the last of June (Fig. 28). Common associates of this species include *Opuntia rafinesquii* and *Tephrosia virginiana.* SP. CIT.: Giant City State Park, *M 171;* three miles south of Elkville, *BS 499.*

FIG. 28. *American Aloe* (Agave virginica) *standing about four feet high with both flowers and fruits in late July. This succulent plant is found on dry sandstone bluffs of the Shawneetown Ridge in Southern Illinois.*

Hypoxis L.

Hypoxis hirsuta (L.) Coville (1:443). One of the first species to bloom in spring on dry sandstone bluffs across Southern Illinois is the bright, yellow-flowered Stargrass. It may be found in flower as early as April 15. SP. CIT.: Giant City State Park, *M 2898;* Midland Hills, *BS 2235.*

Narcissus L.

Narcissus pseudo-narcissus L. (1:443). Daffodil is commonly found around old homesteads. At several locations it has escaped to open woods where it thrives year after year without cultivation. We have collected it in flower as early as February 28. SP. CIT.: Giant City State Park, *M 730.*

30. *Iridaceae* – Iris Family

1. Plants rhizomatous; flowers over 3 cm. across; leaves over 1 cm. wide 2
   2. Flowers orange with purplish spots; seeds black and shining . . . . *Belamcanda*
   2. Flowers blue, lavender, or rust-colored; seeds not black and shining . . . . *Iris*
1. Plants without rhizomes; flowers less than 2 cm. across; leaves less than 1 cm. wide . . . . . . *Sisyrinchium*

FIG. 29. *Blackberry Lily* (Belamcanda chinensis), *a native of Asia and introduced from Europe, has escaped to many moist lowland situations. Photo taken in May.*

Belamcanda ADANS.

Belamcanda chinensis (L.) DC. (1:450). Blackberry Lily is a native of Asia but has sparingly escaped from cultivation in this area (Fig. 29). SP. CIT.: Fountain Bluff, *Cranwill.*

Iris L.

1. Flowers rust-colored; plants of deep swamps . . . *Iris fulva*
1. Flowers blue or lavender as the principal color, often variegated with white or yellow 2
    2. Plants low (to 20 cm.), from slender creeping rootstocks . . . . *Iris cristata*
    2. Plants 40 to 90 cm. tall, from thick rootstocks . . . . *Iris virginica* var. *shrevei*

Iris virginica L. var. *shrevei* (Small) E. Anders. (1:448). Blue Iris is variable in appearance. Leaves are usually, but not always, somewhat glaucous. It is found growing in moist soil throughout Southern Illinois. SP. CIT.: one mile south of Carbondale, *BS 56;* one mile northwest of Oraville, *M. 5252.*

Red Iris (*Iris fulva* Ker) (1:447) is one of the southeastern species characteristic of deep swamps of extreme Southern Illinois (Fig. 30). It flowers in mid-May. We have it from Union and Alexander counties.

Dwarf Iris (*Iris cristata* Ait.) (1:447) is known in Illinois only from Union and Hardin counties where it grows on stream banks in closed woods of oak, beech, and sugar maple (Fig. 31).

Sisyrinchium L.

1. Flowers arising from sessile terminal spathes . . . . *Sisyrinchium albidum*

FIG. 30. *Red Iris* (Iris fulva) *grows sparingly in LaRue Swamp in Union County. Photo taken in May.*

FIG. 31. *Several patches of Dwarf Crested Iris* (Iris cristata) *grow along sandy banks of Rock Creek in Hooven Hollow, Hardin County. Photo taken in mid-May.*

1. Flowers arising from spathes pedunculate from axils of a leaf-like
    bract 2
    2. Pedicels much longer than the inner bract; stems flattened, 2
        to 6 mm. wide . . . . . *Sisyrinchium graminoides*
    2. Pedicels scarcely longer than the inner bract; stems flattened,
        1 to 3 mm. wide . . . . *Sisyrinchium angustifolium*

Sisyrinchium albidum Raf. (1:452). Flowers of this Blue-eyed Grass are white or sometimes purplish. This species grows in moist meadows, in woods, or occasionally in waste ground. SP. CIT.: Giant City State Park, *M 1007;* Midland Hills, *V 666;* north of De Soto, *BS 698.*

Sisyrinchium graminoides Bickn. (1:452). A specimen referable to this species was collected in a rich mesic woods at Lake Murphysboro, May 18, 1954. SP. CIT.: Lake Murphysboro, *M 2454.*

Sisyrinchium angustifolium Mill. (1:452). This species may be found at any site where moisture is plentiful. SP. CIT.: Giant City State Park, *M 193;* south of Carbondale, *M 190.*

31. *Orchidaceae* – Orchid Family

1. Plants with green foliage at flowering time 2
    2. Lip large, yellow, slipper-shaped . . *Cypripedium*
    2. Lip smaller, not yellow, not slipper-shaped 3
        3. Principal leaves all arising from the base of the plant . 4
            4. Flowers with a white lip and purple sepals and pet-
                als; leaves 2 . . . . *Orchis*

4. Flowers without the above color combination .   .   5
    5. Leaf single, long and grass-like; flowers at least
        1 cm. across, rose-purple .   .   .*Calopogon*
    5. Leaves two or more, not decidedly grass-like;
        flowers less than 1 cm. across, white to greenish-
        purple   6
        6. Flowers pedicelled, greenish-purple; leaves
            two, over 2 cm. broad .   .   .*Liparis*
        6. Flowers sessile, white to greenish-white;
            leaves usually more than 2 in number, less
            than 2 cm. broad   7
            7. Leaves green and white; inflorescence
                not twisted   .   .   .   .*Goodyera*
            7. Leaves not variegated; inflorescence of-
                ten twisted   .   .   .   .*Spiranthes*
3. Principal leaves alternate along the stem   8
    8. Perianth with a conspicuous spur   .   .*Habenaria*
    8. Perianth without a spur   .   .   .   .*Triphora*
1. Plants with no green foliage at flowering time (a single withered
    leaf often present in *Aplectrum*)   9
    9. Inflorescence twisted, flowers sessile   .   .   .*Spiranthes*
    9. Inflorescence not twisted, flowers pedicelled or subsessile   .   10
        10. Lip of flower with 1 to 3 ridges   11
            11. A withered oval leaf usually present at flowering
                time; stem arising from one or two solid corms
                .   .   .   .*Aplectrum*
            11. Plants completely saprophytic; stems arising from
                a string of coral-like roots .   .   .*Corallorhiza*
        10. Lip of flower with 5 to 7 ridges .   .   .*Hexalectris*

The orchid family in Southern Illinois is represented for the most
part by rare species. Because of this, orchids should not be picked.
Fifteen species divided among eleven genera occur in Southern Illinois with all of them being found in Jackson County. Two of the
species are restricted to Jackson County. Four locations in the county
are exceptional for orchids. Giant City State Park is the site for seven
species, Little Grand Canyon and Lake Murphysboro for six, and
Carbon Lake for five.

Cypripedium   L.

Cypripedium calceolus L. var. *pubescens* (Willd.) Correll (1:457).
There is no greater excitement among collectors than when one finds
one of the striking Yellow Lady's-slipper Orchids (Fig. 32). Four or

FIG. 32. *Yellow Lady's-slipper Orchid* (Cypripedium calceolus *var.* pubescens) *usually gives a thrill to its discoverer. Photo taken in mid-May.*

five stations are known in Jackson County for this species. An additional plant once seen in the Ava area has since disappeared. The habitat for this species is rich woods on rather steep slopes. At every station Broad Beech Fern (*Dryopteris hexagonoptera*) and Showy Orchis (*Orchis spectabilis*) are in close association with the Lady's-slipper. Elsewhere in Southern Illinois, this species has been collected in a beech-maple woods in Alexander County. The blooming period is between the first and third weeks in May. SP. CIT.: Midland Hills, V *438*; Giant City State Park, *Stewart 83.*

Orchis L.

Orchis spectabilis L. (1:459). Showy Orchis is one of the most beautiful members of the spring flora of Jackson County. All of our collections of this species from the county are from rich oak woods. In Hardin and Pope counties, *Orchis spectabilis* has been found only in beech-maple forests. We have collected it in flower from April 17 to May 9 and in fruit as late as August 17. SP. CIT.: Midland Hills, V *663*; Lake Murphysboro, *M 5393.*

Habenaria WILLD.

Habenaria peramoena Gray (1:463). This is one of the less rare orchids in the area, although it is recorded only from Jackson, Williamson, and Pope counties. In Jackson it grows in low swampy woods at Cave Valley, at Carbon Lake, and west of Crab Orchard Lake. The Fringeless Purple Orchid blooms during the month of July. SP. CIT.: Cave Valley, *V 846;* Carbon Lake, *M 4388;* west of Crab Orchard Lake, *BS 131.*

Calopogon R. BR.

Calopogon pulchellus (Salisb.) R. Br. (1:467). On May 18, 1949, Mr. Julius Swayne, collecting in a relict prairie strip along the Illinois Central Railroad near Elkville, found one plant of the Grass Pink Orchid. Previously this species was known in Illinois only in the northern third of the state. Further search in the area may turn up additional specimens. SP. CIT.: one mile south of Elkville, between Illinois Central Railroad and U.S. highway 51, *BS 399.*

Triphora NUTT.

Triphora trianthophora (Sw.) Rydb. (1:467). This rare orchid is known as Nodding Pogonia. We have only three collections of it from Southern Illinois, and all are in Jackson County. Near Carbon Lake, it grows on a moist wooded slope near a creek, while at Little Grand Canyon, it is found in a rich beech-maple woods. All of our collection dates are either August 6 or 7. SP. CIT.: near Carbondale, *French 2795;* one and one-half miles south of Murphysboro, *Stewart and Hatcher 1155;* Little Grand Canyon, *BS 1688.*

Spiranthes RICH.

Spiranthes tuberosa Raf. var. *grayi* (Ames) Fern. (1:470) [*S. grayi* Ames]. This is distinguished by having one tuberous, thickened root, tiny flowers in one very twisted rank, leaves basal and usually absent at flowering time, and a glabrous rachis and stem. This species was found for the first time in Illinois in 1948 in Pope County by Winterringer (1537). Since then, it has been collected four times, once in Randolph County (*BS 828*) and three times in Jackson County. It usually grows under trees on dry, moss-covered soil. Our collection dates range from August 17 to September 4. SP. CIT.: Little Grand Canyon, *BS 812;* near Carbon Lake, *Hatcher;* Lake Murphysboro, *M 4699.*

Spiranthes gracilis (Bigel.) Beck (1:470). Flowers of this rare graceful orchid are decidedly secund. The stem and rachis are glabrous and the roots are several. In our area, this species is found on the dry Shawneetown Ridge, usually associated with *Krigia dande-lion, Asclepias verticillata,* and *Linum medium.* The basal rosette of leaves, usually absent at flowering time, is quite conspicuous in early spring. Although it has been collected only in Randolph and Jackson counties, it probably occurs in all the counties along the Shawnee-town Ridge. It blooms from late August to October. SP. CIT.: atop sandstone bluff, Giant City State Park, *Mohlenbrock.*

Spiranthes ovalis Lindl. (1:469). This is our rarest Ladies'-tress. It is known from Pulaski, Union, and Jackson counties. The distin-guishing features of this species are the puberulent stem and rachis and the inflorescence less than one centimeter thick. The time of flowering is late September and October. SP. CIT.: Carbondale Res-ervoir, *Stewart.*

Spiranthes cernua (L.) Rich. (1:469). This is the largest and most common Ladies'-tress in Southern Illinois. It grows in low woods, meadows, or along pond margins. Our collection dates are from Sep-tember 24 to October 16. Besides Jackson County, we have it re-corded from Pope and Williamson counties. It is easily separated from all other *Spiranthes* in our area by its pubescent stem and rachis and its inflorescence which is over one centimeter thick. SP. CIT.: east of Carbondale, *Mc 1290;* northern part of Murphysboro, *Hatcher 928;* Thompson's Woods, Carbondale, *V 380;* Giant City State Park, *Mar-berry.*

Goodyera R. BR.

Goodyera pubescens (Willd.) R. Br. (1:471). This curious little orchid, popularly known as Rattlesnake Plantain, is one of the rarest orchids in our flora. It has three stations in Pope County and one in Jackson. In no area may it be considered plentiful. It grows in shaded, moist woods, and seldom grows far from streams. The middle of Au-gust is the time for flowering. SP. CIT.: Little Grand Canyon, *Moh-lenbrock.*

Liparis RICH.

Liparis lilifolia (L.) Rich. (1:473). Twayblade is apparently our most common orchid. It grows in mesophytic woodlands and blooms the last of May. It seems to be found usually in *Acer barbatum* woods. Besides the following cited specimens, this orchid occurs at Giant City State Park and Lake Murphysboro. SP. CIT.: Midland

Hills, V *1224;* Fountain Bluff, *BS 414;* Carbon Lake, *Hatcher and Stewart 782.*

Aplectrum (NUTT.) TORR.

Aplectrum hyemale (Muhl.) Torr. (1:476). Puttyroot Orchid is perhaps most easily detected during winter since its one large leaf is produced in the fall, overwinters, and then withers about the time of flowering. At Giant City State Park a colony of over ninety plants cccurs, while in a woods about six miles west of Murphysboro, a similar number is found. This species blooms during May. It grows with *Valeriana pauciflora* and *Trillium gleasoni* at Giant City State Park. We also have it from Pope and Williamson counties. SP. CIT.: Giant City State Park, *M 806;* four miles north of Carbondale, *BS 387, Marberry 643.*

Corallorhiza CHAT.

Corallorhiza odontorhiza (Willd.) Nutt. (1:474). The Coral-roots are saprophytic plants with purplish stems which make them inconspicuous. This one, Fall Coral-root, is known only from Jackson and Union counties. Both stations are in rich woods. This species is characterized by flowers having an unnotched lip with two purple spots. It blooms from mid-September to mid-October. SP. CIT.: Carbon Lake, *Hatcher 929.*

Corallorhiza wisteriana Conrad (1:474). Wister's Coral-root has a notched lip with several purple spots. It blooms from April 25 to May 25. Rich mesic woods are its favorite haunts. In addition to Jackson, we have it from Union, Hardin, and Pope counties. SP. CIT.:

FIG. 33. *Crested Coral-root Orchid* (Hexalectris spicata) *as it grows on an east-facing slope of oak-hickory woods at Fountain Bluff in Jackson County. Photo taken in July.*

FIG. 34. *A relict of great age is Fountain Bluff rising in the distance from an ancient flood plain of the Mississippi River. The river now flows to the right of the bluff.*

Giant City State Park, *V 1229;* Midland Hills, *V 665, V 1060;* Pomona Natural Bridge, *BS 28;* Lake Murphysboro, *Mohlenbrock.*

Hexalectris RAF.

Hexalectris spicata (Walt.) Bernh. (1:474). The first collection of the saprophytic Crested Coral-root Orchid (Fig. 33) in Illinois was made in 1949 in Randolph County (*Winterringer 2058*). Later, one plant was found at Jackson Hollow in Pope County (*Hatcher and Stewart 1168*). Then on July 20, 1954, accompanied by Mr. David Sanders, the authors found a colony of twelve plants of *Hexalectris* growing on an east-facing slope in an oak-hickory woods at Fountain Bluff (Fig. 34). Two days later, further search in the area turned up five additional plants. SP. CIT.: Fountain Bluff, *Sanders, Voigt, and Mohlenbrock 4330.*

SUBCLASS DICOTYLEDONEAE (JUSS.) DC.

32. *Saururaceae* – Lizard-tail Family

Saururus L.

Saururus cernuus L. (2:3). Lizard-tail is common in low wet ground throughout Southern Illinois. Where it occurs, it is usually very abundant. It often is found in shallow standing water. Flowers appear in mid-June. SP CIT.: *Mc 242; Mc 867; BS 121.*

33. *Salicaceae* – Willow Family

1. Leaves 2 to many times longer than broad . . . .*Salix*
1. Leaves less than twice as long as broad . . . .*Populus*

Salix L.

1. Leaves green on both surfaces 2
   2. Leaves (not of sprouts) up to 20 mm. broad . . . .
   . . . . *Salix nigra*
   2. Leaves (not of sprouts) over 20 mm. broad . . . .
   . . . . *Salix fragilis*
1. Leaves paler on lower surface 3
   3. Leaves glabrous (sometimes minutely pubescent when immature) 4
      4. Leaves sessile or on petioles less than 3 mm. long . .
      . . . . *Salix interior*
      4. Leaves on petioles over 3 mm. long 5
         5. Branches pendulous ("weeping") . . . .
         . . . . *Salix babylonica*
         5. Branches erect or spreading 6
            6. Petioles bearing one or more glands . . .
            . . . . . . *Salix alba* var. *calva*
            6. Petioles without glands 7
               7. Leaves tapering at base . . . .
               . . . . . *Salix amygdaloides*
               7. Leaves rounded or cordate at base . .
               . . . . . . *Salix rigida*
   3. Leaves pubescent at maturity 8
      8. Lower leaf surface with long appressed silvery hairs .
      . . . . *Salix sericea*
      8. Leaves clothed with short ashy-colored hairs . . .
      . . . . *Salix humilis*

Salix nigra L. (2:9). Black Willow, which sometimes attains a height in our area of fifty to sixty feet under favorable conditions, is a common tree along streams and in marshy areas. It occurs throughout Southern Illinois. SP. CIT.: *Mc 316; Mc 709; M 263.*

Salix fragilis L. (2:11). Brittle Willow, so named because of its easily broken branchlets, is a native of Europe. It is known in Southern Illinois only along the Mississippi River four miles south of Cora. At this station, some trees reach a height of thirty-five feet. SP. CIT.: *M 4327.*

Salix interior Rowlee (2:11). Sandbar Willow is common through-out Southern Illinois along rivers and streams. SP. CIT.: *M 4326; M 194.*

Salix alba L. var. *calva* G. F. W. Meyer. White Willow, a native of Europe, has been found growing along a stream one-half mile north of Etherton (Jackson County). SP. CIT.: near Etherton, *M 2351.*

Salix amygdaloides Anders. (2:9). An occasional small tree or shrub in Southern Illinois is Peach-leaved Willow. SP. CIT.: along Mississippi River near Neunert, *Brewer.*

Salix rigida Muhl. (2:11) [*S. cordata* Muhl., not Michx.]. This Wil-low occurs along streams throughout most of Illinois, although we have it only from Saline and Jackson counties. SP. CIT.: *French.*

Salix sericea Marsh. (2:21). Silky Willow is found along rocky streams in Southern Illinois where it is local. We have it from Jack-son, Randolph, Union, and Pope counties. SP. CIT.: *M 764; M 1991; V 853.*

Salix humilis Marsh. (2:23). Along railroads in northern Jackson County is found the Dwarf Prairie Willow. To the south in Illinois, this species is very rare. SP. CIT.: south of Elkville, *BS 672.*

The commonly planted Weeping Willow (*Salix babylonica* L.) (2:11) is seldom spontaneous in our area.

Populus L.

Populus alba L. (2:3) White Poplar is native to Eurasia but is planted quite extensively in Southern Illinois and it sometimes es-capes. The lower surface of leaves is silvery-white. SP. CIT.: *M 168.*

Populus deltoides Marsh. (2:5). The Cottonwood is found in low woods and along rivers and streams throughout Southern Illinois. It flowers in the middle of March. Petioles are flattened in this species. SP. CIT.: *M 23; BS 649.*

Populus heterophylla L. (2:4). Swamp Cottonwood occurs along borders of swamps in all of our southern counties. It, along with *Nyssa aquatica, Fraxinus tomentosa,* and *Taxodium distichum* domi-nates the swampy woods of Alexander, Pulaski, and Massac counties. Petioles are terete in this species. SP. CIT.: swamp three miles west of the junction of Illinois highways 3 and 144, *M 4642.*

34. *Juglandaceae* – Walnut Family

1. Pith separated by small plates .   .   .   .   .   .*Juglans*
1. Pith continuous   .   .   .   .*Carya*

Juglans L.

Juglans cinerea L. (2:27). White Walnut or Butternut is not common in Southern Illinois although we have it recorded from most counties. It grows in moist woodlands. Pith of the twigs is dark brown while the "nut" is much longer than wide. SP. CIT.: Fountain Bluff, *BS 548*.

Juglans nigra L. (2:27). Black Walnut is common in woodlands throughout the state. It may be distinguished from Butternut by its light brown pith and its "nuts" about as wide as long. SP. CIT.: *M 150; BS 704*.

Carya NUTT.

1. Leaflets 5 to 9 (rarely 11 in *C. cordiformis*) 2
2. Leaves usually over 30 cm. long   3
3. Leaflets usually 5, ovate to ovate-lanceolate; leaves 20 to 35 cm. long; bark separating into shaggy strips; fruits with husk nearly round; nut pointed at tip, rounded at bottom   .   .   .   .*Carya ovata*
3. Leaflets 7 (sometimes 5), ovate-lanceolate   .   .   . 4
4. Fruit with husk large (5 to 7.5 cm.); husk thick; nut elongate, somewhat compressed, 2.5 to 4 cm. long; leaves 37 to 50 cm. long .   .*Carya laciniosa*
4. Fruit smaller than above; leaves less than 15 inches long   5
5. Young twigs and rachis of leaves glabrous; fruit obovoid (pear-shaped), 2.5 to 5.0 cm. long   .
   .   .   .   .*Carya glabra*
5. Young twigs and rachis of leaves pubescent or scurfy; fruit rounded .   .   .*Carya tomentosa*
2. Leaves usually under 30 cm. long 6
6. Leaflets ovate, the broadest portion above the middle; fruit pear-shaped to rounded in the husk . *Carya ovalis*
6. Leaflets lanceolate, not broadest above the middle .   . 7
7. Petioles rusty brown in color   .   .   .*Carya texana*
7. Petioles glabrous; buds long and curved, mustard-colored   .   .   .   .   .   .*Carya cordiformis*
1. Leaflets 9 or more (usually at least 13)   8

8. Nut ellipsoid, about twice as long as wide; lower leaf surface
   slightly pubescent .   .   .   .   . .*Carya illinoensis*
8. Nut angled, about as long as wide; leaves glabrous   .   .
   .   .   .   . *Carya aquatica*

Carya ovata (Mill.) K. Koch (2:29). The nuts of Shagbark Hickory are eaten extensively in our area. Trees, which sometimes attain
heights of seventy feet, are common in moist woods in all of our
counties. SP. CIT.: *M 313; M 1491; BS 1807.*

Carya laciniosa (Michx. f.) Loud. (2:29). Shellbark Hickory is
somewhat similar in appearance to Shagbark but the edible nuts are
usually larger. This species is not as common as *C. ovata,* but does
occur locally in moist woods. SP. CIT.: *M 511; BS 1804.*

Carya glabra (Mill.) Sweet (2:31). Pignut is one of our most
abundant hickories. Its usual habitat is dry woods. The nuts are bitter
and therefore are inedible. SP. CIT.: *M 363.*

Carya tomentosa Nutt. (2:29). The fruit of Mockernut Hickory is
very large but after removal of the thick husk, the nut is quite small
as well as being slightly bitter. Most parts of the plant are densely
pubescent. SP. CIT.: *BS 275; French.*

Carya ovalis (Wang.) Sarg. (2:31). Small-fruited Hickory is found
locally in woods in all of our counties. The nut is edible, but very
small so that it is infrequently gathered for food. SP. CIT.: *BS 1810.*

Carya texana Buckl. (2:31). Black Hickory occurs on top of sandstone bluffs of the Shawneetown Ridge. It is not common. The nuts
are very bitter. SP. CIT.: Giant City State Park, *M 244.*

Carya cordiformis (Wang.) K. Koch (2:28). Bitternut Hickory is
our commonest species of *Carya,* occurring in both moist and dry
woods. SP. CIT.: *M 368; M 4032.*

Carya illinoensis (Wang.) K. Koch (2:28). Pecan attains maximum
size in river bottom woods along the Mississippi River between
Neunert and Gorham. SP. CIT.: *BS 2310.*

The rare Water Hickory, *Carya aquatica* (Michx. f.) Nutt. (2:28),
is known from Gallatin, Alexander, Pulaski, and Massac counties
where it occurs in standing water with *Fraxinus tomentosa, Gleditsia
aquatica,* etc.

35. *Betulaceae* – Birch Family

1. Petioles and young twigs with some gland-tipped hairs   .   .
.   .   .   .   *Corylus*
1. Petioles and young twigs pubescent but not glandular .   .   .   2
    2. Leaves with very short single teeth   .   .   . .*Alnus*
    2. Leaves with longer teeth or, if short, then doubly-serrate   .   3
        3. Bark of trunk peeling off in buff to reddish papery strips
.   .   .   . *Betula*
        3. Bark of trunk not peeling off   4
            4. Trunk gray, smooth, with a muscle-like appearance
.   .   .   . *Carpinus*
            4. Trunk with light brown scaly bark .   .   .*Ostrya*

Corylus L.

Corylus americana Walt. (2:33). Hazel-nut or Filbert is common in thickets and along borders of woods. It is one of our earliest flowering plants, beginning to bloom about the second week in February. SP. CIT.: *M 42; M 1802.*

Alnus EHRH.

Alnus serrulata (Ait.) Willd. (2:37). Smooth Alder is very common along streams in southeastern counties, but is rare in southwestern counties. SP. CIT.: three miles south of Boskydell, *BS 303.*

Betula L.

Betula nigra L. (2:35). River Birch is a common and attractive tree found along most watercourses. SP. CIT.: *Welch 315; M 1094.*

Carpinus L.

Carpinus caroliniana Walt. (2:33). Blue Beech is a common small tree of Illinois forests. It generally is found in mesic woods. SP. CIT.: *M 734.*

Ostrya SCOP.

Ostrya virginiana (Mill.) K. Koch (2:33). This species is commonly called Hop Hornbeam or Ironwood although the latter name is sometimes applied to *Carpinus* as well. SP. CIT.: *Mc 911; Mc 972.*

36. *Fagaceae* – Beech Family

1. Leaves entire or lobed . . . . *Quercus*
1. Leaves merely toothed 2
   2. Buds long, tapered, and pointed; bark light in color and very smooth . . . . *Fagus*
   2. Buds not long and pointed; bark rough or, if smoothish, not light in color 3
      3. Leaves usually broadest above middle; fruit an acorn . . . . *Quercus*
      3. Leaves broadest at middle; nutlets surrounded by a prickly bur . . . . *Castanea*

Quercus L.

1. Leaves bristle-tipped 2
   2. Leaves lobed 3
      3. Terminal lobe much longer or broader than others . . 4
         4. Terminal lobe longer than others; acorns pedicelled 5
            5. Leaves with 5 to 7 lobes, terminal one much exceeding the others . . . *Quercus falcata*
            5. Leaves with 3 to 5 lobes, terminal one hardly exceeding the others . . . . . .
              . . . *Quercus falcata* var. *pagodaefolia*
         4. Terminal lobes broader than others; acorns sessile or nearly so . . . . . *Quercus marilandica*
      3. Terminal lobe not longer or broader than others; acorns sessile or nearly so 6
         6. Leaves lobed halfway to middle or less . . . 7
            7. Leaves 7- to 11-lobed, dull green above, pale yellow-green below and smooth . *Quercus rubra*
            7. Leaves 7-lobed, dark green and shining above, brownish green and somewhat pubescent beneath . . . . . . *Quercus velutina*
         6. Leaves lobed deeper than halfway to middle . . 8
            8. Leaves 5- to 9-lobed, dark green and shining above, light green below with tufts of hair along the midrib; acorns 1.5 to 3 cm. long . . . . *Quercus shumardii*
            8. Leaves 7-lobed, lobes broader toward the tip, dark green and shining above, paler and smooth beneath; acorns 1.3 to 2 cm. long . . . 9
               9. Lower branches pendulous; inner bark of trunk yellow or brown . *Quercus palustris*
               9. Lower branches not pendulous; inner bark of trunk reddish . . *Quercus coccinea*

2. Leaves not lobed 10
  10. Leaves more than 5 times longer than broad, smooth
    on lower surface . . . . *Quercus phellos* °
  10. Leaves less than 5 times as long as broad, slightly
    pubescent on lower surface . . *Quercus imbricaria*
1. Leaves not bristle-tipped 11
  11. Leaves dentate or with large rounded teeth or wavy margins 12
    12. Leaves velvety beneath with stellate hairs; acorns on
     peduncles exceeding the petioles . . *Quercus bicolor*
    12. Leaves not velvety beneath although sometimes with
     some stellate hairs; acorns sessile or on peduncles not
     exceeding the petioles 13
      13. Leaves with short erect simple hairs beneath;
       leaves broadest above middle; acorns sessile
       . . . . . . *Quercus michauxii*
      13. Leaves glabrous beneath (except for tufts of
       tomentum in the vein axils) or with stellate hairs;
       acorns sessile or on peduncles 2.5 cm. long . . 14
        14. Leaves broadest at middle; teeth of leaf
         with a callus tip; acorns sessile . . .
         . . . . . *Quercus muhlenbergii*
        14. Leaves broadest above the middle; teeth of
         leaf without a callus tip; acorns on pe-
         duncles 2.5 cm. long . *Quercus montana* °
  11. Leaves lobed 15
    15. Terminal lobes not conspicuously longer or broader
     than the others 16
      16. Leaves wedge-shaped, lowest lobes short; acorn
       with an over-closing cup . . *Quercus lyrata*
      16. Leaves not wedge-shaped, lobes rounded, shallow
       or deep; cup of acorn shallow . . *Quercus alba*
    15. Terminal lobes broad, or broadest part of leaf above
     middle 17
      17. The three upper lobes of the leaves broad and
       equal, forming a cross; acorns small . . .
       . . . . . . *Quercus stellata*
      17. Leaves shallowly lobed; acorns 2.5 to 5 cm.
       broad, the cup with a fringe . . . .
       . . . . . . *Quercus macrocarpa*

Quercus falcata Michx. (2:45). Spanish Oak is an occasional tree of dry woodlands in Southern Illinois. Much less common is the var. *pagodaefolia* (Ell.) Ashe. SP. CIT.: typical: *BS 274; M 804.* Var. *pagodaefolia:* south of Howardton, *Bailey and Hankla 634.*

Quercus marilandica Muench. (2:45). Black Jack Oak is one of the dominant trees of Shawneetown Ridge blufftop woods. It occurs with *Ulmus alata, Quercus stellata,* and *Carya texana.* SP. CIT.: *M 665.*

Quercus rubra L. (2:46). Red Oak is an abundant tree of dry woodlands. It occurs throughout Illinois. SP. CIT.: French; *M 249.*

Quercus velutina Lam. (2:46). Black Oak, often difficult to distinguish from *Quercus rubra,* is a common tree in our area. SP. CIT.: *M 367.*

Quercus shumardii Buckl. (2:47). This attractive tree is found locally in moist woods, often associated with *Q. muhlenbergii.* SP. CIT.: *BS 288.*

Quercus coccinea Muench. (2:47). Scarlet Oak is rare in Southern Illinois where it grows in dry upland woods. We have it from Jackson, Union, and Pulaski counties. SP. CIT.: near Grand Tower, *French 2579.*

Quercus palustris Muench. (2:46). Pin Oak grows abundantly in low swampy woods. Its common associates are Sweet Gum and Honey Locust. SP. CIT.: *BS 2129.*

Quercus imbricaria Michx. (2:44). Shingle Oak is common in woods and along streams. At Lusk Creek on exposed bluffs, this species occurs with leaves about 5 cm. long and 1.3 cm. wide. SP. CIT.: *M 125.* The very rare Willow Oak, *Quercus phellos* L. (2:44), is known in Illinois only from Massac County.

Quercus bicolor Willd. (2:42). Although it is a rather uncommon tree, Swamp White Oak has been collected in all of our southern counties. SP. CIT.: *French 2576; BS 293.*

Quercus michauxii Nutt. (2:43). This species is known as Swamp Chestnut Oak, Cow Oak, or Basket Oak. It is a plant of southeastern states and enters Illinois only in the southernmost counties. SP. CIT.: *M 800.*

Quercus muhlenbergii Engelm. (2:43). Yellow Chestnut or Chinquapin Oak occurs in dry or moist woods of Southern Illinois. SP. CIT.: *M 2347; Evers 12350.*

Rock Chestnut Oak, *Quercus montana* Willd. [*Q. prinus* L.] (2:43), is one of our rarer species of *Quercus.* We have it only from Alexander and Union counties. It was on Atwood Ridge in the latter county that the species was first discovered in Illinois in 1948.

Quercus lyrata Walt. (2:42). Overcup Oak is a plant of river banks and low woods. It is not common. SP. CIT.: near Howardton, *BS 287.*

FIG. 35. *A Post Oak-Black Jack Oak community understoried with blueberry* (Vaccinium) *and Poverty Oat grass* (Danthonia spicata). *Photo taken in June.*

FIG. 36. *Majestic, gray, smooth-barked beech trees* (Fagus grandifolia) *inhabit moist valleys. Photo taken in October.*

Quercus alba L. (2:41). Our most common oak is White Oak. It often is the dominant tree in both moist and dry woods. It is variable in the depth of the leaf sinuses. SP. CIT.: *V 1210; Mc 307; M 4034.*

Quercus stellata Wang. (2:41). This species, called Post Oak because of its great durability and hence usefulness for fence posts, is abundant on dry sandstone bluffs (Fig. 35). SP. CIT.: *M 137; BS 276.*

Quercus macrocarpa Michx. (2:42). Bur Oak, which is our largest species of oak, grows in moist woods. SP. CIT.: *BS 265; M 4831.*

Fagus L.

Fagus grandifolia Ehrh. (2:38). The stately Beech is an inhabitant of rich moist woods where it occurs with *Acer saccharum, Acer barbatum,* and *Nyssa sylvatica* (Fig. 36). The trees attain old age but are of little value commercially because the heartwood usually rots away. Specimens from dry bluffs with very hairy lower leaf surfaces occur in some of our counties. These may be segregated as var. *caroliniana* (Loud.) Fern. forma *mollis* Fern. & Rehd. SP. CIT.: *French; M 4079.*

Castanea MILL.

Castanea dentata (Marsh.) Borkh. (2:38). Chestnut, once abundant in some counties (Pulaski, etc.) has become nearly extinct as a result of Chestnut blight. A large native tree, now dead but with a few living sprouts, occurs in a woods about six miles south of Murphysboro. SP. CIT.: *BS 3235.*

37. *Ulmaceae* – Elm Family

1. Leaves distinctly three-veined from the base; fruit a drupe . . . . *Celtis*
1. Leaves with several prominent veins; fruit not a drupe . . 2
    2. Leaves doubly-toothed; fruit flat and winged . . *Ulmus*
    2. Leaves mostly singly-toothed; fruit a nut and wingless . . . . *Planera* *

Celtis L.

Celtis occidentalis L. (2:51). Hackberries are easily recognized by their warty bark. This species is further distinguished by its sharply toothed leaves which are pale yellow-green beneath and by its purple drupes which are about ten millimeters long. The tree occurs in low woodlands, particularly along streams. Branches often are infested by a fungus, the so-called witch's broom. SP. CIT.: *Mc 314; French; M 4052.*

Celtis laevigata Willd. (2:51) [*C. mississippiensis* Bosc]. Southern or Mississippi Hackberry is a rather small forest tree of low woods. The leaves which bear small teeth or are entire are green on both surfaces and over twice as long as broad. The orange to red drupes are five to eight millimeters long. SP. CIT.: *French; M 738.*

Celtis pumila Pursh [*C. tenuifolia* Nutt.] (2:51). Dwarf Hackberry usually occurs on dry, calcareous bluffs. In these situations, it becomes very gnarled. Its leaves are entire or nearly so, green on both surfaces, and half as wide as long, or broader. The glaucous orange to red drupe is about six millimeters long. SP. CIT.: Giant City State Park, *M 751.*

Ulmus L.

Ulmus americana L. (2:49). American Elm is abundant in woods throughout Illinois although in some sections of the state, it is being decimated by both Dutch Elm Disease and phloem necrosis. Young branches and leaves are glabrous to slightly pubescent. Branches do not bear corky thickenings. SP. CIT.: *Mc 627; M 683; M 4090; M 4796.*

Ulmus rubra Muhl. [*U. fulva* Michx.] (2:49). Slippery Elm is abundant in rich woods throughout our area. Leaves and young twigs are densely pubescent. The branches usually do not bear "wings." The inner bark contains a high percentage of mucilage and in the past was used as a demulcent or soothing agent. SP. CIT.: *M 987; M 1800.*

Ulmus alata Michx. (2:51). Winged Elm is another of the characteristic trees of the Shawneetown Ridge blufftops. At least some of the nearly smooth branches bear corky "wings." SP. CIT.: *Mc 626; M 637; M 1803; M 2173.*

September Elm, *Ulmus serotina* Sarg. (2:51), which blooms in the fall, has been attributed to Illinois from Jackson and Richland counties (Sargent, 1933). However, the Jackson County specimen from Grand Tower, collected by Gleason, is sterile and cannot be placed with certainty under *U. serotina.* Personal correspondence with Dr. Gleason (1954) indicates that unless the species is rediscovered in our area, it should be excluded from the Illinois flora.

Planera   J. F. GMEL.

Water Elm or Planer-tree, *Planera aquatica* (Walt.) J. F. Gmel. (2:51), is a rare tree of swamps in extreme Southern Illinois. It is recorded only from Johnson, Alexander, Pulaski, and Massac counties. It is most abundant along the Cache River near West Vienna.

38. *Moraceae* – Mulberry Family

1. Leaves entire; branches often bearing spines .   .   .*Maclura*
1. Leaves toothed or lobed; branches spineless 2
    2. Leaves velvety pubescent on both surfaces .   .*Broussonetia*
    2. Leaves at most sparsely pubescent or scabrous, not velvety
    .   .   .   .   *Morus*

Maclura   NUTT.

Maclura pomifera (Raf.) Schneid. (2:53). Osage Orange is commonly planted as a windbreak in our area and frequently escapes to roadsides and other waste ground. It is supposedly native only from Arkansas to Oklahoma and Texas. The large Hedge-apples which have a rather pleasing odor are dissappointingly inedible. SP. CIT.: *M 558.*

Broussonetia L'HER.

Broussonetia papyrifera (L.) Vent. (2:53). Paper Mulberry is often cultivated in our area and sometimes escapes to bluffs and other sites. We have it from Randolph, Jackson, and Hardin counties. It is a native of Asia. SP. CIT.: near Fern Rocks, Makanda, *French.*

Morus L.

Morus rubra L. (2:53). Red Mulberry is common in and along the edges of woods. Fruits, which mature in early June, are a delight to birds. Leaves are scabrous above and somewhat downy beneath. SP. CIT.: *French 2544.*

Morus alba L. (2:53). White Mulberry is naturalized from Eurasia. It sometimes escapes from cultivation and persists in waste ground or along fences. Leaves are nearly glabrous on both surfaces, except for some axillary tufts of hairs beneath. SP. CIT.: *M 278.*

39. *Cannabinaceae* – Hemp Family

1. Stems twining; leaves 3- to 5-lobed   .   .   . *.Humulus*
1. Stems upright; leaves with 3 to 7 leaflets   .   . *.Cannabis*

Humulus L.

Humulus americanus Nutt. (2:53). Hop grows in rather dry soil along edges of woods. In our area it is not too common. Although it contains several drugs (lupulin, etc.), its rarity does not make it profitable for collecting in Southern Illinois. The United States specimens may not be distinct from the Old World *Humulus lupulus* L. SP. CIT.: Ava Cave, *M 4654.*

Cannabis L.

Cannabis sativa L. (2:54). This species which is native of Eurasia is the Common Hemp or Marijuana. It contains resins and alkaloids and is used as a narcotic and sedative (Tehon, 1951). It is not common in Southern Illinois.

40. *Urticaceae* – Nettle Family

1. Leaves alternate   2
    2. Plants with stinging bristles   .   .   .   . *.Laportea*

2. Plants without stinging bristles   .   .   . .*Parietaria*
1. Leaves opposite   3
  3. Stems with a few to several stinging bristles (if bristles are absent, then each of the three main veins conspicuously branched)   .   .   .   .*Urtica*
  3. Stems without any stinging bristles; leaves 3-veined   .   . 4
    4. Flowers in axillary spikes; stems opaque .   .*Boehmeria*
    4. Flowers in axillary panicles or racemes; stems translucent
     .   .   .   . *Pilea*

Laportea GAUDICH.

Laportea canadensis (L.) Gaudich. (2:55). Wood Nettle is our most stinging member of the nettle family. It is abundant in low wet woods where it occurs with *Hackelia virginiana, Solidago flexuosa,* and both species of *Impatiens.* It flowers in late July and early August. SP. CIT.: *M 413; M 4028.*

Parietaria L.

Parietaria pennsylvanica Muhl. (2:57). Pennsylvania Pellitory is a rather small and inconspicuous plant of moist or dry woods. It is fairly abundant in sandy soil beneath bluffs. Our records show that it flowers from June 6 to the end of July. SP. CIT.: *French; Welch and Fuller 168; Mc 3605; M 338.*

Urtica L.

Urtica gracilis Ait. (2:55). A collection from Grand Tower, July 11, 1871, by G. H. French, represents the only record of this species in Illinois south of Jersey County. Leaves of this species are over twice as long as broad and bear twenty-five or more pairs of sharp teeth. SP. CIT.: Grand Tower, *French.*

Urtica chamaedryoides Pursh (2:55). This curious nettle has been found in Illinois only from Alexander and Jackson counties. In both localities it occurs along the bases of bluffs. Leaves are about as broad as long and bear ten to twenty pairs of rounded teeth. SP. CIT.: near Grand Tower, *French;* base of limestone bluff north of Grand Tower, *BS 584.*

Boehmeria JACQ.

Boehmeria cylindrica (L.) Sw. (2:57). False Nettle is common in moist woods in all southern counties. It begins to flower in July (our first date is July 23) and continues to bloom until the end of August. Petioles exceed one inch in length and leaves are rather thin and nearly glabrous or slightly scabrous above. SP. CIT.: *Mc 959.*

Boehmeria drummondiana Wedd. [*B. cylindrica* (L.) Sw. var. *drummondiana* Wedd.]. This is a rare plant for Illinois, known only from a peat bog in Lake County and from a marsh in Jackson County. Petioles are less than one inch long and the thick rather coriaceous leaves are pubescent and harshly scabrous. SP. CIT.: marshy soil, one mile north of Murphysboro *M 567.*

Pilea LINDLEY

Pilea pumila (L.) Gray (2:57). This species is called Clearweed because of its translucent stems. It grows in wet soil in woods or in waste ground. It flowers from July to October. SP. CIT.: *Bell; M 1719.*

41. *Santalaceae* – Sandalwood Family

Comandra NUTT.

Comandra umbellata (L.) Nutt. (2:59). Our only member of this family is False Toadflax. It grows in sandy soil, usually along edges of bluffs but sometimes along roadsides. We have it only from Randolph and Jackson counties. Along Rock Castle Creek (Randolph County) it grows with *Ranunculus harveyi* (Fig. 37). It flowers from mid-April to mid-May. SP. CIT.: *Mc 744; BS 5351.*

FIG. 37. *A delightful reflecting pool at Rock Castle Creek in Randolph County.*

42. *Loranthaceae* – Mistletoe Family

Phoradendron   NUTT.

Phoradendron flavescens (Pursh) Nutt. (2:59). Mistletoe is a parasitic shrub on several species of deciduous trees. It is abundant in the extreme southern part of Illinois, but is found only sparingly in Jackson and Randolph counties. SP. CIT.: near Murphysboro, *Mohlenbrock.*

43. *Aristolochiaceae* – Birthwort Family

1. Plants with alternate leaves borne along the stems; twining or
    erect   .   .   .   .   *Aristolochia*
1. Plants without aerial stems, but with basal leaves   .   .*Asarum*

Aristolochia   L.

Aristolochia tomentosa Sims (2:63). Dutchman's Pipe, a woody climber, is known only from Devil's Bake Oven near Grand Tower (Fig. 38). The peculiar flowers are produced in late May while the tomentose fruits mature through the middle of August. SP. CIT.: Devil's Bake Oven near Grand Tower, *M 2546.*

Aristolochia serpentaria L. (2:63). Virginia Snakeroot is usually an inconspicuous plant of mesic woods (Fig. 39). The roots have an odor of turpentine. According to Tehon (1951), it contains drugs,

FIG. 38.   *Dutchman's Pipe* (Aristolochia tomentosa) *is rare in Southern Illinois. It is found near Grand Tower in Jackson County. Photo taken June, 1954.*

FIG. 39.   *Virginia Snakeroot* (Aristolochia serpentaria) *grows inconspicuously among fallen leaves.*

volatile oils, and alkaloids and is used as an aromatic bitter stimulant. It flowers during June. SP. CIT.: *M 4081; M 4312; M 4716. A. hastata* Nutt., having long and narrow leaves, has been collected in Pulaski County in a cypress swamp by H. E. Ahles (No. 6959).

Asarum L.

Asarum reflexum Bickn. [*A. canadense* L. var. *reflexum* (Bickn.) B. L. Robins.] (2:61). Wild Ginger is a common plant of mesic woods. Along with Dutchman's-breeches, Squirrel-corn, Dog-tooth Violets, Jack-in-the-Pulpit, and many others, it makes our early spring flora attractive and interesting. SP. CIT.: *M 37; M 4048; V 507.*

44. *Polygonaceae* – Buckwheat Family

1. Plants climbing by tendrils; stipules absent . .*Brunnichia* °
1. Plants erect (if climbing, not by tendrils); leaves with sheathing stipules 2
  2. Outer sepals erect, inner ones without tubercles . *Polygonum*
  2. Outer sepals spreading, inner ones with tubercles or at least winged . . . . *Rumex*

Brunnichia BANKS

The Buckwheat Vine, *Brunnichia cirrhosa* Banks (2:85), is a high climber in the low swampy woods of Johnson, Pope, Alexander, Pulaski, and Massac counties.

Polygonum L.

1. Flowers borne in small axillary clusters 2
  2. Leaves somewhat folded and bearing tiny spinulose teeth . . . . *Polygonum tenue*
  2. Leaves flat, mostly entire, at least not spinulose-toothed . 3
    3. Sheaths with bristles over 5 mm. long 4
      4. Plants annual; achenes well exserted from the calyx . . . . *Polygonum exsertum*
      4. Plants perennial; achenes not well exserted from the calyx . . . . *Polygonum ramosissimum*
    3. Sheaths with bristles less than 5 mm. long . . . 5
      5. Stems erect . . . .*Polygonum erectum*
      5. Stems prostrate . . .*Polygonum aviculare*
1. Flowers borne in terminal and sometimes axillary spikes or racemes 6
  6. Stems trailing, twining, or climbing 7

7. Stems with reflexed prickles . . . .
    . . . . *Polygonum sagittatum*
7. Stems without reflexed prickles 8
    8. Flowers distant, about 1 cm. apart; plants annual
        . . . . . . *Polygonum convolvulus*
    8. Flowers close together (less than 1 cm. apart);
        plants perennial 9
        9. Leaves ovate, usually over 5 cm. long; mature
            fruit 7 to 10 mm. long . . . . .
            . . . . . *Polygonum dumetorum*
        9. Leaves broad, usually less than 5 cm. long;
            mature fruit 10 to 15 mm. long . . .
            . . . . . .*Polygonum scandens*
6. Stems erect or decumbent, not trailing, twining, or climbing 10
    10. Panicles or racemes terminal 11
        11. Inflorescence interrupted; flowers white . .
            . . . . . .*Polygonum virginianum*
        11. Inflorescence crowded; flowers scarlet to pink
            . . . . . .*Polygonum coccineum*
    10. Panicles or racemes terminal and axillary . . . 12
        12. Sheathing stipules entire, or without bristles . 13
            13. Racemes nodding . . . .
                . . . . *Polygonum lapathifolium*
            13. Racemes erect, or only slightly spreading .
                . . . .*Polygonum pennsylvanicum*
        12. Sheathing stipules with bristles 14
            14. Leaves 3 to 6 inches broad, finely pubescent
                throughout . . .*Polygonum orientale*
            14. Leaves mostly under 3 inches broad, not
                finely pubescent throughout 15
                15. Flowers pink or rose 16
                    16. Leaves with a red blotch; ra-
                        cemes over 7 mm. thick . .
                        . . .*Polygonum persicaria*
                    16. Leaves without a red blotch; ra-
                        cemes mostly slender . . .
                        . .*Polygonum hydropiperoides*
                15. Flowers white or greenish . . . 17
                    17. Stipular sheaths with bristles over
                        5 mm. long 18
                        18. Leaves strigose on both sur-
                            faces . . . . .
                            . .*Polygonum setaceum*
                        18. Leaves not as above . . 19
                            19. Achenes over 2 mm.

long; leaves lanceolate
.   .   . *Polygonum*
hydropiperoides

19. Achenes less than 2
mm. long; leaves lin-
ear-lanceolate   .   .
.   .   .*Polygonum*
opelousanum

17. Stipular sheaths with bristles less
than 5 mm. long 20

20. Some of the flowers distant
(7 to 10 mm. apart) .   . 21

21. Perennial; leaves deep
green .   .   .   .
Polygonum punctatum

21. Annual; leaves yellow-
green .   .   .   .
Polygonum punctatum
var.   *leptostachyum*

20. Flowers crowded   .   .
. *Polygonum hydropiper* °

Polygonum tenue Michx. (2:74). Atop dry, exposed sandstone bluffs with *Ruellia humilis, Opuntia rafinesquii, Agave virginica, Stylosanthes biflora*, etc.; not too common. SP. CIT.: BS 863; M 4720.

Polygonum exsertum Small. Sandy beeches along Mississippi River; rare; Jackson and Union counties. SP. CIT.: (see Jones *et al.*, 1955).

Polygonum ramosissimum Michx. (2:74). Moist woods and wet open places; occasional; often growing with *Carex muskingumensis, Hibiscus lasiocarpos*, etc.; Jackson and Union counties. SP. CIT.: BS 2881.

Polygonum erectum L. (2:75). Waste ground; common. SP. CIT.: M 451.

Polygonum aviculare L. (2:76). Waste ground, particularly a weed of lawns; very common; naturalized from Europe. SP. CIT.: M 450.

Polygonum sagittatum L. (2:83). Tear Thumb grows in marshes and along streams; not common; Jackson, Union, and Pope counties. Begins to flower in May. SP. CIT.: M 408; V 1012.

Polygonum convolvulus L. (2:83). Fields and waste places; native of Eurasia; Jackson, Union, and Saline counties. Begins to flower in May. SP. CIT.: M 342.

Polygonum dumetorum L. [*P. cristatum* Engelm. & Gray]. Waste places; Jackson and Union counties. Begins to flower in August. SP. CIT.: *M 508; M 4721; BS 575.*

Polygonum scandens L. (2:85). Edges of woods; common. Flowers in August and September. SP. CIT.: *M 543; Mc 395; Mc 431.*

Polygonum virginianum L. (2:83). [*Tovara virginiana* (L.) Raf.]. Virginia Knotweed grows in moist woods; very common. SP. CIT.: *Mc 436; Mc 989; Mc 1253; M 517; M 4039.*

Polygonum coccineum Muhl. (2:77). Wet ground, particularly around lakes; Jackson and Union counties. SP. CIT.: *M 409; BS 3104.*

Polygonum lapathifolium L. (2:78). Moist ground; very common. SP. CIT.: *BS 540, BS 3227.*

Polygonum pennsylvanicum L. (2:78). Moist ground; very common. SP. CIT.: *Mc 1128; Mc 440; M 400; Bell.*

Polygonum orientale L. (2:79). Prince's Feather; native of India, frequently cultivated, often escapes into moist ground; Randolph and Jackson counties. SP. CIT.: *V 1169; M 4621.*

Polygonum persicaria L. (2:81). Native to Europe; waste places, even sometimes in woodlands; common. SP. CIT.: *Mc 140.*

Polygonum hydropiperoides Michx. (2:83). Wet ground, often growing in water; common. SP. CIT.: *M 275; Mc 140; BS 3105.*

Polygonum setaceum Baldw. (2:83). Swamps and frequently flooded open ground; not common; Jackson, Saline, Union, Alexander, and Pulaski counties. SP. CIT.: north of Oraville, July 5, 1955, *M 5469.*

Polygonum opelousanum Riddell (2:83). Wet ground, marshes, edges of swamps; rare; Jackson, Massac, Pulaski, and Saline counties. SP. CIT.: *V 1005.*

Polygonum punctatum Ell. (2:80). Moist roadside ditches; common. An attractive species. SP. CIT.: *M 394; Bell.*

Polygonum punctatum Ell. var. *leptostachyum* (Meisn.) Small. Occurs with the species in a roadside ditch along Illinois highway 3 near junction of Illinois highways 3 and 144. SP. CIT.: *M 4335.*

One species of *Polygonum* occurs in Southern Illinois which has not been recorded from Jackson County. It is *Polygonum hydropiper* L. (2:80) which grows in roadside ditches in Saline and Union counties.

Rumex L.

1. Leaves with a sour taste, hastate . . . *Rumex acetosella*
1. Leaves not as above 2
  2. Sepals or valves with spiny teeth at their margins . . . 3
    3. Pedicels as long as or shorter than the fruit; flowers very densely clustered . . . . . *Rumex fueginus*
    3. Pedicels longer than fruit but not more than twice as long; leaf blades obtuse . . *Rumex obtusifolius*
  2. Sepals or valves without spiny teeth 4
    4. Sepals or valves bearing tubercles 5
      5. Leaves flat 6
        6. Pedicels about 3 times as long as fruit . . . . *Rumex verticillatus*
        6. Pedicels the length of fruit or shorter . . . . *Rumex triangulivalvis*
      5. Leaves wavy-margined; pedicels 1 to 2 times as long as fruit . . . . . *Rumex crispus*
    4. Sepals or valves not conspicuously tuberculate, or only one valve bearing a tubercle 7
      7. Leaves flat, light green, pointed at each end . . . . *Rumex altissimus*
      7. Leaves wavy-margined, dark green . . . . *Rumex patientia*

Rumex acetosella L. (2:66). This common species of waste ground is known as Field Sorrel or Sour Dock. It is abundant in every county. SP. CIT.: *Mc 17; M 628; V 1227.*

Rumex fueginus Phil. [*R. maritimus* L. var. *fueginus* (Phil.) Dusen] (2:71). This species is rare in Southern Illinois where it grows on sandy shores of the Mississippi River. It is recorded from only eight Illinois counties with Jackson and Randolph being the only two in Southern Illinois. SP. CIT.: Grand Tower slough, August 20, 1954, *M 4814.*

Rumex obtusifolius L. (2:71). Bitter Dock is a common weed of moist ground. It is naturalized from Europe. SP. CIT.: *M 372; M 4683; Bell.*

Rumex verticillatus L. (2:68). Verticillate or Swamp Dock with its pendulous inflorescence is common in wet ground throughout Illinois. SP. CIT.: *M 2437; M 5422; M 5341.*

Rumex triangulivalvis (Danser) Rech. f. [*R. mexicanus* Meisn.] (2:69). This species is uncommon in Southern Illinois, being known

only from Jackson and Pulaski counties where it grows along highways and railroads. SP. CIT.: north of Oraville, May 15, 1955, *M
5243.*

Rumex crispus L. (2:68). Curly Dock is our most common species
of *Rumex*, occurring in all kinds of waste ground. SP. CIT.: *M 1278.*

Rumex altissimus Wood (2:69). Another common weed of roadsides and other waste ground is Narrow-leaved or Pale Dock. SP.
CIT.: *M 228; M 2331.*

Rumex patientia L. (2:68). Patience Dock, naturalized from Europe, is known in our area only from a waste area in Giant City
State Park. SP. CIT.: *M 313.*

45. *Chenopodiaceae* – Goosefoot Family

1. Calyx in fruit winged; low bushy plants of sandy habitat   .   .
     .   .   .   .*Cycloloma*
1. Calyx in fruit not winged   2
    2. Leaves broad or narrowly rhomboid or linear, dentate (often
       coarsely so except in *Chenopodium polyspermum* and *C.
       leptophylla*)   3
        3. Leaves dentate or if entire, not silvery-mealy on both
           surfaces   .   .   .   .*Chenopodium*
        3. Leaves hastately toothed, silvery-mealy on both surfaces
           .   .   .   .   *Atriplex* °
    2. Leaves linear, entire, plant somewhat bushy branched .   .
       .   .   .   .   *Kochia*

Cycloloma   MOQ.

Cycloloma atriplicifolia (Spreng.) Coult. (2:87). Winged Pigweed
is an inhabitant of pure sandy areas. We have it from Jackson and
Union counties. SP. CIT.: Grand Tower slough, August 20, 1954, *M
4631.*

Chenopodium   L.

1. Leaves and stems with resinous glands; leaves sinuate-pinnatifid   2
    2. Flowers in dense glomerules   .   *Chenopodium ambrosioides*
    2. Flowers in loose panicles   .   .   .*Chenopodium botrys*
1. Leaves and stems not glandular   3
    3. Leaves (except sometimes the lowermost) entire .   .   .   4
        4. Leaves green on both sides   5
            5. Flowers few in glomerules along branches   .   .   6

6. Leaves rhombic-lanceolate (the lower sometimes sinuate-dentate) . . . . .
. . . . . *Chenopodium boscianum*

6. Leaves elliptical, cuspidate . . . .
. . . . .*Chenopodium berlandieri* °

5. Flowers in cymes or spikes, numerous . . .
. . . . . *Chenopodium polyspermum*

4. Leaves white-mealy beneath 7

7. Leaves elliptical, cuspidate . . . . .
. . . . . *Chenopodium berlandieri* °

7. Leaves narrow, 5 to 6 times longer than broad .
. . . . . *Chenopodium leptophyllum*

3. Leaves sinuate-dentate or hastate 8

8. Leaves green on both sides or only slightly mealy beneath 9

9. Leaves large toothed (1 to 4 on each side), leaves
nearly as broad as long . . . .
. .*Chenopodium hybridum* var. *gigantospermum*

9. Leaves (at least some of them) hastately toothed
. . . . . .*Chenopodium berlandieri* °

8. Leaves white-mealy beneath 10

10. Leaves broadly wedge-shaped at base, rhombic-
ovate or lanceolate, toothed . *Chenopodium album*

10. Leaves not broadly wedge-shaped at base, 2 to 4
low teeth on each side of leaf . . . . .
. . . . . .*Chenopodium glaucum* °

Chenopodium ambrosioides L. (2:89). This unpleasantly aromatic
plant, called Mexican Tea or Stinkweed, is abundant in waste ground
where it often attains a height of three feet. SP. CIT.: *M 449; BS
182; French.*

Chenopodium botrys L. (2:89). This species, which also has an
unpleasant odor, usually reaches a height of one to one and one-half
feet. It is not as common as the preceding species. SP. CIT.: along
railroad south of Elkville, *J. Garrison.*

Chenopodium boscianum Moq. [*C. standleyanum* Aellen] (2:90).
This occasional species grows in open woods. SP. CIT.: *M 4745.*

Chenopodium berlandieri Moq. has been found in dry soil in Alexander and Pulaski counties. It is often placed under *Chenopodium
album.*

Chenopodium polyspermum L. (2:91). This species has been found
growing along the Gulf, Mobile and Ohio Railroad in Murphysboro.
Our specimen is the only one we have seen for Illinois. SP. CIT.:

along Gulf, Mobile and Ohio Railroad, Murphysboro, June 18, 1955, *Mohlenbrock.*

Chenopodium leptophyllum Nutt. [*C. pratericola* Rydb.] (2:90). This is another species which we have found only along railroads. Its nearest station to the Jackson County site is in Sangamon County. SP. CIT.: along Illinois Central Railroad, Carbondale, August 17, 1954, *M 4826.*

Chenopodium hybridum L. var. *gigantospermum* (Aellen) Rouleau (2:93). This plant is naturalized from Europe and has escaped into dry woods in a few of our counties. In specimens from Fountain Bluff, some leaves measure over four inches long and broad. SP. CIT.: *M 4358.*

Chenopodium album L. (2:89). Lamb's Quarters is very abundant in waste ground throughout Illinois. SP. CIT.: *BS 181.*

Oak-leaved Goosefoot, *Chenopodium glaucum* L. (2:91), an adventive from Europe, is found in waste ground in Randolph and Union counties. It undoubtedly is more common than our records show.

Atriplex L.

Two species of *Atriplex* occur in Southern Illinois but neither has been found in Jackson County. *Atriplex hastata* L., with triangular hastate leaves, is recorded from Gallatin and Pope counties, while *Atriplex patula* L. (2:97) is known from Cottonwood Pond in Saline County. The latter species has linear or lanceolate, essentially entire, leaves.

Kochia ROTH

Kochia scoparia (L.) Roth (2:95). This is a common species in railroad switching yards although in our area we have it only from Jackson County. SP. CIT.: *M 4746.* Variety *culta* Farw., the Burning-bush, is a common cultivar.

46. Amaranthaceae – Amaranth Family

1. Leaves alternate; plants green 2
    2. Calyx present in both staminate and pistillate flowers . .
    . . . . *Amaranthus*
    2. Calyx present in only staminate flowers . . .*Acnida*

1. Leaves opposite; plants white woolly or green and smooth . . 3
    3. Plants woolly; flowers in dense spikes on long peduncles .
        . . . . *Froelichia*
    3. Plants smooth; inflorescence paniculate . . . *Iresine* *

Amaranthus L.

1. Flowers terminal *and* axillary in elongated spikes or panicles . 2
    2. Stems with paired axillary spines at base of leaf . . .
        . . . . *Amaranthus spinosus*
    2. Stems without spines 3
        3. Spikes 2 to 3 cm. long, 4 to 6 mm. thick . .
            . . . . *Amaranthus hybridus*
        3. Spikes 3 to 5 cm. long, 7 to 15 mm. thick . . .
            . . . . . . *Amaranthus retroflexus*
1. Flowers in small axillary clusters . . *Amaranthus albus*

Amaranthus spinosus L. (2:104). Spiny Pigweed is common in waste ground. It is native in tropical America. SP. CIT.: *M 332*.

Amaranthus hybridus L. (2:105). Green Amaranth is found in waste ground and fields throughout Illinois. SP. CIT.: *M 801*.

Amaranthus retroflexus L. (2:104). Like other species of *Amaranthus* in our area, Rough Pigweed is very common. SP. CIT.: *M 465; BS 251*.

Amaranthus albus L. (2:102). Tumbleweed is the least abundant of our Amaranths, being recorded in our study area only from Jackson County. SP. CIT.: *M 4510; M 321*.

Acnida L.

Acnida subnuda (S. Wats.) Standl. [*A. altissima* Riddell var. *subnuda* (S. Wats.) Fern. (2:107). This species of Water Hemp grows in moist soil in roadside ditches in Jackson, Union, Alexander, and Massac counties. The staminate flowers have acute sepals. Leaves are broadest above the middle. SP. CIT.: near Gorham, *Mohlenbrock*.

Acnida altissima Riddell (2:107). Tall Water Hemp occurs along streams only locally. We have it from Jackson, Randolph, Union, Pulaski, and Massac counties. The staminate flowers have acute sepals. Leaves are broadest below the middle. SP. CIT.: *BS 3228; BS 3206*.

A third species, *Acnida tamariscina* (Nutt.) Wood (2:106), with the staminate flowers having long subulate sepals, is known from waste places in Randolph, Union, Pulaski, and Massac counties.

Froelichia MOENCH

Froelichia gracilis (Hook.) Moq. (2:109). One of the most common species of railroad waifs is Woolly Froelichia. It is adventive from the western United States. SP. CIT.: *Hardy 115; BS 557; M 4358.*

Iresine P.BR.

A specimen collected by Fricke from Pulaski County is *Iresine rhizomatosa* Standl. (2:109). We have not seen this species in our field work.

47. *Phytolaccaceae* – Pokeweed Family

Phytolacca L.

Phytolacca americana L. (2:113). Pokeweed is abundant in all kinds of habitats in our area. The purple berries contain a bitter alkaloid which in the past was used as a purgative. SP. CIT.: *M 311.*

48. *Nyctaginaceae* – Four-o'clock Family

Mirabilis L.

Mirabilis nyctaginea (Michx.) MacM. (2:110). Wild Four-o'clock, or Umbrella-wort, is abundant along railroads throughout our area. It flowers from mid-April through May. SP. CIT.: *Mc 3501; M 2190; BS 403.*

49. *Illecebraceae* – Whitlow-wort Family

Paronychia ADANS.

Paronychia canadensis (L.) Wood (2:120). Forked Chickweed is found sparingly in dry soil of woods or semi-waste areas in Southern Illinois. The stems are completely glabrous. SP. CIT.: Ava Cave, August 26, 1954, *M 4652.*

Paronychia fastigiata (Raf.) Fern. (2:120). This species is somewhat more common than the preceding where it is found in similar situations. The stems are puberulent. SP. CIT.: *Mc 475; M 783.*

50. *Aizoaceae* – Carpetweed Family

Mollugo L.

Mollugo verticillata L. (2:113). Carpetweed is abundant in moist waste ground and along streams. It is naturalized from the South. SP. CIT.: *M 333; M 4395; Mc 219.*

51. *Portulacaceae* – Purslane Family

1. Plants annual; leaves obovate-spatulate or, if somewhat terete, then cauline . . . . *.Portulaca*
1. Plants perennial; leaves linear-lanceolate or terete and basal . 2
   2. Leaves terete and basal; flowers rose or bright pink . .
   . . . . *Talinum* °
   2. Leaves linear-lanceolate or sometimes broader, cauline; flowers white or pale pink . . . . *.Claytonia*

Portulaca L.

Portulaca oleracea L. (2:115). Common Garden Purslane is a weed in lawns and cultivated areas. The leaves are obovate-spatulate, the flowers yellow, and the stamens six to ten. SP. CIT.: *M 441.*

Portulaca grandiflora Hook. (2:115). Rose Moss or Moss Pink is a garden plant which occasionally escapes but seldom persists. Flowers are showy, leaves are subterete and cauline, and the stamens number about forty.

Talinum ADANS.

Two species of the genus *Talinum* occur in Southern Illinois. Both are rare and neither is known from Jackson County. On dry sandstone bluffs in Union, Johnson, and Pope counties is found tiny *Talinum parviflorum* Nutt. (2:117). Flowers appear in June or July and bear four to eight stamens. The petals are five to seven millimeters long. The even rarer *Talinum calycinum* Engelm. (2:117) has been collected only at Castle Rock near Leanderville in Randolph County. It was first discovered September 25, 1954 (*M 4911*), growing along the edge of a sandstone bluff with *Cyperus filiculmis, Polygonum tenue,* and *Isanthus brachiatus.* Petals are over ten millimeters long and there are twenty to many stamens. Flowers open about two o'clock in the afternoon. The Randolph County station marks the eastern limit of this species.

Claytonia L.

Claytonia virginica L. (2:116). Spring Beauty is perhaps our most common spring woodland wild flower. It blooms as early as February 26 and continues well into May. The flower is variable. Robust specimens with leaves over one and one-half centimeters broad may be segregated as forma *robusta* (Somes) Palmer & Steyerm. SP. CIT.: typical: *M 1891; M 2186; M 4792; Mc 630; Mc 686.* Forma *robusta: M 5257.*

52. *Caryophyllaceae* – Pink Family

The Pink Family is represented in Southern Illinois by twenty species distributed among ten genera of which fourteen species are adventive. Nine genera and fifteen species occur in Jackson County.

1. Sepals separate or free to base 2
  2. Flowers in cymes; petals sometimes notched . . . 3
    3. Petals deeply 2-cleft 4
      4. Styles 3; capsules exceeding calyx . .*Cerastium*
      4. Styles 5; capsules about equalling calyx . *Stellaria*
    3. Petals entire or absent 5
      5. Leaves (in our species) 3 to 5 mm. broad . .
       . . . . *Arenaria*
      5. Leaves minute, 1 mm. broad or less . .*Sagina*
  2. Flowers in peduncled umbels; petals not deeply notched .
   . . . . *Holosteum* °
1. Sepals united at least part way to form a calyx tube . . . 6
  6. Flowers bright pink to rose, the petals about 1 cm. long .
   . . . . *Dianthus*
  6. Flowers white, lavender, or bright red, if pinkish, then with petals over 1 cm. long 7
    7. Calyx with 5 ribs; flowers whitish-pink . . *Saponaria*
    7. Calyx with 10 ribs 8
      8. Calyx teeth longer than calyx tube; flowers lavender . . . .*Agrostemma*
      8. Calyx teeth shorter than calyx tube; flowers white or bright red 9
        9. Flowers white; styles 5; flowers usually opening in the evening . . . .*Lychnis*
        9. Flowers white or bright red; styles 3 to 4; flowers blooming during the day . .*Silene*

Cerastium L.

1. Petals over 1 cm. long   .   . *Cerastium arvense* var. *villosum*
1. Petals less than 1 cm. long   2
    2. Petals equalling or shorter than sepals   3
        3. Sepals long-bearded at tip . *Cerastium brachypetalum* °
        3. Sepals pubescent but not long-bearded . . .   4
            4. Perennial; pedicels usually somewhat longer than
                sepals   .   .   .   .   .*Cerastium vulgatum*
            4. Annual; pedicels shorter than or equalling sepals   .
                .   .   .   .   .   . *Cerastium viscosum*
    2. Petals longer than sepals   5
        5. Pedicels over 15 mm. long .   . .*Cerastium nutans*
        5. Pedicels less than 15 mm. long . . . . . .
            .   .   .   .   .   . *Cerastium brachypodum*

Cerastium vulgatum L. (2:129). Common Mouse-ear Chickweed
is abundant in fields and waste ground throughout our area. SP. CIT.:
BS 1410; Mc 3506.

Cerastium nutans Raf. (2:129). Another common plant of waste
ground is Nodding Mouse-ear Chickweed. SP. CIT.: *M 59; M 2130;
M 1989; BS 1337.*

Cerastium brachypodum (Engelm.) Robins. (2:129). This adventive from western United States is known from Jackson, Williamson,
Union, and Alexander counties. SP. CIT.: *Mc 13; M 746.*

Cerastium viscosum L. (2:129). While not as abundant as most of
our naturalized species of *Cerastium*, this one is recorded from most
of our counties. SP. CIT.: *M 454; BS 1340.*

Two additional species of *Cerastium* occur in Southern Illinois.
The attractive and native *C. arvense* L. var. *villosum* (Muhl.) Hol-
lick & Britt. [*C. velutinum* Raf.] (2:129) has been found on lime-
stone bluffs in Gallatin, Pope, and Saline counties. A specimen of
C. brachypetalum Pers. was collected along a road north of Ullin on
May 7, 1951, by H. E. Ahles (No. 7988). It is a native of Europe.

Stellaria L.

Stellaria media (L.) Vill. (2:124). Common Chickweed is abundant in lawns and waste ground. It is of European origin and is
naturalized here. Leaves of this species are never more than twice
as long as broad. SP. CIT.: *Mc 612; BS 1338; M 731.*

Stellaria longifolia Muhl. (2:127). Long-leaved Chickweed is known in Southern Illinois from only Jackson County where it grows west of the Crab Orchard Lake dam. It flowers from mid-May to mid-June. The sepals are less than four and one-half millimeters long. SP. CIT.: *BS 449.*

Stellaria graminea L. (2:127). Grass-leaved Chickweed, native of Europe, grows on the campus of Southern Illinois University. The sepals are over four and one-half millimeters long. SP. CIT.: June 17, 1955, *VM 5418.*

Arenaria L.

Thyme-leaved Sandwort, *Arenaria serpyllifolia* L. (2:132), an adventive from Europe, has been collected in Massac and Jackson counties.

Sagina L.

Sagina decumbens (Ell.) Torr. & Gray (2:132). The tiny Pearlwort grows in moist waste ground locally throughout our area. SP. CIT.: *BS 1339; M 692.*

Holosteum L.

Jagged Chickweed, *Holosteum umbellatum* L. (2:129), has been collected along a roadside near West Vienna, Johnson County, by H. E. Ahles (No. 3556) and G. S. Winterringer (No. 9005).

Dianthus L.

Dianthus armeria L. (2:144). Deptford Pink is an attractive plant which has escaped to stream banks and woods over most of Southern Illinois. SP. CIT.: *BS 718.*

Saponaria L.

Saponaria officinalis L. (2:143). Bouncing Bet is an abundant escape in waste ground. Color of the petals is variable. SP. CIT.: *M 371.*

Saponaria vaccaria L. (2:143). This plant is known as Cow-herb. It is known in Southern Illinois only from near Carbondale, Jackson County. SP. CIT.: May 28, 1940, *Mc 82.*

Agrostemma L.

Agrostemma githago L. (2:136). Corn Cockle is a common species in wheat fields and other areas of open ground. The seeds are poisonous. SP. CIT.: *V 874; M 177; M 2519.*

Lychnis L.

Lychnis alba Mill. (2:136). Evening Campion has been collected in Southern Illinois only along the road at the base of Cedar Creek bluff about nine miles south of Murphysboro. SP. CIT.: September 5, 1954, *M 4825.*

Silene L.

Silene antirrhina L. (2:139). This common plant, known as Sleepy Catchfly, is found along roads and railroads. It flowers from late April through May. The white petals are minute and fringeless. SP. CIT.: *BS 1012; M 2341; Stewart 4.*

Silene stellata (L.) Ait. f. (2:140). This attractive species is found in woods throughout Southern Illinois. It is one of the few showy woodland wild flowers blooming in mid-summer. The white fringed petals are about 10 mm. long. The leaves are whorled. SP. CIT.: *Mc 813; M 280; M 4080; M 4391; Voigt.*

One of the most beautiful and striking of wild flowers is Firepink, *Silene virginica* L. (2:142). The bright red petals are produced by late April. We have this species from moist woods in Johnson and Pope counties.

53. *Magnoliaceae* – Magnolia Family

1. Leaves with four broad lobes   .   .   .   .   .*Liriodendron*
1. Leaves entire, unlobed   .   .   .   .*Magnolia*

Liriodendron L.

Liriodendron tulipifera L. (2:154). Tulip Tree is a stately tree of mesic woods. It is excellent for lumber. Some in our area attain a height of one hundred feet. Flowers are produced during May. SP. CIT.: *BS 390; BS 2191.*

Magnolia L.

Magnolia acuminata L. (2:153). Cucumber Tree is a southern species which reaches north in Illinois only to Jackson County where it occurs in moist woods near Grassy Knob. It flowers in May. SP. CIT.: *BS 285.*

54. *Annonaceae* – Custard-apple Family

Asimina ADANS.

Asimina triloba (L.) Dunal. (2:154). Pawpaw is the only member of its family to occur as far north as this. The interesting maroon flowers are produced in April while the banana-like fruits ripen in September. They are eaten by raccoons, opossums, and some people. SP. CIT.: *BS 341; BS 2158.*

55. *Ranunculaceae* – Buttercup Family

The Buttercup Family is represented here chiefly by native wood-land species with attractive flowers. Thirteen genera and thirty-two species occur in Southern Illinois with twelve genera and thirty species reported from Jackson County. All species except *Cimicifuga cordata* and some species of *Clematis* bloom from March to the end of June.

1. Leaves grass-like; flowers yellow, minute, on a scape; plants up to 15 cm. tall . . . . *Myosurus*
1. Leaves not grass-like 2
 2. One or more of the petals spurred 3
  3. All five petals spurred; flowers yellow inside, red outside . . . . *Aquilegia*
  3. Only one petal spurred; flowers without the above color combinations . . . . . . *Delphinium*
 2. None of the petals spurred 4
  4. Stems climbing; leaves opposite . . *Clematis*
  4. Stems not climbing 5
   5. Flowers pale to bright yellow. . *Ranunculus*
   5. Flowers greenish, whitish, red, blue, or purple . 6
    6. Leaves simple 7
     7. Leaves broadly 3-lobed or cleft, coriaceous . . . . *Hepatica*
     7. Leaves palmately lobed or cleft; rhizome yellow . . . . . *Hydrastis*
    6. Leaves compound 8
     8. Leaves of stem opposite or whorled, or leaves arising from the ground . . . 9
      9. Inflorescence of panicles or paniculate racemes 10

> 10. Flowers perfect; leaflets broad and maple-leaf shaped . *Cimicifuga* *
> 10. Flowers unisexual (plants dioecious or polygamous); leaflets at most 4 cm. long, ternately lobed . . . . . *Thalictrum*
> 9. Inflorescence an umbel of 3 to 4 flowers or flowers solitary 11
>> 11. Flowers 3 to 4 in an umbel, short-peduncled . . .*Anemonella*
>> 11. Flowers solitary, long-penduncled . . . . . .*Anemone*
> 8. Leaves alternate on stem 12
>> 12. Leaflets pinnately veined, serrate; flowers borne in a raceme . *Actaea*
>> 12. Leaflets palmate-reticulate veined, apiculate, not serrate; flowers usually solitary . . . . .*Isopyrum*

Myosurus L.

Myosurus minimus L. (2:179). Mousetail is found in moist ground in fields or woods or sometimes (at Giant City State Park) on wet sandstone bluffs. It is often overlooked perhaps because of its small size. We have collected it in flower as early as March 20. SP. CIT.: *M 34; Voigt.*

Aquilegia L.

Aquilegia canadensis L. (2:165). Wild Columbine is one of the most handsome of spring wild flowers. It grows in crevices of moist bluffs and in wooded ravines throughout Southern Illinois. It flowers from mid-May to mid-June. SP. CIT.: *BS 17; M 421.*

Delphinium L.

Delphinium tricorne Michx. (2:167). Wild Larkspur is a common species of moist woods where it occurs with *Asarum reflexum, Hybanthus concolor,* and other spring wild flowers. It flowers during April. The flower contains three pistils. Occasional specimens are white-flowered. These may be segregated as forma *albiflora* Millsp. SP. CIT.: *BS 6; M 38; V 504.*

Delphinium ajacis L. (2:167). The cultivated Larkspur often escapes to roadsides, and may become established in other places as well. Flowers are unicarpellate and variable in color. SP. CIT.: *BS 1709.*

Clematis L.

1. Sepals thick and leathery 2
    2. Sepals up to 25 mm. long . . . .*Clematis pitcheri*
    2. Sepals over 30 mm. long . . . .*Clematis crispa* °
1. Sepals thin, not leathery 3
    3. Leaflets usually 3; sepals with long white hairs . . .
        . . . .*Clematis virginiana*
    3. Leaflets usually 5; sepals tomentose . . . . . .
        . . . .*Clematis dioscoreifolia*

Clematis pitcheri Torr. & Gray (2:187). Leather Flower is not common in Southern Illinois where it grows in moist woods or along railroads. It begins to flower in mid-July. SP. CIT.: *BS 136.*

Clematis virginiana L. (2:185). Virgin's Bower occurs in moist soil locally throughout our area. It blooms in August and September. SP. CIT.: *Mc 449; M 4707.*

Clematis dioscoreifolia Levl. & Vaniot (2:185). This Japanese species is commonly cultivated in our area where it has often escaped into woods. SP. CIT.: near Ava, *Mohlenbrock.*

In Alexander and Pulaski counties in low swampy woods is found *Clematis crispa* L. (2:187). This is a species of southern United States which just enters our range.

Ranunculus L.

1. Aquatic plants 2
    2. Petals longer than sepals; leaves finely dissected . . .
        . . . .*Ranunculus flabellaris*
    2. Petals about equalling the sepals or shorter; leaves not lobed or dissected 3
        3. Petals pale yellow, 1 to 2 mm. long . . . . .
            . . . .*Rununculus pusillus*
        3. Petals bright yellow, 3 to 6 mm. long . . . .
            . . . .*Rununculus laxicaulis*
1. Land plants (may be in mud, but most of life cycle is on land) 4
    4. Stems or petioles pubescent (*R. abortivus* might be looked for here) 5
        5. Beak of achene about one-half the length of body of achene or longer 6
            6. Petals longer than sepals 7

7. Leaves prominently 3-lobed, the lobes broad . 8
    8. Terminal lobe of leaf stalked; plants usually smooth . .*Ranunculus septentrionalis*
    8. Terminal lobe of leaf not stalked; plants hispid . . . .*Ranunculus hipidus*
7. Leaves 3-lobed, each lobe again divided (i.e., the lobes not broad) . *Ranunculus fascicularis*
  6. Petals and sepals about equal in length, petals pale yellow . . . . .*Ranunculus recurvatus*
5. Beak of achene less than one-half the length of body of achene 9
  9. Petals longer than sepals 10
    10. Lower leaves undivided . . . . .
      . . . *Ranunculus harveyi* forma *pilosus*
    10. Lower leaves divided . *Ranunculus sardous*
  9. Petals and sepals equal or petals less than length of sepals 11
    11. Achenes hooked or spiny (under 10× magnification) . . . . *Ranunculus parviflorus*
    11. Achenes smooth . . *Ranunculus micranthus*
4. Stems or petioles glabrous (lower stems in some specimens of *R. abortivus* may be appressed pubescent) . . . 12
12. Basal leaves cordate or reniform 13
    13. Petals longer than sepals; flower 1 to 2 cm. broad . . . .*Ranunculus harveyi*
    13. Petals shorter or no longer than sepals; flowers less than 1 cm. broad . .*Ranunculus abortivus*
12. Basal leaves denticulate or crenate (or if entire, not cordate or reniform) 14
    14. Sepals and petals nearly equal or petals not exceeding sepals; leaves mostly entire . . . 15
      15. Petals bright yellow, 3 to 6 mm. long . .
      . . . . .*Ranunculus laxicaulis*
      15. Petals pale yellow, 1 to 2 mm. long . .
      . . . . . *Ranunculus pusillus*
    14. Petals longer than sepals; leaves deeply lobed
    . . . . . .*Ranunculus sceleratus*

Ranunculus flabellaris Raf. [*R. delphinifolius* Torr.] (2:174). Dissected-leaved Buttercup grows in swamps and ponds in only a few southern counties. It is abundant in LaRue Swamp of Union County. SP. CIT.: *French.*

Ranunculus pusillus Poir. (2:172). Small Buttercup occurs in shallow water or very wet roadside ditches. It is particularly common along Illinois highway 3 east of Gorham where it grows with *Ranunculus laxicaulis, Phyllanthus caroliniensis, Ammannia coccinea, Cyperus acuminatus*, and others. It is known from Jackson, Saline, and Union counties. It flowers from mid-May into June. SP. CIT.: *BS 57; V 1378; M 5295*.

Ranunculus laxicaulis (Torr. & Gray) Darby (2:172). This species occurs in low woods and wet roadside ditches in Saline, Jackson, and Union counties. SP. CIT.: *BS 411; M 5351*.

Ranunculus septentrionalis Poir. (2:177). Marsh Buttercup is common in moist ground and rich woods all over Illinois. It flowers as early as March 20. SP. CIT.: *BS 24; M 791*.

Ranunculus hispidus Michx. (2:177). Bristly Buttercup is common in dry woods or along roads. Our collection dates show it to bloom from April 12 to May 9. SP. CIT.: *V 852; M 627; M 1952*. Specimens in which leaves are deeply triple cleft and the stem only sparsely pubescent occur with the species. This is var. *falsus* Fern.

Ranunculus fascicularis Muhl. (2:177). Early Buttercup is common in open grassy places atop sandstone bluffs where it often forms a beautiful meadowy appearance with *Nothoscordum bivalve*. Specimens with leaf segments nearly entire and broadest well above the middle are var. *apricus* (Greene) Fern. SP. CIT.: typical: Giant City State Park, March 13, 1953, *M 635*. Var. *apricus*: Peter's Cave, April 5, 1954, *M 1897*.

Ranunculus recurvatus Poir. (2:175). This species grows in mesic woods locally but probably is in every southern county. It does not begin to flower until May. SP. CIT.: *V 667; V 1210; M 2426*.

Ranunculus sardous Crantz (2:179). This European species is spreading rapidly in moist fields in Southern Illinois. We have it from Jackson, Union, Pope, and Pulaski counties. SP. CIT.: *Bailey and Hankla 635; BS 3200; V 1455*.

Ranunculus parviflorus L. (2:179). This rare species has been collected at Saltpeter Cave (Jackson County). SP. CIT.: May 13, 1954, *M 2437*.

Ranunculus abortivus L. (2:173). Small-flowered Buttercup is very common in waste ground and in woods throughout Illinois. It flowers during April and early May. SP. CIT.: *Mc 641; Mc 678; M 1938; BS 14*.

Ranunculus sceleratus L. (2:175). This species grows in wet, marshy places about one mile northwest of Oraville with *Acorus calamus* and *Sagittaria latifolia*. Otherwise, it is not known south of St. Clair County in Illinois. SP. CIT.: near Oraville, May 7, 1955, *M 5267*.

A recently discovered plant for Illinois is Harvey's Buttercup, *Ranunculus harveyi* (Gray) Britt. (2:174) (Fig. 40). It was first found at Piney Creek (Randolph County) (Fig. 41), on rocky slopes, April 24, 1954 (*M 4911*). In Jackson County it is abundant on Jones's farm near Ava. In spring of 1955, forma *pilosus* Palmer and Steyerm. was found along Rock Castle Creek (Randolph County) about three miles north of the Piney Creek Station (*M 5000*).

Hepatica MILL.

Hepatica acutiloba DC. (2:183). Liverleaf is one of the first spring wild flowers to bloom, flowering as early as the first few days in March. It grows in rich, shaded woods. We have it from Jackson, Hardin, Pope, and Gallatin counties. SP. CIT.: *BS 316; Mohlenbrock*.

Hydrastis L.

Hydrastis canadensis L. (2:157). The attractive Goldenseal produces its curious flowers as the leaf unfolds during mid-April and develops its scarlet fruits during the last part of July (Fig. 42). The golden-yellow rhizome was previously collected for its demulsifying properties. As a result of this extensive collecting Goldenseal is now only occasionally seen. SP. CIT.: *V 1058*.

FIG. 40. *Harvey's Buttercup* (Ranunculus harveyi) *occurs frequently in Randolph County. This buttercup, only recently discovered in Illinois, has Ozark affinities.*

FIG. 41. *A sluice-like run of water in sandstone at Piney Creek in Randolph County.*

FIG. 42. *Goldenseal* (Hydrastis canadensis) *in flower.*

Cimicifuga L.

A rare native species is Black Cohosh, *Cimicifuga cordifolia* Pursh (2:157). It is known only from deep wooded ravines in Pope County (Jackson Hollow and Belle Smith Springs). It flowers during August.

Thalictrum L.

1. Each lobe of the 3-lobed leaflets with 1 to 3 teeth   .   .   .
    .   .   .   .*Thalictrum dioicum*
1. Leaflets entire or with 2 to 3 entire lobes 2
      2. Leaflets minutely glandular beneath   .   .   .   .   .
          .   .   .   .*Thalictrum   revolutum*
      2. Leaflets without glandular hairs beneath although sometimes
          pubescent   3
            3. Leaflets puberulent beneath . *Thalictrum dasycarpum*
            3. Leaflets glabrous beneath   . *Thalictrum hypoglaucum*

Thalictrum dioicum L. (2:160). Early Meadow-Rue is an occasional species of rich woods in a few southern counties. SP. CIT.: *Mc 58*.

Thalictrum revolutum DC. (2:160). Waxy Meadow-Rue is an occasional species of rich woods. SP. CIT.: (see Jones *et al.*, 1955).

Thalictrum dasycarpum Fisch. & Lall. (2:160). This is the Purplish Meadow-Rue and it has been collected only in Jackson and Johnson counties where it grows in rich woods. SP. CIT.: Giant City State Park, June 11, 1953, *M 312;* Midland Hills, May 16, 1951, *V 493.*

Thalictrum hypoglaucum Rydb. This species attains a height of nearly four feet along a stream north of Oraville. We have it from Jackson, Massac, and Saline counties. SP. CIT.: *M 232; M 5244; Mc 232.*

Anemonella SPACH

Anemonella thalictroides (L.) Spach (2:185). Rue-Anemone is a local species of dry, open woods. It is sometimes confused with *Isopyrum biternatum,* a plant of rich woods, but the latter species lacks whorled leaves beneath the inflorescence. SP. CIT.: Giant City State Park, May 15, 1953, *M 760;* Little Grand Canyon, May 17, 1952, *BS 1892.*

Anemone L.

Anemone virginiana L. (2:181). Common Anemone is found in moist or dry woods throughout the southern counties. It flowers from late May to early September. Leaves of the involucre are petioled. SP. CIT.: *Bell; French.*

Anemone canadensis L. (2:182). This species, Meadow Anemone, has its only Southern Illinois station in a woods at Riverside Park in Murphysboro. Leaves of the involucre are sessile. SP. CIT.: Riverside Park, April 18, 1951, *M 536.*

Actaea L.

Actaea alba (L.) Mill. (2:159). Doll's-eyes is one of our most interesting spring, moist-woodland, wild flowers. Bright white berries with "eye-markings" mature during August. This species is probably in every southern county. SP. CIT.: *M 99; M 2418.*

Isopyrum L.

Isopyrum biternatum (Raf.) Torr. & Gray. (2:168). False Rue-Anemone is common in mesophytic woods. It flowers as early as February 28 in favorable situations. SP. CIT.: *Mc 622; M 636.*

56. *Cabombaceae* – Watershield Family

1. Leaves palmately dissected . . . . . *Cabomba* °
1. Leaves entire or nearly so . . . . . *Brasenia* °

Members of this water-plant family are rare in Southern Illinois and none is known from Jackson County. *Cabomba caroliniana* Gray (2:147) is abundant in Wolf Lake (Union County). This is also our only station in the twelve-county area for Watershield, *Brasenia schreberi* Gmel. (2:148).

57. *Nelumbonaceae* – Lotus Family

Nelumbo ADANS.

Nelumbo lutea (Willd.) Pers. (2:148). Giant Lotus Lily or American Lotus with its round leaves and large showy pale yellow flowers occurs in most large lakes in our area. It flowers from late June to mid-August. SP. CIT.: BS 2591; BS 2719.

58. *Nymphaeaceae* – Waterlily Family

1. Flowers yellow, 3 to 4 cm. broad . . . . *Nuphar*
1. Flowers white, 7 to 12 cm. broad . . . *Nymphaea* °

Nuphar SMITH

Nuphar advena (Ait.) Ait. f. (2:150). Yellow Pond Lily is locally abundant in shallow water in a few of our counties. SP. CIT.: Campbell's Lake, near Elkville, June 10, 1941, *Mc 830*.

Nymphaea L.

The beautiful Fragrant Water Lily, *Nymphaea odorata* Ait. (2:151), has been collected in Union and Johnson counties. It flowers from late July through August.

59. *Ceratophyllaceae* – Hornwort Family

Ceratophyllum L.

Ceratophyllum demersum L. (2:147). This species is commonly called Coontail. It has been found in quiet water in most of our counties. SP. CIT.: BS 2677.

60. *Berberidaceae* – Barberry Family

1. Flowers solitary on each plant 2
    2. Flowers borne in axils of a pair of leaves . .*Podophyllum*
    2. Flowers on a naked scape; leaves all basal . .*Jeffersonia*
1. Flowers several on each plant . . . . .*Caulophyllum*

Podophyllum L.

Podophyllum peltatum L. (2:189). Mayapple, or Mandrake, as it is sometimes called, is abundant in moist woods. Rhizomes, leaves, and seeds are poisonous, although the flesh of the fruit may be eaten. SP. CIT.: *MV 2076; MV 2793; M 2420.*

Jeffersonia BART.

Jeffersonia diphylla (L.) Pers. (2:189). Twinleaf is very rare in Southern Illinois where it is reported from rich woods in Jackson, Pope, Pulaski, and Massac counties. Flowers are borne in late April or early May. SP. CIT.: (see Jones *et al.*, 1955).

Caulophyllum MICHX.

Caulophyllum thalictroides (L.) Michx. (2:189). Blue Cohosh is a plant of moist woods. Inconspicuous greenish-yellow flowers appear in late April while the large five to eight millimeters blue fruits mature in August. SP. CIT.: Giant City State Park, *V 669.*

61. *Menispermaceae* – Moonvine Family

1. Leaves peltate along the margin . . . .*Menispermum*
1. Leaves not peltate 2
    2. Leaves entire or 3-lobed; petals 6; fruit red . .*Cocculus*
    2. Leaves 3- to 7-lobed; petals none; fruit black . . .
        . . . . *Calycocarpum*

Menispermum L.

Menispermum canadense L. (2:191). Moonseed is a relatively common plant in woods and thickets throughout Southern Illinois. It is sometimes confused with *Passiflora lutea* but differs from it in lacking tendrils. SP. CIT.: Fountain Bluff, *BS 2318.*

Cocculus DC.

Cocculus carolinus (L.) DC. (2:192). Carolina Snailseed produces brilliant scarlet waxy-looking drupes in October. The seeds are curved like a snail. It is not common in this area, but it is often abundant where it does occur. SP. CIT.: *BS 252; BS 2882; M 322; French.*

Calycocarpum NUTT.

Calycocarpum lyoni (Pursh) Gray (2:192). Cupseed is rare in Illinois where it occurs in rich woods generally along rivers in only a few counties. We have it from near Grand Tower along the Mississippi River. SP. CIT.: *Mohlenbrock.*

62. *Lauraceae* – Laurel Family

1. Leaves unlobed, pinnately veined . . . . . *.Lindera*
1. Leaves, or some of them, usually lobed, palmately veined . .
   . . . . *Sassafras*

Lindera THUNB.

Lindera benzoin (L.) Blume (2:193). Spicebush (Fig. 43) is one of our most attractive shrubs of moist woods. It produces yellow flowers in March and bright red fruits in autumn. SP. CIT.: *M 632; BS 652.*

Sassafras NEES & EBERM.

Sassafras albidum (Nutt.) Nees (2:193). Sassafras is abundant along edges of woods, in waste ground, or along fences. Specimens with hairy leaves are var. *molle* (Raf.) Fern. These latter are some-

FIG. 43. *An early blooming shrub is Spicebush* (Lindera benzoin). *It is named for its pleasing aromatic odor.*

times called Red Sassafras. Roots are often collected and used in the making of sassafras tea. SP. CIT.: typical; *M 32; Mc 309; M 4083; Evers 12347.* Var. *molle: M 2161.*

63. *Papaveraceae* – Poppy Family

1. Flowers white . . . .*Sanguinaria*
1. Flowers yellow or red 2
   2. Flowers red . . . .*Papaver*
   2. Flowers yellow 3
      3. Petals 2 cm. long or longer . . . .*Stylophorum*
      3. Petals about 1 cm. long . . . .*Chelidonium*

Sanguinaria L.

Sanguinaria canadensis L. (2:197). Bloodroot is one of the first spring flowers to open in rich woods. We have it as early as March 13. The petals soon drop off. Rhizomes contain a blood-red sap. SP. CIT.: *Mc 635; M 48.*

Papaver L.

Papaver rhoeas L. (2:198). Corn Poppy often escapes from flower gardens into waste ground where it may persist for a few years. It is a native of Europe.

Stylophorum NUTT.

Stylophorum diphyllum (Michx.) Nutt. (2:197). Celandine Poppy is an unusually beautiful spring wild flower which flowers during April. It usually occurs on valley floors of rich ravines. SP. CIT.: *M 656; BS 3165; BS 335.*

Chelidonium L.

Chelidonium majus L. (2:197). Cultivated Celandine Poppy is occasionally found in waste ground where it has escaped from cultivation. SP. CIT.: Crab Orchard Lake, *Mc 693.*

64. *Fumariaceae* – Fumitory Family

1. Flowers yellow . . . .*Corydalis*
1. Flowers white . . . .*Dicentra*

Corydalis MEDIC.

Corydalis micrantha (Engelm.) Gray (2:202). Small-flowered Corydalis occurs in rich woods in Jackson, Pope, and Alexander counties. It is distinguished from the following species by the spur over two and one-half millimeters long which is straight or upwardly curved. SP. CIT.: Giant City State Park, April 25, 1953, *M 421*.

Corydalis flavula Raf. (2:202). The more common Pale Corydalis is locally abundant in moist woods where it grows with *Dicentra cucullata, Viola* spp., *Hydrophyllum appendiculatum,* etc. The spur is less than two and one-half millimeters long and is incurved toward the pedicel. It flowers from mid-March to mid-April. SP. CIT.: *Mc 633; BS 319; M 634; M 1972; M 1896.*

Dicentra BERNH.

Dicentra cucullaria (L.) Bernh. (2:201). The fascinating Dutchman's-breeches begins to bloom as early as March 13 and continues until mid-April. It is found in rich woods near bluffs. The spurs are pointed and widely spreading. SP. CIT.: *Mc 634; BS 318; M 46; M 1932; M 1879.*

Dicentra canadensis (Goldie) Walp. (2:201). Squirrel-corn is for the most part rare in Southern Illinois, being known only from Jackson, Union, and Williamson counties. It begins to bloom somewhat later than the preceding (March 20). The spurs are rounded and not spreading. It usually is found in the same stations as Dutchman's-breeches. SP. CIT.: *BS 1990; M 24; M 5113.*

65. *Cruciferae* – Mustard Family

1. Petals white, pink, or pale purple 2
    2. Leaves palmately compound, three in a whorl . . .
        . . . . *Dentaria laciniata*
    2. Leaves not palmately compound 3
        3. None of the leaves pinnatifid, pinnately compound, or auriculate or clasping at base 4
            4. Plants glabrous 5
                5. Some cauline leaves as broad as long . .
                    . . . . . . *Cardamine bulbosa*
                5. Leaves 2 to several times longer than broad .
                    . . . . . . *Lepidium virginicum*
            4. Plants with some kind of pubescence . . . 6

6. Stems with simple hairs at base; leaves denticulate 7
   7. Some cauline leaves over 5 mm. broad . . . . *Arabis canadensis*
   7. Cauline leaves 1.0 to 4.5 mm. broad . . . . *Lepidium virginicum*
6. Stellate hairs present; leaves entire . . . 8
   8. All or most leaves basal 9
      9. Petals 2-cleft . . *Draba verna*
      9. Petals entire 10
         10. Stem with usually 2 or 3 widely separated cauline leaves . . . . . *Arabidopsis thaliana*
         10. Cauline leaves none or 2 or 3 crowded toward base of stem . . . . . *Draba reptans*
   8. Numerous cauline leaves present . . . . . . *Draba brachycarpa*
3. Leaves either pinnatifid, pinnately compound, or simple *and* auriculate or clasping 11
   11. Plants glabrous 12
      12. Some leaves clasping 13
         13. All leaves simple, unlobed . . . 14
            14. Leaves entire, glaucous . . . . . . *Conringia orientalis*
            14. Leaves denticulate 15
               15. Lower leaves on petioles 3 cm. long or longer . . . . *Armoracia rusticana*
               15. Lower leaves sessile or on petioles less than 3 cm. long 16
                  16. Plants glaucous; most leaves sessile . . . . *Arabis laevigata*
                  16. Plants not glaucous; some basal leaves on short petioles . . . . *Thlaspi arvense*
         13. At least lower leaves deeply pinnatifid 17
            17. Petals 10 mm. long or longer . . . . . *Iodanthus pinnatifidus*
            17. Petals 6 to 8 mm. long . . . . . . *Armoracia rusticana*
      12. None of the leaves clasping 18
         18. Plants aquatic; lower leaves with threadlike divisions . *Neobeckia aquatica*

18. Lower leaves not with thread-like divisions 19

    19. Usually in water; petals 4 to 5 mm. long; leaflets nearly entire . . . . . *Nasturtium officinale*

    19. Land plants; petals up to 4 mm. long; leaflets entire or lobed . . 20

        20. Cauline leaves simple, usually incised 21

            21. Leaves up to 5 mm. broad . . . . . *Lepidium virginicum*

            21. Leaves 10 to 35 mm. broad . . . . . *Iodanthus pinnatifidus*

        20. Cauline leaves pinnately compound or pinnatifid . . . . . . *Cardamine*

11. Plants pubescent 21

    21. Some leaves clasping or auriculate . . 22

        22. Hairs forked 23

            23. Pedicels hirsute . *Arabis shortii*

            23. Pedicels glabrous or minutely pubescent . *Capsella bursa-pastoris*

        22. Hairs simple 24

            24. Cauline leaves entire or essentially so . . . *Arabis pycnocarpa*

            24. Cauline leaves denticulate . . 25

                25. Petals 3 to 4 mm. long . . . . . *Cardaria draba*

                25. Petals up to 2 mm. long . . *Lepidium campestre*

    21. None of the leaves clasping or auriculate . 26

        26. Pubescence of stellate hairs . . . . *Capsella bursa-pastoris*

        26. Pubescence of simple hairs . . . 27

            27. Petals pale purple, 10 to 15 mm. long . . . *Raphanus sativus*

            27. Petals white, up to 5 mm. long . 28

                28. Cauline leaves simple, usually incised; stamens 2 or 4 . . . . *Lepidium virginicum*

                28. Cauline leaves pinnately compound or pinnatifid . . 29

                    29. Terminal leaflet much

broader than lateral
ones   .   .   . *Car-*
damine   pennsylvanica

      29. Terminal leaflet not
much broader than lat-
eral ones   .   .   .
.   . *Arabis virginica*

1. Petals yellow   30

    30. Leaves simple, coarsely dentate, denticulate, or entire; hairs
branched or none (basal leaves in *Erysimum* often pinnatifid
but rarely deeply so)   31

        31. Cauline leaves auriculate or clasping   32

            32. Plants glabrous and glaucous; leaves entire .   .
.   .   .   .   .   . *Conringia orientalis*

            32. Plants with branched hairs; leaves denticulate
.   .   .   .   .   . *Camelina microcarpa*

        31. Cauline leaves not auriculate or clasping   .   .   .
.   .   .   .   .   . *Erysimum repandum*

    30. Some leaves deeply pinnatifid or pinnately compound .   . 33

        33. Pubescence of stellate hairs; leaves with finely cut
divisions 34

            34. Cauline leaves pinnately compound .   .   .
.   .   .   .   . *Descurainia brachycarpa*

            34. Cauline leaves entire or dentate .   .   .   .
.   .   .   .   .   . *Erysimum repandum*

        33. Pubescence of simple hairs or none   35

            35. Lower flowers of the raceme with pinnatifid
leafy bracts .   .   .   . *Erucastrum gallicum*

            35. Lower flowers of the raceme without pinnatifid
bracts (except occasionally the very lowest one) 36

                36. Petals 1 cm. long or longer   .   . *Brassica*

                36. Petals up to 1 cm. long 37

                    37. Plants glabrous at maturity   .   . 38

                        38. Cauline leaves subentire, not
lobed, glaucous .   .   .   .
.   .   .   . *Brassica rapa*

                        38. Cauline leaves usually lobed or
coarsely toothed 39

                            39. Flowers 6 to 10 mm. wide;
fruits several times longer
than broad .   . *Barbarea*

                            39. Flowers 1.5 to 5.0 mm.
wide; fruits about 3 to 4
times longer than broad
.   .   .   . *Rorippa*

37. Plants at maturity hirsute, at least
    near the base   40
    40. Flowers about 3 mm. broad
                .   .   *.Sisymbrium officinale*
    40. Flowers 6 to 10 mm. broad   .   41
         41. At least upper leaves with
               linear divisions .   .   .
               .   *Sisymbrium altissimum*
         41. Divisions of leaves not
               linear   .   .   *.Brassica*

Arabidopsis HEYNH.

Arabidopsis thaliana (L.) Heynh. (2:247). Mouse-ear Cress is common in fields and waste ground. It flowers from mid-March to mid-May. It is distinguished from species of *Draba* by its elongated pods. SP. CIT.: *Mc 628; V 1331; M 1883; M 1437; M 2183.*

Arabis L.

1. Pedicels and fruits erect and appressed .   *.Arabis pycnocarpa*
1. Pedicels and fruits spreading or pendulous, not appressed .   .   2
    2. Stem leaves with ear-like flaps at base   3
        3. Pedicels less than 6 mm. long; basal and all but very
           uppermost cauline leaves pinnatifid .   *.Arabis virginica*
        3. Pedicels 7 mm. long or longer; only basal leaves some-
           times pinnatifid   4
            4. Stem leaves hairy .   .   .   *.Arabis canadensis*
            4. Stem leaves glabrous and usually glaucous   .   .
               .   .   .   . *Arabis laevigata*
    2. Stem leaves without ear-like flaps at base   5
        5. Leaves pubescent .   .   .   .   .   *.Arabis shortii*
        5. Leaves glabrous and usually glaucous   .   .   .   .
           .   .   .   . *.Arabis laevigata*

Arabis pycnocarpa Hopkins [*A. hirsuta* (L.) Scop. var. *pycnocarpa* (Hopkins) Rollins] (2:234). This rare species has been collected in Southern Illinois only on the limestone base of Devil's Bake Oven along the Mississippi River. Its nearest known station is in DeWitt County, more than two hundred miles to the north. SP. CIT.: near Grand Tower, May 30, 1954, *M 2550.*

Arabis virginica (L.) Poir. [*Sibara virginica* (L.) Rollins] (2:235). This small species occurs in rocky woods or occasionally along roads in most of the southern counties. We have found it flowering as early as March 13. SP. CIT.: *M 792.*

Arabis canadensis L. (2:235). Sicklepod occurs sparingly on dry wooded slopes where it grows with *Lespedeza stuevei, Danthonia spicata,* and others. We have it from Randolph, Jackson, Johnson, and Pope counties.

Arabis laevigata (Muhl.) Poir. (2:235). Smooth Rockcress grows commonly in moist woods or on moist sandstone bluffs. It sometimes attains a height of over two feet. Flowers appear in early April. Occasional nonglaucous specimens occur which resemble *Arabis viridis* Harger but *A. viridis* may be distinguished by petals which about equal the length of sepals. SP. CIT.: *M 30; M 1884.*

Arabis shortii (Fern.) Gleason [*A. dentata* Torr. & Gray] (2:236). This species has been found in Southern Illinois only in moist woods in Jackson and Alexander counties. SP. CIT.: Midland Hills, May 9, 1951, V *850.*

Armoracia GAERTN.

Armoracia rusticana (Lam.) Gaertn. (2:227). Horse-radish, a native of Europe, often escapes from cultivation to roadside ditches and other waste areas. SP. CIT.: Giant City State Park, *English.*

Barbarea R. BR.

Barbarea vulgaris R. Br. (2:241). Wintercress is a common plant of roadsides, fields, and waste places. It is a native of Europe. The basal leaves have up to four pairs of leaflets. SP. CIT.: *M 134; M 2180.* A second species, *Barbarea verna* (Mill.) Aschers. (2:241), has been collected by H. E. Ahles (No. *7997*) from Johnson County. This species has basal leaves with 4 to 10 pairs of leaflets.

Brassica L.

1. Some leaves clasping . . . . . . *Brassica rapa*
1. None of the leaves clasping 2
    2. Pedicels 10 mm. long or longer . . *Brassica juncea*
    2. Pedicels 3 to 7 mm. long 3
        3. Base of beak of fruit with one seed . *Brassica kaber*
        3. Base of beak of fruit without seeds . *Brassica nigra*

Brassica rapa L. Turnip sometimes escapes into waste ground in our area. SP. CIT.: *M 334.*

Brassica kaber (DC.) L. Wheeler (2:207). This species is called

Field Mustard or Charlock. It is a native of Europe and has become established in waste places. SP. CIT.: *M 2324.*

Brassica nigra (L.) Koch. (2:207). Black Mustard has been found in this area only in Jackson and Williamson counties where it occurs in waste ground. SP. CIT.: *Welch 103.*

Indian Mustard, *Brassica juncea* (L.) Cosson (2:207), is known from Union, Alexander, and Pulaski counties. It is a native of Europe.

Camelina CRANTZ

Camelina microcarpa Andrz. (2:247). This species occurs along railroads in Jackson and Randolph counties. It has been naturalized from Europe. SP. CIT.: *V 1351; M 2152.*

Capsella MEDIC.

Capsella bursa-pastoris (L.) Medic. (2:217). Shepherd's-purse, another naturalized plant of European origin, is very abundant in lawns and waste areas throughout Illinois. SP. CIT.: *Voigt; M 33.*

Cardamine L.

1. Stem leaves lobed, often deeply pinnatifid 2
    2. Basal leaves few or absent 3
        3. Terminal leaflet 1 to 2 cm. broad . . . . .
            . . . . *Cardamine pennsylvanica*
        3. Terminal leaflet less than 1 cm. broad . . . .
            . . . . *Cardamine arenicola*
    2. Basal leaves numerous . . . . *Cardamine hirsuta* °
1. Stem leaves simple, or sometimes with two lobes . . . . .
        . . . . *Cardamine bulbosa*

Cardamine pennsylvanica Muhl. (2:231). Pennsylvania Bitter-cress is local in moist ground throughout the southern counties. It flowers during April. SP. CIT.: *M 112; M 2135; M 1885.*

Cardamine arenicola Britt. [*C. parviflora* L. var. *arenicola* (Britt.) O. E. Schulz] (2:231). This species is common in moist soil in woods or open ground. SP. CIT.: *M 20; M 2198; BS 355; Welch and Fuller 150.*

Cardamine bulbosa (Schreb.) BSP. (2:229). This species is local in moist ravines. It flowers during early May. SP. CIT.: *M 2406.*

A fourth species, *Cardamine hirsuta* L. (2:231), occurs in waste ground in Pope and Hardin counties. It is adventive from Europe.

Cardaria DESV.

Hoary Cress, *Cardaria draba* (L.) Desv. (2:215), occurs in waste ground in Alexander County. It probably has been overlooked in the other southern counties.

Conringia ADANS.

Conringia orientalis (L.) Dumort. (2:211). Hare's-ear Mustard has been found along railroads in Jackson and Union counties. It is a native of Europe. It flowers in late April. SP. CIT.: four miles south of Murphysboro, railroad siding, *M 2326*.

Dentaria L.

Dentaria laciniata Muhl. (2:231). Toothwort or Pepperroot is a native woodland crucifer. It is variable in flower color and leaf shape. It is abundant in moist woods where it begins to flower as early as mid-February. SP. CIT.: *M 1479; BS 317; Mc 621.*

Descurainia WEBB & BERTH.

Descurainia brachycarpa (Richards) O. E. Schulz (2:247). Tansy Mustard is found occasionally in our area where it grows most often on limestone. It is abundant in talus at Pine Hills. SP. CIT.: *M 14; BS 359.*

Draba L.

Draba brachycarpa Nutt. (2:222). Short-fruited Whitlow-grass is abundant in waste ground and at edges of woods. It is one of the first species to flower, often blooming early in February. It is our only species of *Draba* with more than one or two pairs of leaves on the stem. SP. CIT.: *M 423; M 2167; M 4797.*

Draba reptans (Lam.) Fern. (2:222). This small plant is found on limestone bluff prairies along the Mississippi River. It is variable in degree of pubescence. The petals are not cleft and the stems generally have one to two pairs of leaves. SP. CIT.: *BS 2079.*

Draba verna L. (2:222). Mouse-ear Whitlow-grass, a native of Europe, grows in lawns and fields. It flowers in March. Petals are deeply cleft. All leaves are in a basal rosette. SP. CIT.: *M 633; BS 302.*

Erucastrum (DC.) PRESL

Erucastrum gallicum (Willd.) O. E. Schulz (2:209). Our station for this European plant is along a railroad near Pomona. The nearest Illinois station is in Champaign County. SP. CIT.: one mile north of Pomona, July 5, 1951, *BS 1515*.

Erysimum L.

Erysimum repandum L. (2:243). This European species has escaped along roads in only a few of our counties (Randolph and Jackson). SP. CIT.: Giant City State Park, April 25, 1953, *M 16;* one mile north of Etherton, April 25, 1954, *M 2328*.

Iodanthus TORR. & GRAY

Along streams in Jackson, Randolph, Alexander, Pope, Hardin, and Massac counties *Iodanthus pinnatifidus* (Michx.) Steud. (2:241) has been found. It flowers during May and June. SP. CIT.: "Granny" woods west of Murphysboro, *Mohlenbrock*.

Lepidium L.

Lepidium virginicum L. (2:212). Common Peppergrass is abundant in waste ground throughout Illinois. Leaves of the stem are not auriculate. SP. CIT.: *Voigt; M 96*.

Lepidium campestre (L.) R. Br. (2:212). Field Peppergrass occurs along railroads. We have it from Randolph, Jackson, and Union counties. Leaves of the stem have ear-like flaps at the base. SP. CIT.: four miles south of Murphysboro, railroad siding, April 30, 1954, *M 2346*.

Nasturtium R. BR.

Nasturtium officinale R. Br. (2:239). Watercress is often collected for salads. It occurs in great abundance in LaRue Swamp (Union County). We have it in Jackson County from the Fountain Bluff spring. SP. CIT.: *BS 429*.

Neobeckia GREENE

Another crucifer which grows in water is *Neobeckia aquatica* (A. Eaton) Greene (2:227). The submerged leaves resemble *Ceratophyllum demersum*. It is known in our area only from LaRue Swamp. It flowers during June.

Raphanus L.

Radish, *Raphanus sativus* L. (2:211), is an occasional escape from cultivation but seldom persists.

Rorippa scop.

1. Pedicels at maturity less than 3 mm. long . *Rorippa sessiliflora*
1. Pedicels at maturity 3 to 10 mm. long 2
    2. Petals 1 to 2 mm. long; annual . . .*Rorippa islandica*
    2. Petals 3 mm. long or longer; perennial 3
        3. Leaves with ear-like flaps at base . .*Rorippa sinuata* °
        3. Leaves without ear-like flaps at base . . . .
        . . . . *Rorippa sylvestris* °

Rorippa sessiliflora (Nutt.) Hitchc. Short-stalked Yellow Cress is common in moist soil where it grows with *Phyla lanceolota, Gratiola neglecta,* and others. It flowers from August through October. SP. CIT.: Bell; *M 593.*

Rorippa islandica (Oeder) Borbas [*R. palustris* (L.) Besser; *R. islandica* (Oeder) Borbas var. *fernaldiana* Butters & Abbe] (2:239). This species, Marsh Yellow Cress, is common in moist soil or sometimes in water. It begins to flower in April and continues into August. SP. CIT.: *Mc 300; M 594; M 2184.*

Two additional species of *Rorippa* occur in Southern Illinois. *R. sinuata* (Nutt.) Hitchc. (2:239) occurs on river banks in Randolph, Pope, and Massac counties while *R. sylvestris* (L.) Besser (2:239) grows in moist soil in Pope, Hardin, and Massac counties. The latter is an introduction from Europe.

Sisymbrium L.

Sisymbrium altissimum L. (2:245). Tumble Mustard is an occasional waif along railroads and highways. It flowers during April and May and is an introduced species from Europe. The leaf divisions are thread-like. SP. CIT.: four miles south of Murphysboro, railroad siding, *M 2325.*

Sisymbrium officinale (L.) Scop. (2:245). Common Hedge Mustard occurs in waste ground in a few southern counties. It flowers from May to September. Leaf divisions are not thread-like. SP. CIT.: *BS 484.*

Thlaspi L.

Thlaspi arvense L. (2:217). Penny Cress is common in fields, along railroads, and in other waste areas throughout Southern Illinois. Young plants often resemble *T. perfoliatum* which does not occur in Illinois. SP. CIT.: *BS 267; M 1988.*

66. *Capparidaceae* – Caper Family

1. Petals notched at summit; stamens more than 6 .   .*Polanisia*
1. Petals not notched at summit; stamens 6 .   .   . .*Cleome*

Polanisia RAF.

Polanisia graveolens Raf. (2:249). Clammyweed is a common railroad waif which blooms from mid-July through early September. Petals are 4 to 8 mm. long. SP. CIT.: Fountain Bluff, *BS* 553. A second species, *P. trachysperma* Torr. & Gray (2:249), has been collected in Union County. The petals of this species are about 1 cm. long.

Cleome L.

Cleome speciosissima Deppe (2:249). This species is a cultigen which has escaped into low woods in Jackson County. SP. CIT.: low wet soil, Little Muddy River bottom, three miles east of Elkville, July 11, 1940, *Mc* 269.

67. *Crassulaceae* – Stonecrop Family

1. Leaves succulent .   .   .   .*Sedum*
1. Leaves not succulent .   .   .   .   .   .*Penthorum*

Sedum L.

1. Leaves terete   2
    2. Flowers pink or white   .   .   .*Sedum pulchellum*
    2. Flowers yellow   .   .   .   .   .*Sedum acre*
1. Leaves flat   3
    3. Flowers white   .   .   .   .*Sedum ternatum*
    3. Flowers pink or reddish   4
        4. Flowers pale pink .   .   .*Sedum telephioides* °
        4. Flowers reddish-purple   .   .*Sedum triphyllum*

Sedum pulchellum Michx. (2:256). Stonecrop grows on very dry exposed sandstone bluffs across the Shawneetown Ridge where it is associated with *Opuntia rafinesquii, Agave virginica, Tephrosia virginiana, Oenothera linifolia, Ruellia humilis,* and others. It flowers during May and June. Although the Appalachian species *Sedum nevii* Gray is periodically attributed to Southern Illinois (Gleason, 1952, and others), the specimens on which this statement is based are thought to be only juvenile stages of *Sedum pulchellum* (Baldwin, 1942). SP. CIT.: Giant City State Park, *M 95*.

Sedum acre L. (2:258). Yellow Mossy Stonecrop is a common plant in cultivation which sometimes escapes to rocky places. It is a native of Europe.

Sedum ternatum Michx. (2:256). Three-leaved Stonecrop is found only locally on moist bluffs in Illinois. It has a very interrupted distribution in Illinois. Flowers are produced in May. SP. CIT.: *BS 665; BS 2154; BS 2336.*

A plant with a most peculiar range is *Sedum telephioides* Michx. (2:257). Its range, according to Gleason (1952), is "s. Pa. to Va., W. Va., and Ga.; s. Ill." In our area, it occurs on steep bluffs in southeastern counties. It flowers in August and September.

Sedum triphyllum (Haw.) S. F. Gray [*S. telephium* L.] (2:257). Live-for-ever or Orpine is an escape from gardens. It occurs abundantly on Devil's Bake Oven near Grand Tower where it blooms during August. SP. CIT.: *M 620.*

Penthorum   L.

Penthorum sedoides L. (2:254). Ditch Stonecrop is a common plant of moist ground. It almost invariably is found with *Phyla lanceolata*. It flowers from July through September. Fruits turn a brilliant reddish-brown. SP. CIT.: *Bell; M 568; Mc 459; Mc 886; V 999; BS 2674.*

68. *Saxifragaceae* – Saxifrage Family

1. Leaves unlobed, longer than broad   .   .   .   .*Saxifraga*
1. Leaves lobed, about as wide as long   2
   2. Petals white, fringed; stamens 10 .   .   .   .   .*Mitella*
   2. Petals yellow to greenish, not fringed; stamens 5 . *Heuchera*

Saxifraga   L.

Saxifraga pennsylvanica L. (2:264). Pennsylvania Saxifrage has been collected in Southern Illinois only from wet woods in Jackson County (see Jones *et al.*, 1955). Leaves are sparsely hairy above and the two beaks of the fruit are only slightly spreading.

Saxifraga forbesii Vasey (2:265). Forbes' Saxifrage is named for Stephen Alfred Forbes who first discovered this species at Makanda in Jackson County in about 1870 (undoubtedly on the moist dripping sandstone bluffs at Giant City State Park). This species closely resembles S. *pennsylvanica* but differs from it in its very hairy leaves and in fruits which have the two beaks widely spreading. It is known in Illinois from Jackson, Williamson, Pope, and Gallatin counties. It flowers during April and May. SP. CIT.: Giant City State Park, *M 60;* Little Grand Canyon, *BS 2434.*

Virginia Saxifrage, *Saxifraga virginiensis* Michx. (2:264), occurs in moist rocky woods in Hardin County at Blind Hollow and Hooven Hollow where it is rather common. It begins to flower in late April.

Mitella L.

Mitella diphylla L. (2:267). The attractive Bishop's Cap (Fig. 44) is found on moist sandstone bluffs in Jackson, Williamson, and Hardin counties. It is more abundant in northern Illinois. It has occurred at Giant City State Park at least since 1871. SP. CIT.: Giant City State Park, May 1, 1951, *M 926;* Little Grand Canyon, May 30, 1952, *BS 2435.*

Heuchera L.

Heuchera parviflora Bartl. var. *rugelii* (Shuttlw.) Rosend., Butt., & Lak. (2:268). Small-flowered Alumroot grows on moist sandstone bluffs in a few southern counties. It flowers from late June until early November. The base of the flower is covered with long white hairs, none of which are gland-tipped. The plant is of southeastern origin. SP. CIT.: *Mc 509; Mc 428; BS 3164; BS 279.*

FIG. 44. *Bishop's Cap* (Mitella diphylla), *a delicate plant of northern affinity, grows sparsely on a large moist boulder in Giant City State Park. Photo taken May 1, 1956, by Marvin Rensing.*

FIG. 45. *Virginia Willow* (Itea virginica), *a handsome shrub of low swampy woods, has Coastal Plain affinities.*

Heuchera americana L. var. *interior* Rosend., Butt., & Lak. (2:269). This plant is found occasionally in dry woods. The base of the flowers has gland-tipped hairs, the petals are one to one and one-half millimeters long, and the calyx tube is nearly regular. SP. CIT.: Giant City State Park, May 9, 1953, *M 136;* near Etherton, April 19, 1954, *M 2168.*

Heuchera americana L. var. *hirsuticaulis* (Wheelock) Rosend., Butt., & Lak. (2:269) [*H. hirsuticaulis* (Wheelock) Rydb.]. This plant is found in similar situations as the preceding but is more common. The base of the flowers has gland-tipped hairs, the petals are two to two and one-half millimeters long, and the calyx tube is very irregular. SP. CIT.: *M 2397; M 4064; Mc 746; Mc 717.*

69. *Escalloniaceae* – Escallonia Family

Itea L.

Virginia Willow, *Itea virginica* L. (Fig. 45) (2:274), a Coastal Plain species, is a characteristic shrub of swampy woods of southern counties. We have it from Union, Johnson, Pulaski, Pope, and Alexander counties. It flowers in late May and matures its fruits during July.

70. *Hydrangeaceae* – Hydrangea Family

1. Stamens about 10 . . . .*Hydrangea*
1. Stamens 20 or more . . . .*Philadelphus*

Hydrangea L.

Hydrangea arborescens L. (2:274). Wild Hydrangea is found in moist rocky woods throughout Southern Illinois. Many marginal flowers are very showy but sterile. Specimens with densely pubescent leaves are var. *deamii* St. John. They are not as abundant as the typical variety. SP. CIT.: typical: *Mc 866; M 2136; M 4073.* Var. *deamii: Mc 424; M 381; Bell.*

Philadelphus L.

Two species of Mock Orange have been collected in Southern Illinois. One, *Philadelphus inodorus* L., an escape from cultivation, was found in an orchard in Union County. Flowers are in clusters of one to three. On rocky limestone bluffs of the Ohio River near Golconda is the native *Philadelphus pubescens* Loisel. (2:273). It was collected by E. J. Palmer in 1919 and 1920. Flowers are in clusters of five to seven.

71. *Grossulariaceae* – Gooseberry Family

Ribes L.

Ribes missouriensis Nutt. (2:276). Missouri Gooseberry occurs along streams in southern counties which border the Mississippi River. Stamens greatly exceed the length of petals and sepals. SP. CIT.: near Carbondale, July 16, 1941, *Mc 922.*

Ribes cynosbati L. (2:276). Pasture Gooseberry occurs in similar situations as the preceding. Stamens only about equal the length of petals and sepals. SP. CIT.: Fountain Bluff, July 12, 1871, *French; BS 315.*

72. *Hamamelidaceae* – Witch-hazel Family

Liquidambar L.

Liquidambar styraciflua L. (2:280). Sweet Gum is one of our most stately forest trees. It occurs commonly with Pin Oak, Tulip Tree,

and Shag-bark Hickory. It is a popular shade tree. SP. CIT.: *M 247;*
M 2175; M 4025.

73. *Platanaceae* – Plane-tree Family

Platanus L.

Platanus occidentalis L. (2:281). Sycamore is abundant along
streams and moist woods. In an apparently relict habitat atop the
dry limestone Devil's Backbone, sycamore occurs with cottonwood,
sea-oats (*Uniola latifolia*), and other lowland plants. SP. CIT.:
Mc 308.

74. *Rosaceae* – Rose Family

1. Plants woody 2
    2. Leaves compound 3
        3. Leaves palmately compound with 3 to 5 leaflets . *Rubus*
        3. Leaves pinnately compound with 3 to many leaflets
            . . . . *Rosa*
    2. Leaves simple 4
        4. Petals oblanceolate, never broadly rounded at summit;
            small trees without thorns . . . . *Amelanchier*
        4. Petals obovate, broadly rounded at summit; trees with or
            without thorns 5
            5. Style 1; leaves not lobed; plants usually without
                sharp curved spines . . . . . *Prunus*
            5. Styles 2 or more; leaves often lobed; plants often
                with spines 6
                6. Thorns none 7
                    7. Flowers perigynous, often double; fruit a
                        folicle or dehiscent capsule . *Spiraea*
                    7. Flowers perigynous with adnate hypanthium
                        (Plate 2); flowers not double . . . 8
                        8. Anthers pink or red . . *Pyrus*
                        8. Anthers white or yellow . *Malus*
                6. Thorns present 9
                    9. Styles united at base . . . *Pyrus*
                    9. Styles free to base . . *Crataegus*
1. Plants not woody 10
    10. Leaves 2 to 3 pinnate; flowers unisexual with petals about
         1 mm. long . . . . *Aruncus*
    10. Leaves once compound; petals more than 1 mm. long . . 11
        11. Flowers white 12

Rubus L.

The individuals of the genera *Rubus, Rosa,* and *Crataegus* are very variable and many hundreds of species have been proposed in these genera. Since they are based mainly on tenuous characters, we are following Gleason's treatment of *Rubus* by recognizing only taxa which can be regarded as collective species. While no doubt many of our specimens may be classified into microspecies, we are not attempting to employ such segregations.

7. Terminal leaflet widest at middle   .   .   .
.   .   .   . *Rubus ostryifolius*
7. Terminal leaflet widest well below middle .   .
.   .   .   .   . *Rubus pennsylvanicus*

Rubus occidentalis L. Black Raspberry is common in moist soil although it is infrequently collected. SP. CIT.: *M 264; BS 713; BS 39.*

Rubus flagellaris Willd. (2:310). Fruits of Dewberry are in great demand in our area. This species is common along railroads and edges of fields and woods. SP. CIT.: *BS 54; M 1509; M 2176.*

Southern Dewberry, *Rubus trivialis* Michx. (2:306), only recently discovered in Southern Illinois, grows along roadsides or edges of woods in Randolph, Union, Massac, and Jackson counties. SP. CIT.: *Mohlenbrock.*

Rubus allegheniensis Porter (2:310). Allegheny Blackberry is an occasional species in dry woods, although we have collected it in three counties. SP. CIT.: Giant City State Park, *M 77.*

Rubus orarius Blanchard (2:311) [*Rubus alumnus* Bailey]. Only one station is known in Illinois for this species. SP. CIT.: along Illinois Central Railroad, Carbondale, *BS 62.*

Rubus ostryifolius Rydb. (2:312) (includes *Rubus schneckii* Bailey). High-bush Blackberry is a common plant in a wide variety of habitats. SP. CIT.: *M 84; BS 519.*

Rubus pennsylvanicus Porr. (2:314) (includes *Rubus frondosus* Biegl.). This is our most common species of Blackberry. Fruits are often collected for eating purposes. SP. CIT.: *M 63.*

The more eastern *Rubus enslenii* Tratt (2:311) has been collected once in Illinois along Piney Creek in Randolph County.

Rosa L.

1. Styles united, protruding beyond hypanthium   2
   2. Leaflets 7 to 9 .   .   .   .   .   . *Rosa multiflora*
   2. Leaflets 3 to 5 .   .   .   .   .   . *Rosa setigera*
1. Styles free from each other, barely or not at all protruding beyond hypanthium   3
   3. Hypanthium and pedicels without gland-tipped hairs .   .
   .   .   .   . *Rosa blanda* °
   3. Hypanthium and pedicels with gland-tipped hairs   .   . 4
   4. Leaflets with teeth over 1 mm. long   .   .   .   .
   .   .   .   . *Rosa carolina*

4. Leaflets with teeth less than 1 mm. long .   .   .   .
    .   .   .   . *Rosa palustris*

Rosa multiflora Thunb. (2:323). Multiflora Rose, a native of Japan and China, is often planted as a fence row in our area. SP. CIT.: *Voigt; Evers 16750.*

Rosa setigera Michx. (2:323). Climbing Rose is common along edges of woods or in pastures. Our specimens belong to var. *tomentosa* Torr. & Gray with the lower surfaces of leaves being tomentose. SP. CIT.: Giant City State Park, June 26, 1947, *BS 98.*

Meadow Rose, *Rosa blanda* Ait. (2:326), was discovered growing along the edge of a woods in the Union County part of Giant City State Park. This is the only station for this species in Illinois south of Peoria County. SP. CIT.: Giant City State Park, wood's edge, May 15, 1953, *M 66,* and May 22, 1953, *M 742.*

Rosa carolina L. (2:326). Pasture Rose is our most common species of rose. It occurs in waste ground or in woods. SP. CIT.: *Mc 189; M 398; M 270; M 2413.*

Rosa palustris Marsh. (2:325). Swamp Rose, which is common in marshes and swamps throughout our area, sometimes attains a height of eight feet and becomes very bushy. SP. CIT.: "marsh," Murphysboro, *Mohlenbrock.*

Amelanchier MEDIC.

Amelanchier arborea (Michx. f.) Fern. (2:377). Shadbush, or Service-berry, is found along edges of sandstone bluffs. It is a most attractive plant when in flower during late March and early April. Fruits are edible. SP. CIT.: *M 51; M 2185.*

Prunus L.

1. Leaves softly and densely pubescent beneath   2
    2. Petioles with 2 to several glands; leaves broadly rounded at
        the base   .   .   .   .   .   . *Prunus mexicana*
    2. Petioles without glands; leaves tapering to the base   .   .
        .   .   .   . *Prunus lanata*
1. Leaves glabrous or minutely pubescent beneath   3
    3. Flowers pink; plants escaped from cultivation   .   .   .
        .   .   .   . *Prunus persica*
    3. Flowers white; native species   4
        4. Petioles without glands   .   .   . *Prunus americana*

4. Petioles with 1 to 6 glands 5
    5. Leaf teeth not gland-tipped; inflorescence a many-flowered raceme . . . . *Prunus serotina*
    5. Leaf teeth gland-tipped or with scars where glands have been; flowers few, solitary, or in umbels or corymbs 6
        6. Leaf teeth sharp, bearing the gland terminally . . . . *Prunus hortulana*
        6. Leaf teeth blunt and low, bearing the gland next to the sinus 7
            7. Glands of teeth red; mid-vein pubescent beneath . . . *Prunus munsoniana*
            7. Glands of teeth not red; mid-vein glabrous beneath . . . *Prunus angustifolia*

Prunus persica (L.) Batsch. (2:329). Peach has occasionally escaped to roadsides throughout our area. SP. CIT.: *M 938*.

Prunus serotina Ehrh. (2:329). Wild Black Cherry is a common tree of woods, thickets, and waste ground. The attractive inflorescences are produced in May. The fruits are a delight to birds. SP. CIT.: *Mc 317; BS 366; M 4625*.

Prunus americana Marsh. (2:332). American Plum is common in moist woods or semiwaste ground in all southern counties. SP. CIT.: *BS 314; Mc 679*.

Prunus lanata (Sudw.) Mack. & Bush (2:332). Woolly Wild Plum is local on wooded bluffs in a few southern counties. SP. CIT.: Fountain Bluff, *BS 2737*.

Prunus mexicana Wats. (2:332). Big Tree Plum has been found in Illinois at the base of a steep limestone bluff one mile north of Pine Hills in Jackson County. Its previous range was southern Missouri to Texas. SP. CIT.: north of Pine Hills, July 5, 1955, *Mohlenbrock and Stewart 5468*.

Prunus angustifolia Marsh. (2:333). Chickasaw Plum occurs locally in Southern Illinois, chiefly in sandy soil. SP. CIT.: one mile north of Makanda, July 4, 1940, *G. N. Jones 19948*.

Prunus hortulana Bailey (2:333). Wild Goose Plum occurs primarily along edges of dry oak-hickory woods where it is fairly common. SP. CIT.: *M 52; M 1898*.

Munson Wild Plum, *Prunus munsoniana* Wight and Hedrick (2:333), is known from Gallatin, Johnson, Pope, Jackson, and Alexander counties. SP. CIT.: *Mohlenbrock*.

Pyrus L.

Pyrus communis L. (2:335). Pear is frequently found along road-ways where it has escaped from nearby orchards. Leaves are smooth at maturity and anthers are red. SP. CIT.: *M 44*.

Malus MILL.

1. Plants without thorns; species escaped from cultivation; leaves
   pubescent at maturity; anthers yellow; fruit an apple .   .   .
   .   .   .   . *Malus pumila*
1. Plants usually with thorns; native species 2
   2. Leaves pubescent beneath at maturity .   .   .   .   .
      .   .   .   . *Malus ioensis*
   2. Leaves glabrous at maturity .   .   .   . *Malus coronaria*

Malus pumila Mill. (2:335) [*Pyrus malus* L.]. Like the preceding, Apple is a common escape into waste ground. SP. CIT.: *M 45*.

Malus ioensis (Wood) Britt. (2:336) [*Pyrus ioensis* (Wood) Car-ruth.]. Iowa Crab Apple is the only woody plant which occurs in limestone hill prairies in our area. It also occurs along edges of woods. SP. CIT.: *BS 2736; Mc 311; M 677*.

Malus coronaria Mill. (2:335) [*Pyrus coronaria* L.]. Wild Sweet Crab Apple is less common than the preceding but does occur locally along edges of woods in a few of our counties. SP. CIT.: (see Jones *et al.*, 1955).

Spiraea L.

Spiraea prunifolia Sieb. and Zucc. (2:285). This is one of the culti-vated Spiraeas which often escapes. SP. CIT.: rocky hillside, south-east quarter of section 33, near Makanda, *Mc 680*.

Crataegus L.

1. Leaves broadly rounded at summit, broadest above middle, un-lobed, glossy 2
   2. Leaves glabrous, mostly over 2 cm. broad .   .   .   .
      .   .   .   . *Crataegus crus-galli*
   2. Leaves pubescent, up to 2 cm. broad .   .   .   .   .
      .   .   .   . *Crataegus engelmanni*
1. Leaves more or less acute at summit, broadest at or below mid-dle, often lobed, usually dull above 3
   3. Leaves soft pubescent beneath 4

    4. Leaves widest at middle; petioles winged .   .   .
        .   .   .   . *Crataegus calpodendron*

    4. Leaves widest near base; petioles usually not winged .
        .   .   .   . *Crataegus mollis*

  3. Leaves glabrous beneath or with tufts of tomentum in the
    vein axils   5

    5. Leaves broadest near middle; leaves mostly cuneate .
        .   .   .   . *Crataegus viridis*

    5. Leaves broadest near base; leaves mostly cordate or
    truncate .   .   .   .
        .   .   .   . *Crataegus pruinosa*

Crataegus crus-galli L. (2:344). Cock-spur Thorn is common in woods and open ground in all southern counties. SP. CIT.: *M 124; M 5255.*

Crataegus engelmanni Sarg. (2:348). Barberry Thorn, so named because the foliage suggests that of species of *Berberis,* occurs rarely on dry sandstone or limestone bluffs. We have it from Jackson, Union, Pope, and Hardin counties. SP. CIT.: *M 5342.*

Crataegus calpodendron (Ehrh.) Medic. (2:372). This Hawthorn is local in dry woods. It is reported in our area only from Jackson, Johnson, and Pope counties. SP. CIT.: *M 499.*

Crataegus mollis (Torr. & Gray) Scheele (2:369). Red Haw is the most common species of Hawthorn in Southern Illinois. The fruits are edible. SP. CIT.: *M 3.*

Crataegus viridis L. (2:348). This species is rather common in low wet woods. In the Oakwood Bottoms east of Gorham it comprises a great percentage of the shrubby layer species. SP. CIT.: *M 4779.*

Crataegus pruinosa (Wendl.) K. Koch (2:363). Although infrequently collected in our area, this species is rather common in dry woods. SP. CIT.: *M 127.*

Aruncus ADANS.

Aruncus dioica (Walt.) Fern. (2:287). Goat's-beard is local in deep, wooded ravines where it occurs with *Aralia racemosa, Athyrium pycnocarpon, Mitchella repens,* and others. It flowers during early June. SP. CIT.: *BS 2451; Mc Cree, Welch and Fuller 183.*

Gillenia MOENCH

Gillenia stipulacea (Muhl.) Trel. (2:288). Indian Physic, or American Ipecac, grows in dry oak-hickory woods. It is one of the notable summer flowering plants. SP. CIT.: *M 184*.

Fragaria L.

Fragaria virginiana Duch. (2:289). Wild Strawberry occurs occasionally along roads and in fields. It blooms in April and May. SP. CIT.: *Mc 90*.

Fragaria chiloensis Duch. var. *ananassa* Bailey (2:290). Garden Strawberry commonly escapes to roadsides and waste ground where it persists for several years. SP. CIT.: *M 1953; M 1900*.

Geum L.

Geum vernum (Raf.) Torr. & Gray (2:301). Spring Avens is abundant in moist woods where it flowers in April and May. The bright yellow petals are longer than the sepals. SP. CIT.: *M 773; M 2434*.

Geum canadense Jacq. (2:301). One of the most common summer flowering plants of moist woods is White Avens. The white petals are longer than the sepals. SP. CIT.: *M 1342; Mc 885; Mc 958*.

Geum virginianum L. (2:302). The rather rare Virginia Avens is found in dry woods where it is never abundant. We have it from Randolph, Jackson, Williamson, Union, Pope, and Gallatin counties. Pale yellow or cream petals are about as long as, or somewhat shorter than, the sepals. SP. CIT.: Lake Murphysboro, July 7, 1955, *M 5503*.

Agrimonia L.

Agrimonia parviflora Ait. (2:320). Small-flowered Agrimony is a very common species of moist woods. Leaves are softly pubescent and with gland-tipped hairs in the inflorescence. Flowers appear from mid-July through September. SP. CIT.: *Mc 935; M 436; M 4092; Mc 404; BS 748*.

Agrimonia rostellata Wallr. (2:319). This is an occasional species of moist woods. The leaves are nearly glabrous (at least not softly pubescent) and the inflorescence bears gland-tipped hairs. SP. CIT.: *M 578; BS 527*.

Agrimonia pubescens Wallr. (2:320). This Agrimony grows in dry woods. It begins to flower about one week earlier than the other species of *Agrimonia* in our area. The inflorescence lacks gland-tipped hairs. SP. CIT.: *M 274; M 4397*.

Potentilla L.

Potentilla paradoxa Nutt. (2:295). The rare River Cinquefoil is known in Illinois only from Jackson, Randolph, and St. Clair counties. Two of its common associates are *Hemicarpha micrantha* and *Cyperus acuminatus*. SP. CIT.: near Grand Tower, August 20, 1954, *M 4637;* Cora cutoff, August 24, 1954, *M 4651.*

Potentilla simplex Michx. (2:293). This is the Common Cinquefoil which may be found in a wide variety of open habitats. It flowers from April to June. Specimens with glabrous stems and petioles are var. *calvescens* Fern. SP. CIT.: typical: *Mc 9; M 123; M 1978; Voigt.* Var. *calvescens: M* 2134.

Potentilla recta L. (2:295). This handsome roadside plant which is a native of Europe is common in most of the southern counties. It blooms from early May into August. SP. CIT.: *M 2600; BS 1514.*

Potentilla monspeliensis L. (2:294) (including *P. norvegica* L.). Rough Cinquefoil is common about bodies of water, being particularly abundant around Walker Hill Pond just east of Grand Tower. SP. CIT.: *BS 415; BS 723; M 4832.*

75. *Leguminosae* – Pea Family

6. Stems either thorny, spiny, or densely hispid . 7
  7. Flowers pink, stems densely hispid . .
    . . . . *Robinia*
  7. Flowers not entirely, if at all, pink; stems
    with spines or thorns 8
    8. Stipules modified into short, sharp
      spines . . . . . *Robinia*
    8. Not as above; spines often branched
      . . . . *Gleditsia*
6. Stems without thorns and not densely hispid . 9
  9. Leaves once-pinnate; leaflets ovate, 2.5 to
    7.0 cm. broad . . . .*Cladrastis* *
  9. Leaves often twice-pinnate; leaflets up to
    2.5 cm. broad 10
    10. Leaflets entire . .*Gymnocladus*
    10. Leaflets with small teeth near apex .
      . . . . . . *Gleditsia*
1. Plants herbaceous 11
  11. Petals bright yellow or orange 12
    12. Leaves simple . . . . . .*Crotalaria*
    12. Leaves compound 13
      13. Leaves 3-foliolate . . . .*Stylosanthes*
      13. Leaves 4- to many-foliolate . . . *Cassia*
  11. Petals some other color than bright yellow or orange . . 14
    14. Leaflets 3 15
      15. Leaflets with small teeth 16
        16. Terminal leaflet sessile or subsessile . .
        . . . . *Trifolium*
        16. Terminal leaflet on a distinct stalk . . 17
          17. Flowers blue or purple .*Medicago*
          17. Flowers yellow or white . . . 18
            18. Flowers in elongated racemes .
            . . . . . *Melilotus*
            18. Flowers in small rounded heads,
            yellow 19
              19. Calyx 2-lipped .*Trifolium*
              19. Calyx with nearly equal
              teeth . . .*Medicago*
      15. Leaflets entire 20
        20. Flowers pale yellow or cream-colored or
        sometimes with a purplish spot . . . 21
          21. Flowers over 15 mm. long .*Baptisia*
          21. Flowers less than 12 mm. long . .
          . . . . . . *Lespedeza*
        20. Flowers white, blue, pink, or purple . . 22
          22. Stems upright or prostrate but never
          climbing or twining 23

32. Leaves with an uneven number of leaflets .   35
   35. Plants twining .   .   . *.Apios*
   35. Plants upright   36
      36. Flowers orchid and yellow; leaf-
         lets densely covered with gray-
         ish-white hairs .   *.Tephrosia*
      36. Not as above   37
         37. Leaves less than 1 cm.
            long, usually retuse *.Dalea*
         37. Leaves over 1 cm. long,
            acute or obtuse *.Astragalus*

Cercis L.

Cercis canadensis L. (2:383). Redbud is a very handsome small tree when flowering during April. It is common in rich woods or on wooded slopes (Fig. 46). Specimens with completely glabrous leaves may be segregated as var. *glabrifolia* Fern. SP. CIT.: typical: *M 651, BS 325.* Var. *glabrifolia: Welch and Fuller 186.*

Puereria DC.

Puereria lobata (Willd.) Ohwi (2:452). Kudzu-vine is becoming an aggressive weed in some southern counties where it often kills many of the trees on which it climbs (Fig. 47). SP. CIT.: *M 4710.*

Wisteria NUTT.

Wisteria floribunda (Willd.) DC. (2:412). The cultivated Wisteria is a native of Japan which sometimes escapes to roadsides and railways. It has fifteen or more leaflets and pubescent pods. SP. CIT.: along railroad, Murphysboro, August 14, 1954, *M 4624.* The native Kentucky Wisteria (*Wisteria macrostachya* Nutt.) (2:412) occurs in

FIG. 46. *Early spring blooming of Red Bud* (Cercis canadensis) *in a deciduous forest near Harrisburg creates a layering effect.*

low swampy woods in the most southern counties. It has up to fifteen leaflets and glabrous pods. It flowers during May.

Amorpha L.

Amorpha canescens Pursh (2:409). Lead-plant (Fig. 48) belongs
to the prairie flora and has been collected in prairie soil along the
Illinois Central Railroad south of Elkville. This is the only Southern
Illinois station south of St. Clair County. The leaves are densely
canescent. SP. CIT.: near Elkville, *Joe Garrison.*

Amorpha fruticosa L. (2:411). The typical variety with leaflets
about half as broad as long occurs on the limestone Devil's Bake
Oven north of Grand Tower. Variety *tennesseensis* (Shuttlw.) Palmer
with narrower leaflets occurs in low moist ground. Both of these may
be distinguished from other species of *Amorpha* in our area by their
pubescent petioles and absence of canescence on leaves. SP. CIT.:
typical; Devil's Bake Oven, June 20, 1955, *M 5427.* Var *tennesseensis,*
Turkey Bayou, *BS 2884.* On June 5, 1919, E. J. Palmer collected a
specimen of *Amorpha nitens* from Pope County along the Ohio River
near Golconda (No. *15371*). This species lacks canescence and is
glabrous on the petioles.

Robinia L.

Robinia pseudoacacia L. (2:414). Black Locust is native in Illinois
only in southeastern counties although it is a common tree throughout Illinois. The petioles are not hispid and flowers are white and
fragrant. SP. CIT.: *M 366; BS 37.*

Robinia hispida L. (2:414). Bristly Locust, a native of the southeastern states, has been planted occasionally along roads in Southern

FIG. 47. *In southwestern Jackson County introduced Kudzu-vine* (Puereria lobata) *smothers young trees.*

FIG. 48. *A shrubby legume is Prairie Lead Plant* (Amorpha canescens) *found growing along a railroad right-of-way near Elkville in northern Jackson County.*

Illinois. Petioles are densely hispid and the flowers are rose or pink.
SP. CIT.: east of Murphysboro, *Mohlenbrock.*

Gleditsia L.

Gleditsia triacanthos L. (2:383). Honey Locust is a common forest
tree of bottomland and swampy woods. Pods are usually over fifteen
centimeters long and contain several seeds. A form completely lacking thorns is forma *inermis* (L.) Zabel. SP. CIT.: *M 261; Mc 310.*
The One-seeded Locust occurs in swamps of several southern counties where it is found with Water Hickory, Pumpkin Ash, Bald Cypress, Swamp Cottonwood, and others. Pods usually do not exceed
ten centimeters and contain only one to three seeds.

Cladrastis RAF.

The handsome Yellowwood (*Cladrastis lutea* (Michx. f.) K. Koch)
(2:392) is known only from three stations in Alexander County near
Olive Branch. The beautiful flowers are produced during late May.

Gymnocladus LAM.

Gymnocladus dioicus (L.) K. Koch (2:383). Kentucky Coffee Tree
is a local species of woods, often being found at the base of steep
bluffs. This is one of the last woody species to put forth its leaves in
spring. SP. CIT.: Fountain Bluff, *BS 650;* Giant City State Park, *M 568.*

Crotalaria L.

Crotalaria sagittalis L. (2:394). Rattlebox is our only herbaceous
legume with simple leaves. The mature legumes contain loose seeds
which rattle when brushed against. This plant generally occurs in dry
open soil, seldom being found in the woods. Flowers are produced
from mid-June to the end of August. The leaves are nearly sessile.
SP. CIT.: *M 1226; BS 462; McCree and Wilson 1140. Crotalaria
spectabilis* Roth (2:395) has been found only once in Illinois in
Alexander County at the edge of a soybean field near Olive Branch
on August 20, 1949, by H. M. Franklin (No. *34*).

Stylosanthes SW.

Stylosanthes biflora (L.) Taub. (2:439). Pencil-flower has one of
the longest blooming periods of any of our species, beginning to
flower in mid-May and continuing well into October. It grows in dry

oak woods atop sandstone bluffs, although it occasionally has been found along stream banks. It is variable in degree of pubescence from rarely completely glabrous to a thin line of pubescence on the stem to somewhat hispid in the inflorescence. SP. CIT.: *BS 86; M 4721; M 4051.*

Cassia L.

1. Leaflets over 3 cm. long 2
    2. A black gland between lowest pair of leaflets . *Cassia tora* °
    2. A black gland near base of petiole 3
        3. Gland on a stalk; stipules setaceous . *Cassia hebecarpa*
        3. Gland not stalked; stipules linear-lanceolate . . .
            . . . . *Cassia marilandica*
1. Leaflets less than 2.5 cm. long 4
    4. Petals 1 cm. long or longer; stamens 10 . *Cassia fasciculata*
    4. Petals less than 1 cm. long; stamens 5 . . *Cassia nictitans*

Cassia hebecarpa Fern. (2:384). This species, Wild Senna, grows in rich woods where it has been found in Jackson, Union, Alexander, and Pulaski counties. At the Jackson County station it is found with *Thaspium trifoliatum, Elephantopus carolinianus,* and others. SP. CIT.: *M 4663.*

Cassia marilandica L. (2:384). This species of Senna occurs mostly along roadways which remain fairly moist throughout the year. It flowers in late July or early August. SP. CIT.: *M 4354; M 563; Welch 337.* A specimen of *Cassia tora* L. (2:386) collected by Fricke in Alexander County is in the University of Illinois Herbarium.

Cassia fasciculata Michx. (2:386). Partridge Pea is very abundant in waste ground all over Illinois. It often occurs in dense stands. It flowers in August. SP. CIT.: *BS 2709; M 163; Mc 896.*

Cassia nictitans L. (2:386). This is the Wild Sensitive Plant. In the evenings, leaves of this and the preceding species fold up. This species is not common in our area. SP. CIT.: *M 432.*

Trifolium L.

1. Flowers bright crimson . . . . . *Trifolium incarnatum*
1. Flowers white, pink, or yellow 2
    2. Flowers yellow 3
        3. Terminal leaflet stalked 4
            4. Heads with 20 or more flowers . . . . .
                . . . . *Trifolium procumbens*

    4. Heads up to 15-flowered . . *Trifolium dubium*
  3. Terminal leaflet sessile or subsessile . *Trifolium agrarium*
2. Flowers not yellow   5
    5. Each flower in the head with a distinct pedicel . . 6
      6. Calyx lobes over twice as long as the calyx tube .
        . . . . *Trifolium reflexum*
      6. Calyx lobes less than twice as long as the calyx tube   7
        7. Stems upright . . . *Trifolium hybridum*
        7. Stems creeping . . . *Trifolium repens*
    5. Each flower in the head without a distinct pedicel . . 8
      8. Plants erect   9
        9. Calyx puberulent or glabrous; flowers over 10
          mm. long; leaflets usually about half as wide as
          broad . . . . . *Trifolium pratense*
        9. Calyx villous; flowers at most 7.5 mm. long;
          leaflets usually much less than half as wide as
          broad . . . . . *Trifolium arvense* °
      8. Plants creeping; heads globose, 10 to 15 mm. in di-
        ameter; calyx labiate . . *Trifolium fragiferum*

Trifolium incarnatum L. (2:400). Crimson Clover. Along roads; not common; occasionally planted. SP. CIT.: *M 765; M 5137.*

Trifolium procumbens L. (2:403). Low Hop Clover. Fields and waste places; common; native of Europe. SP. CIT.: *M 234; Mc 69.*

Trifolium dubium Sibth. (2:403). Little Hop Clover. Fields and waste places; not as common as the preceding; restricted in Illinois to southern counties; native of Europe. SP. CIT.: *M 141.*

Trifolium agrarium L. (2:402). Yellow Hop Clover. The only Southern Illinois specimens are from Union County (F. S. Earle in 1879) and Jackson County (near Carbondale, July 16, 1878, *French 543*).

Trifolium reflexum L. (2:402). Buffalo Clover. Waste ground, and along streams; rare. SP. CIT.: *French,* from Fern Rocks, 1871.

Trifolium hybridum L. (2:401). Alsike Clover. Waste ground; common. SP. CIT.: *M 81; M 2348; Mc 31; Voigt.*

Trifolium repens L. (2:401). White Clover. Fields, lawns, waste ground; very abundant; native of Europe.

Trifolium pratense L. (2:400). Red Clover. Fields and waste ground; very common. SP. CIT.: *M 85; BS 51.*

Rabbit-foot Clover, *Trifolium arvense* L. (2:400), has been found in waste ground in a few of our counties. It is yet to be discovered in Jackson County.

Trifolium fragiferum L. (2:401). Strawberry Clover. Found growing east of Illinois Avenue near College Street, Carbondale, Illinois, Jackson County, *Joe E. Garrison.*

Medicago L.

Medicago sativa L. (2:405). Alfalfa is commonly cultivated in our area where it readily escapes. The flowers are blue. SP. CIT.: *M 358; Mc 3534.*

Medicago lupulina L. (2:405). Black Medic, with yellow flowers up to four millimeters long, is abundant in fields. SP. CIT.: *M 131; M 2516.*

Medicago arabica (L.) Huds. (2:405). This species is known from only one station in Illinois. The yellow flowers are about six millimeters long. SP. CIT.: orchard near Carbondale, May 15, 1940, *G. H. Boewe.*

Melilotus MILL.

Melilotus alba Desr. (2:403). White Sweet Clover is very common in all parts of Illinois. It has a somewhat longer blooming period than the Yellow Sweet Clover. SP. CIT.: *M 251; M 4625.*

Melilotus officinalis (L.) Lam. (2:403). Yellow Sweet Clover occurs abundantly in waste ground and fields. SP. CIT.: *V 1352; M 180; M 2332.*

Baptisia VENT.

Baptisia leucantha Torr. & Gray (2:392). Wild Indigo grows along railroad-strips or along edges of woods. The bracts are less than one centimeter long. The showy flowers open in June or July. SP. CIT.: *BS 476; MV 1587.*

The smaller *Baptisia leucophaea* Nutt. (2:392) occurs in prairies mostly to the north of our area although it has been recorded from Randolph County. The bracts are one to three centimeters long.

Clitoria L.

Clitoria mariana L. (2:447). Butterfly-Pea is found in dry woods or on sandstone bluffs. The large, attractive flower is produced in June and July. SP. CIT.: Giant City State Park, June 26, 1947, *BS 104.*

Petalostemum   MICHX.

Petalostemum candidum (Willd.) Michx. (2:411). White Prairie Clover is a characteristic species of hill prairies (Fig. 49). It usually is less common than Purple Prairie Clover. SP. CIT.: near Carbon Lake, June 20, 1951, *Hardy.*

Petalostemum purpureum (Vent.) Rydb. (2:411). Purple Prairie Clover (Fig. 49) occurs on hill prairies where it begins to flower just a little after the preceding species. SP. CIT.: north of Pine Hills, *M 4760.*

Desmodium   DESV.

1. Flowers borne on a leafless stalk arising directly from the ground
    .   .   .   . *Desmodium nudiflorum*
1. Not as above   2
    2. Flowers borne in a terminal panicle which arises from a cluster of leaves at the summit of the stem .   .   .   .   .
        .   .   .   . *Desmodium glutinosum*
    2. Not as above   3
        3. Flowers white; calyx nearly regular .   .   .   .   .
            .   .   .   . *Desmodium pauciflorum*
        3. Flowers pink or purple; calyx 2-lipped   4
            4. Leaflets one-quarter or less as wide as long; petioles 1 to 5 mm. long .   .   . *Desmodium sessilifolium*
            4. Leaflets over one-quarter to nearly as wide as long; petioles mostly over 5 mm. long   5
                5. Leaflets bearing hooked hairs so that they stick to clothing   6

FIG. 49. *A pleasing summer scene from a hill prairie situation includes White Prairie Clover* (Petalostemum candidum) *and Purple Prairie Clover* (Petalostemum purpureum).

6. Stipules glabrous or merely ciliate; leaflets
   not conspicuously net-veined . . .
   . . . . . *Desmodium canescens*
6. Stipules pubescent; leaves conspicuously
   net-veined . . . *Desmodium illinoense*
5. Leaflets without hooked hairs although often
   pubescent 7
   7. Leaflets glabrous or nearly so . . . 8
      8. Leaflets never half as wide as long .
         . . . *Desmodium paniculatum*
      8. Some leaflets one-half or more as wide
         as long 9
         9. Petioles mostly over 5 cm. long .
            . . *Desmodium cuspidatum* °
         9. Petioles mostly 1 to 5 cm. long . 10
            10. At least the terminal leaflet 4
                cm. long . . . . .
                . . *Desmodium laevigatum*
            10. None of the leaflets 4 cm. long
                . *Desmodium marilandicum*
   7. Leaflets with some kind of pubescence . 11
      11. All leaflets 3 to 4 times longer than
          broad . . *Desmodium paniculatum*
      11. Some leaflets broader than above . 12
          12. Leaflets as wide as long or wider,
              obtuse at the apex; stems usually
              prostrate . . . . .
              . . *Desmodium rotundifolium*
          12. Leaflets usually never as wide as
              long, often acute or acuminate;
              stems erect or ascending . . 13
              13. Terminal leaflet at most
                  averaging about 2.5 mm.
                  long . *Desmodium ciliare*
              13. Terminal leaflet averaging
                  more than 3 cm. long . . 14
                  14. Stipules lanceolate,
                      over 8 mm. long .
                      . . . . *Des-
                      modium longifolium*
                  14. Stipules narrower,
                      never over 8 mm. long 15
      15. Leaflets velvety beneath . . .
          . . . . *Desmodium nuttallii*
      15. Leaflets not velvety beneath although
          they may be densely pubescent . . 16

16. Flowers 6 mm. long or less; calyx
     3 mm. long or less   .   .   .
          .   .   . *Desmodium rigidum*
16. Flowers 7 to 13 mm. long; calyx
     over 3 mm. long   17
     17. Petioles 3 to 18 mm. long
          . *Desmodium canadense*
     17. Petioles over 20 mm. long
          .   . *Desmodium dillenii*

Desmodium nudiflorum (L.) DC. (2:427). Moist or dry woods; common. SP. CIT.: *M 448; Mc 338; Mc 910; BS 513.*

Desmodium glutinosum (Muhl.) Wood (2:427). Moist woods; common. SP. CIT.: *M 380.*

Desmodium pauciflorum (Nutt.) DC. (2:427). Woods; known in Illinois only from seven southern counties. SP. CIT.: *French; Mc 987;* Lake Murphysboro, August 3, 1954, *M 4731.*

Desmodium sessilifolium (Torr.) Torr. & Gray (2:428). Woods; not very common; we have it only from Jackson, Williamson, Saline, and Pope counties. SP. CIT.: *Mc 1126; BS 777;* Giant City State Park, August 6, 1953, *M 598.*

Desmodium canescens (L.) DC. (2:428). Dry woods; occasional. SP. CIT.: *M 519; M 4627; French 624.*

Desmodium illinoense Gray (2:429). Dry open ground; occasional. SP. CIT.: *M 425;* Devil's Backbone, August 24, 1948, *BS 589.*

Desmodium paniculatum (L.) DC. (2:430). Woods; not uncommon. SP. CIT.: *Mc 446; Mc 1020;* Riverside Park, Murphysboro, August 8, 1954, *M 4742.*

From Union and Johnson counties we have recorded *Desmodium cuspidatum* (Muhl.) Loud. (2:431). This species grows in open woods.

Desmodium laevigatum (Nutt.) DC. (2:431). This species is uncommon in our area, being known only from Jackson, Williamson, and Union counties where it grows in rich woods. SP. CIT.: *M 4661; Bell.*

Desmodium marilandicum (L.) DC. (2:430). This northern species occurs in Southern Illinois at Giant City State Park. SP. CIT.: Giant City State Park, August 28, 1953, *M 546.*

Desmodium rotundifolium DC. (2:427). Round-leaf Beggar's-tick is local in woods in some southern counties. SP. CIT.: *French;* near Ava Cave, August 26, 1954, *M 4656*.

Desmodium ciliare (Muhl.) DC. (2:430). This species is rare in Southern Illinois where we have it from Jackson and Union counties. It occurs in open dry woods. SP. CIT.: *M 536;* two miles south of Elkville, August 11, 1948, *BS 556*.

Desmodium longifolium (Torr. & Gray) Smyth (2:431) [*D. cuspidatum* (Muhl.) Loud. var. *longifolium* (Torr. & Gray) Schubert]. This rather uncommon species is found in dry open woods in Jackson, Alexander, and Massac counties. SP. CIT.: *French 615;* Clear Springs, September 11, 1952, *BS 2979*.

Desmodium nuttallii (Schindl.) Schubert (2:431) [*D. viridiflorum* (L.) DC.]. Locally in dry woods in a few counties. SP. CIT.: Giant City State Park, September 6, 1953, *M 539.*

Desmodium rigidum (Ell.) DC. (2:429) is reported from sandy soil in Williamson and Pope counties.

Desmodium canadense (L.) DC. (2:429). This species grows along railroads near Elkville. It is also listed from Williamson, Union, and Pope counties. SP. CIT.: near Elkville, July 19, 1948, *BS 522*.

Desmodium dillenii Darl. (2:431) [*D. perplexum* Schubert; *D. glabellum* (Michx.) DC.]. So far this species is known in Southern Illinois only from southwestern counties where it is found in dry, open woods. SP. CIT.: Giant City State Park, August 31, 1953, *M 497.*

Psoralea L.

Psoralea onobrychis Nutt. (2:406). This rather rare species grows along streams. It has been collected in our area in Jackson and Williamson counties. The mature leaflets are over two and one-half centimeters broad. SP. CIT.: June 24, 1941, *Mc 850*.

Psoralea psoralioides (Walt.) Cory var. *eglandulosa* (Ell.) F. L. Freeman (2:406). Scurf-pea occurs on dry wooded slopes in southern counties of Illinois. Leaflets do not exceed two centimeters in width. SP. CIT.: Giant City State Park, May 24, 1953, *M 181;* near Boskydell, May 29, 1949, *Evers 16795*.

Phaseolus L.

Phaseolus polystachyus (L.) BSP. (2:451). Wild Kidney Bean is an occasional species of woods and edges of woods. Flowers appear from late July through August. SP. CIT.: Giant City State Park, July 23, 1948, *BS 529*.

Galactia P. BR.

Galactia volubilis (L.) Britt. (2:451). Milk Pea grows in dry oak woods. In Illinois, it is confined to six southern counties. Wherever we have seen it, it has always been associated with *Scutellaria incana*. SP. CIT.: Lake Murphysboro, August 17, 1955, *M 56*.

Strophostyles ELL.

Strophostyles helvola (L.) Britt. (2:453). Wild Bean is a common species along roads and edges of woods. It may be distinguished by its more or less lobed leaflets, its glabrous calyx tube, and its pointed bracteoles. It blooms from mid-July through August. SP. CIT.: *M 506; BS 156*.

Strophostyles umbellata (Muhl.) Britt. (2:453). Umbellate Wild Bean, our rarest species of *Strophostyles*, occurs along roads, often growing with *Strophostyles helvola*. The leaflets are unlobed, the calyx tube is glabrous, and the bracteoles are blunt. SP. CIT.: *M 1341; McCree & Wilson 1145*.

Strophostyles leiosperma (Torr. & Gray) Piper (2:453). This species is abundant in dry woods and around strip mines and other dry open ground. The leaflets are unlobed and the calyx tube is pubescent. SP. CIT.: *M 435*.

Amphicarpa ELL.

Amphicarpa bracteata (L.) Fern. (2:451). Hog-peanut is local in open woods where it commonly grows with other twining legumes (*Galactia, Phaseolus*, and others). The terminal leaflet is thinly pubescent and seldom exceeds six centimeters in length. SP. CIT.: Fountain Bluffs, September 14, 1940, *Mc 450*.

Amphicarpa comosa (L.) G. Don (2:451) [*A. bracteata* (L.) Fern. var. *comosa* (L.) Fern.]. This species apparently is less common than the preceding. In our area it is recorded only from Jackson and Hardin counties. The terminal leaflet is densely pubescent and usually exceeds six centimeters in length. SP. CIT.: *BS 561; McCree & Wilson 1131*.

Desmanthus WILLD.

Desmanthus illinoensis (Michx.) MacM. (2:380). The Illinois Mimosa is very common along rivers in Southern Illinois. It generally is found in sandy soil. SP. CIT.: *Mc 1014*.

Pisum L.

Pisum sativum L. Garden Pea is cultivated and occasionally escapes into waste places. SP. CIT.: Carbondale, May 14, 1941, *Mc 734*.

Lathyrus L.

Lathyrus odoratus L. The cultivated Sweet Pea is spontaneous along roads.

Vicia L.

1. Flowers single or in pairs	2
2. Flowers over 2 cm. long . . . . . *Vicia sativa*	
2. Flowers less than 2 cm. long . . . *Vicia angustifolia*	
1. Racemes bearing 15 or more flowers	3
3. Calyx with a swollen area at base; flowers over 13 mm. long	4
4. Axis of raceme with long spreading hairs . *Vicia villosa*	
4. Axis of raceme with short appressed hairs . . .	
. . . . *Vicia dasycarpa*	
3. Calyx without a swollen area at base; flowers less than 13 mm. long . . . . *Vicia cracca*	

Vicia sativa L. (2:441). Spring Vetch is a native of Europe which has become established locally in Illinois. SP. CIT.: near Etherton, May 11, 1954, *M 2409*.

Vicia angustifolia Reich. (2:441). Common Vetch, another native of Europe, is an escape along roads and in waste ground in most of our counties. SP. CIT.: *M 233*. Specimens in which the leaflets of the upper leaf are emarginate are var. *segetalis* (Thuill.) W. D. J. Koch. SP. CIT.: near Etherton, *Mohlenbrock*.

Vicia villosa Roth (2:443). Winter Vetch is a common introduction along railroads and in waste ground. It is a native of Europe. SP. CIT.: *M 235; Mc 102; M 2408*.

Vicia dasycarpa Ten. (2:443). This species has been found in Illinois in only a few counties. SP. CIT.: *M 2329*.

Vicia cracca L. (2:443). Tufted Vetch is an occasional waif of roadsides and waste ground. It has been naturalized from Europe. SP. CIT.: *Mc 66.*

Apios MEDIC.

Apios americana Medic. (2:448). The species of this genus are our only twining legumes with five leaflets. The flowers are reddish. Hairy specimens may be segregated as forma *pilosa* Steyerm. Specimens with loose lanceolate-attenuate racemes belong to var. *turrigera* Fern. SP. CIT.: typical: *M 331; Mc 407; V 883.* F. *pilosa:* the "marsh," one mile north of Murphysboro, July 14, 1954, *M 4416.* Var. *turrigera:* one and one-half miles north of Etherton, July 15, 1941, *Mc 910.*

Apios priceana Robinson differs from the preceding in having pale rose-colored flowers. The standard has a fleshy knob on the apex. This species is very rare in moist woods or thickets. SP. CIT.: Wolf Lake, September 8, 1941, *G. D. Fuller 664.*

Tephrosia PERS.

Tephrosia virginiana (L.) Pers. (2:413). Goat's-rue is found on dry sandstone bluffs of the Shawneetown Ridge. Our specimens belong to var. *holosericea* (Nutt.) Torr. & Gray. SP. CIT.: Giant City State Park, May 24, 1953, *M 297; BS 2568.*

Dalea WILLD.

Dalea alopecuroides Willd. (2:408). This plant grows in moist waste ground. In Jackson County we have it only along the Mississippi River where it grows with *Boltonia interior, Mimulus alatus,* and others. SP. CIT.: *M 4647.*

Astragalus L.

Astragalus canadensis L. (2:420). Canadian Milk-Vetch is found along river banks. We have it only from Jackson, Hardin, and Pulaski counties. SP. CIT.: near Neunert, along the Mississippi River, September 5, 1954, *Brewer.*

76. *Geraniaceae* – Geranium Family

Geranium carolinianum L. (2:459). Carolina Cranesbill is a common species of waste ground. It flowers during April and May. SP. CIT.: *Mc 758; Stewart 1; M 241; Voigt.*

Geranium maculatum L. (2:458). Wild Geranium is abundant in rich woods. SP. CIT.: *M 2399; Mc 61; BS 11; M 1957.*

77. *Oxalidaceae* – Wood-sorrel Family

1. Flowers violet; plants with a bulb . . . . *Oxalis violacea*
1. Flowers white or yellow or greenish 2
   2. Flowers white; plants with a bulb . *Oxalis violacea* f. *albida*
   2. Flowers yellow or greenish; no bulbs 3
      3. Flowering stems creeping, rooting at nodes . . .
         . . . . *Oxalis corniculata*
      3. Flowering stems erect or ascending, not rooting at nodes 4
         4. Pedicels and stems with spreading hairs . . .
            . . . . *Oxalis stricta*
         4. Pedicels and stems with appressed hairs . . .
            . . . . *Oxalis dillenii*

Oxalis violacea L. (2:456). The attractive Violet Wood-sorrel is found on dry sandstone bluffs, in moist woods, or in waste ground. At the former habitat, it occurs with *Spiranthes gracilis, Krigia dandelion,* and others. White-flowered forms occur which are known as forma *albida* Fassett. SP. CIT.: *M 2403; M 1962; M 1895; BS 379.*

Oxalis corniculata L. (2:455). Creeping Wood-sorrel, native of Europe, has been collected in Illinois only in Champaign, Cook, and Jackson counties. SP. CIT.: waste ground around buildings, Giant City State Park, May 6, 1953, *M 175.*

Oxalis stricta L. (2:455) [*O. cymosa* Small]. The variable Common Wood-sorrel occurs throughout Illinois in waste ground. SP. CIT.: *Mc 873; V 1226; BS 423; M 2427.* Sometimes greenish double-flowered specimens occur.

Oxalis dillenii Jacq. (2:455) [*O. stricta* L.]. Upright Yellow Wood-sorrel is abundant in waste ground and woods in all of our counties. SP. CIT.: *M 1; Mc 712; Voigt.*

78. *Linaceae* – Flax Family

Linum L.

1. Flowers blue . . . . *Linum usitatissimum*
1. Flowers yellow 2
   2. Leaves with a pair of small glands at base . *Linum sulcatum*
   2. Leaves without glands at base 3

3. Most leaves opposite; plants of wet ground . . .
   . . . . Linum striatum
3. Most leaves alternate; plants of dry soil 4
   4. Inner sepals with gland-tipped cilia . Linum medium
   4. Inner sepals without gland-tipped cilia . . .
      . . . . Linum virginianum °

Linum usitatissimum L. (2:464). The only station for this species in Southern Illinois is at Fountain Bluff. SP. CIT.: Fountain Bluff, July 17, 1917, Cranwill.

Linum sulcatum Riddell (2:463). We have recorded this species from Jackson, Union, Pope, and Pulaski counties. It grows in dry rocky woods. SP. CIT.: near Carbondale, July 10, 1871, French.

Linum striatum Walt. (2:462). This Flax is rare in Illinois where it is known only from Jackson, Johnson, and Pope counties. At the Jackson County station, it was growing with Fimbristylis autumnalis and Hypericum mutilum. SP. CIT.: along stream near Ava Cave, September 1, 1954, M 4820.

Linum medium (Planch.) Britt. (2:463). This species grows in dry situations atop sandstone bluffs across Southern Illinois. It is the commonest species of flax. Our specimens belong to var. texanum (Planch.) Fern. SP. CIT.: Giant City State Park, June 11, 1953, M 309.

A fifth species, Linum virginianum L. (2:463), which grows in similar situations as Linum medium, has been collected in Union, Pope, Alexander, and Pulaski counties.

79. Balsaminaceae – Balsam Family

Impatiens L.

Impatiens biflora Walt. (2:513) [I. capensis Meerb.]. The Orange-spotted Touch-me-not is abundant in low flood-plain woods and other moist areas throughout Illinois. It is more common than the following species. The flowering period extends from June to September. SP. CIT.: Mc 430; Mc 825; M 4806.

Impatiens pallida Nutt. (2:513). Pale Touch-me-not is common in moist woods, often being found with Impatiens biflora. SP. CIT.: Mc 438; M 502.

80. *Zygophyllaceae* – Caltrop Family

Puncture-weed, *Tribulus terrestris* L. (2:467), has been collected in waste ground in Union County.

81. *Rutaceae* – Rue Family

1. Plants herbaceous . . . . *Ruta*
1. Plants woody 2
    2. Leaflets 3 . . . . *Ptelea*
    2. Leaflets 5 to 9 . . . . *Xanthoxylum*

Ruta L.

Ruta graveolens L. (2:469). Rue, sometimes cultivated, rarely escapes into waste ground. SP. CIT.: Giant City State Park, June, 1871, *French.*

Ptelea L.

Ptelea trifoliata L. (2:469). Hop-tree or Wafer-ash is a rare species in Southern Illinois where it grows along edges of woods. We have it from Randolph, Jackson, and Union counties. SP. CIT.: *French,* in 1870.

Zanthoxylum L.

Zanthoxylum americanum Mill. (2:467). Prickly Ash, or Toothache Tree, is known in Southern Illinois only from a single specimen collected in 1878 by French "in rocky woods and banks" and from two collections in 1919 by E. J. Palmer in Jackson and Johnson counties.

82. *Simarubaceae* – Quassia Family

Ailanthus DESF.

Ailanthus altissima (Mill.) Swingle (2:469). Tree-of-Heaven is often planted and frequently escapes. SP. CIT.: *Mc 975.*

83. *Polygalaceae* – Milkwort Family

Polygala L.

Polygala sanguinea L. (2:472). Red Milkwort is an occasional species of open woods or fields. It flowers from July to September. Racemes are five to fifteen millimeters thick. SP. CIT.: *M 406; Fuller, Welch, & Marberry 487.*

Polygala verticillata L. (2:473). Whorled Milkwort is a rare plant of dry soil in Southern Illinois. We have it only from Jackson County where it was collected in prairie soil along the railroad seven miles north of Murphysboro. The raceme is less than five millimeters thick and the leaves are opposite or whorled. SP. CIT.: *French.*

Polygala ambigua Nutt. (2:473) [*P. verticillata* var. *ambigua* (Nutt.) Wood]. We have recorded this rare species from Jackson and Pope counties where it grows in fields or open woods. The raceme is less than five millimeters thick and the leaves are alternate. SP. CIT.: *M 5378.*

84. *Euphorbiaceae* – Spurge Family

1. Milky juice present . . . . *Euphorbia*
1. Milky juice absent 2
  2. Plants with stellate hairs or scales 3
    3. Leaves 5 to 7 mm. broad, pointed at apex . *Crotonopsis*
    3. Leaves usually broader or, if only 5 to 7 mm. broad, then obtuse at apex . . . . *Croton*
  2. Plants lacking stellate hairs or scales 4
    4. Leaves entire; flowers minute and axillary . *Phyllanthus*
    4. Leaves serrate or denticulate 5
      5. Some of the flowers in terminal spikes . *Acalypha*
      5. Flowers either in lateral racemes or axillary . . 6
        6. Flowers in lateral racemes; leaves heart-shaped . . . . *Tragia* °
        6. Flowers axillary, subtended by palmately cleft bracts; leaves not heart-shaped . . *Acalypha*

Euphorbia L.

1. Leaves asymmetrical at base 2
  2. Leaves entire . . . . . . *Euphorbia serpens* °
  2. Leaves toothed (at least near the apex) 3
    3. Stems at maturity glabrous or glabrate . . . . *Euphorbia maculata*
    3. Stems at maturity villous 4
      4. Leaves hairy beneath . . . *Euphorbia supina*
      4. Leaves glabrous . . . *Euphorbia humistrata*
1. Leaves symmetrical at base 5

5. Leaves with white borders . . . *Euphorbia marginata*
5. Leaves not as above 6
  6. Plants with conspicuous white petal-like glands . .
    . . . . *Euphorbia corollata*
  6. Plants without conspicuous white petal-like glands . . 7
    7. Leaves 1 to 3 mm. wide, entire . . . . .
     . . . . *Euphorbia cyparissias*
    7. Leaves broader, serrate 8
      8. Leaves mostly opposite, pubescent . . .
       . . . . . . *Euphorbia dentata*
      8. Leaves mostly alternate, usually glabrous . . 9
        9. Some upper leaves lobed and often spotted
         with red or white at the base . . .
         . . . . *Euphorbia heterophylla* °
        9. None of the leaves lobed or spotted near
         base . . . . *Euphorbia obtusata* °

Euphorbia maculata L. (2:484) [*Chamaesyce maculata* (L.) Small; *Euphorbia preslii* Guss.]. Nodding Spurge is a common weed of waste ground. The habit of the plant is usually erect, although sometimes it may be prostrate. SP. CIT.: *Mc 415*.

Euphorbia supina Raf. [*Chamaesyce supina* (Raf.) Moldenke]. Milk Spurge is abundant in waste ground throughout Illinois. It commonly grows in cracks of sidewalks. SP. CIT.: *M 349; BS 587*.

Euphorbia humistrata Engelm. (2:484) [*Chamaesyce humistrata* (Engelm.) Small]. This is a rather common species which occurs along rivers and streams. It is usually found in sandy situations. SP. CIT.: *BS 2877*. Another species which grows in sand along our larger rivers is *Euphorbia serpens* HBK. [*Chamaesyce serpens* (HBK.) Small]. We have it from Randolph and Union counties.

Euphorbia marginata Pursh (2:485). Snow-on-the-Mountain often escapes from cultivation into waste ground where it persists. It is native in western United States (Fig. 50). SP. CIT.: *M 4626*.

Euphorbia corollata L. (2:486). Flowering Spurge may be found in dry woods, prairie areas, or waste ground throughout Illinois. On limestone hill prairies of Jackson County, the soft pubescent var. *mollis* Millsp. occurs. SP. CIT.: typical: *M 1484; M 4060; M 1588; M 2499*. Var. *mollis*: *M 1208*.

Euphorbia cyparissias L. (2:490). Cypress Spurge is most commonly found in cemeteries. It is naturalized from Europe. SP. CIT.: *BS 1006*.

FIG. 50. *Snow-on-the-Mountain* (Euphorbia marginata) *is a cultivated plant sometimes occurring as a weed in waste ground.*

Euphorbia dentata Michx. (2:487) [*Poinsettia dentata* (Michx.) Small]. Wild Poinsettia is a common plant of waste ground, although it sometimes may be found in woods or in the Pine Hills area along limestone bluffs. SP. CIT.: *M 4364; M 4625; BS 585.*

Lobed-leaf Wild Poinsettia, *Euphorbia heterophylla* L. (2:487) [*Poinsettia heterophylla* (L.) Klotzsch & Garcke], has been collected in Union County.

In talus at the base of steep limestone bluffs in Pine Hills of Union County has been found Bluntleaved Milk Spurge, *Euphorbia obtusata* Pursh (2:487). It flowers during May.

Crotonopsis MICHX.

Crotonopsis elliptica Willd. (2:477). This species may be found on sandstone bluffs, in fields, or sometimes in dry woods in most of our counties. SP. CIT.: Giant City State Park, July 14, 1953, *M 1492.*

Croton L.

Croton glandulosus L. (2:475). Sand Croton, adventive from southern United States, is an occasional plant of waste ground. Our plants belong to var. *septentrionalis* Muell.-Arg. The leaves are coarsely toothed. SP. CIT.: *M 4824.*

Croton capitatus Michx. (2:477). Capitate Croton occurs in sandy soils, often being found atop sandstone bluffs of the Shawneetown

Ridge (Fig. 51). The three deeply cleft styles bear twelve to twenty-four stigmas. SP. CIT.: *M 4723; Wilson & McCree 1178.*

Croton monanthogynus Michx. (2:475). This species is likewise found on sandstone bluffs or in sandy waste ground throughout our area. The two bifid styles bear four stigmas. SP. CIT.: *M 347; M 4512; M 4626.*

Phyllanthus   L.

Phyllanthus caroliniensis Walt. (2:475). These small plants grow in moist soil. This species is commonly associated with *Ammannia coccinea, Gratiola neglecta, Lindernia dubia,* and others. It is not common. SP. CIT.: ditch at junction of Illinois highways 3 and 144, *Mohlenbrock.*

Acalypha   L.

1. Leaves heart-shaped at base; inflorescence of terminal racemes
    .   .   .   .   *Acalypha ostryaefolia*
1. Leaves narrowed to base; inflorescence axillary   2
    2.   Petioles more than half as long as blades   .   .   .   .
         .   .   .   .   *Acalypha rhomboidea*
    2.   Petioles at most half as long as blades, usually less .   .   .   3
         3.   Stem with some spreading hairs .   . *Acalypha virginica*
         3.   Stem with appressed or incurved hairs   .   .   .   .
              .   .   .   .   *Acalypha gracilens*

Acalypha ostryaefolia Riddell (2:477). This rather uncommon species occurs along roads or in woods in a few of our counties. SP. CIT.:

FIG. 51. *Thin soil and exposed sandstone of uplands above Panther's Den in Williamson County.* Croton capitatus *is found in this habitat.*

grassy roadside between Crane and Cora, October 13, 1951, *H. E. Ahles 5654.*

Acalypha rhomboidea Raf. (2:477). This species is confined mostly to woodlands in Southern Illinois. SP. CIT.: *M 158; BS 821.*

Acalypha virginica L. (2:479). Virginia Three-seeded Mercury occurs in woods, along streams, or in waste ground throughout our area. SP. CIT.: *M 335; M 4340.*

Acalypha gracilens Gray (2:479). Slender Three-seeded Mercury, confined in Illinois mostly to southern counties, occurs in woods, fields, or open ground. SP. CIT.: *M 350.*

Tragia L.

The very rare *Tragia cordata* Michx. (2:479) is known in Illinois only from banks of the Ohio River at Golconda, Pope County, where it has been collected by S. A. Forbes, and later by E. J. Palmer.

85. *Celastraceae* – Staff-tree Family

1. Leaves opposite; plants not climbing or twining .   . *Euonymus*
1. Leaves alternate; plants twining   .   .   .   .   . *Celastrus*

Euonymus L.

Euonymus atropurpureus Jacq. (2:503). Wahoo, or Burning-bush, is a small tree occurring in woods, usually near streams. It flowers during May. The bright pink or scarlet fruit matures during the latter part of August. The flowers have four petals. SP. CIT.: *M 396; M 2503; M 4071.*

Euonymus obovatus Nutt. (2:503). Running-strawberry-bush is found in woods where it occurs with *Aplectrum hyemale* and other rather rare species. We have it from Jackson, Gallatin, Pope, and Hardin counties. The petals are five and the leaves are petioled. SP. CIT.: *M 98; M 4046.*

The very rare *Euonymus americanus* L. (2:503) has been collected in a woodland two miles west of Pulaski, Pulaski County, September 24, 1931, by *H. S. Pepoon* and *E. G. Barrett* (No. 505). Petals of this species are five and the leaves are sessile.

Celastrus L.

Celastrus scandens L. (2:503). Fruits of Climbing Bitter-sweet are sought during the winter season for their decorative value. While

FIG. 52. *Round-leafed White Bittersweet* (Celastrus orbiculatus), *a handsome ornamental climber, displays glossy leaves and fruit in July.*

Bitter-sweet is still rather common in our area, it is rapidly becoming scarce. SP. CIT.: *Mc 711; BS 795.*

Celastrus orbiculatus Thunb. (2:503). Round-leaved White Bitter-sweet (Fig. 52) sometimes escapes into woodlands. In Southern Illinois, we have it from Union and Jackson counties. Fruits are borne in clusters of one to three and the leaves are round. SP. CIT.: *M 291.*

86. *Aquifoliaceae* – Holly Family

Ilex L.

Ilex decidua Walt. (2:499). Swamp Holly has a peculiar habitat range in Southern Illinois. While it commonly grows in swampy woods or along streams, it may be found with surprising regularity atop dry limestone or sandstone bluffs. We have not noticed any habitat-induced changes in the plants. The deciduous leaves are glabrous or pubescent on only the veins beneath. SP. CIT.: Fountain Bluff, August 24, 1878, *French 1784.*

Winterberry, *Ilex verticillata* (L.) Gray (2:500), chiefly a species of northern Illinois, has been found in rich, moist woods in Pope and Union counties. The deciduous leaves are appressed pubescent beneath. The discovery of a tree of *Ilex opaca* Ait. (2:499) south of Alto Pass in Union County by Leon Minckler adds a new woody species to the Illinois flora. While this species has been attributed to

Southern Illinois for years, the Union County specimen (in Rambarger Hollow) is the only authentic report. It is thought that this tree is native in Union County (Voigt, 1955).

87. *Anacardiaceae* – Sumac Family

1. Leaves trifoliolate 2
    2. Leaves very aromatic; flowers appear before leaves; berries red, pubescent . . . . *Rhus aromatica*
    2. Leaves poisonous to the touch; flowers appear after leaves; berries whitish, glabrous . . . . . *Rhus radicans*
1. Leaves with 5 to many leaflets 3
    3. Leaf rachis winged . . . . . . *Rhus copallina*
    3. Leaf rachis not winged 4
        4. Twigs and leaves densely pubescent . . *Rhus typhina*
        4. Twigs and leaves glabrous . . . *Rhus glabra*

Rhus aromatica Ait. (2:497). Fragrant Sumac is an attractive shrub of dry woodlands. Flowers are formed during the latter part of February and in March. Red berries are mature about the last of August. Typical specimens have leaflets acute at the apex and somewhat wedge-shaped at the base. Specimens with leaflets rounded at the apex and with strongly wedge-shaped base are var. *serotina* (Greene) Rehd. Plants with velvety leaflets and tomentose twigs are var. *illinoensis* (Greene) Fern. SP. CIT.: typical: *M 107; M 1886*. Var. *serotina: V 1212*. Var. *illinoensis: Mc 763*.

Rhus radicans L. (2:494). The variable Poison Ivy is a most common plant of our area. Trifoliolate leaves borne alternately distinguish it from any other vine in our area. This plant also occurs in shrub form. One variety and two forms have been recorded from our area. Poison Oak, *Rhus toxicodendron* L., does not occur in Illinois. SP. CIT.: *Mc 93; Mc 767; M 147; Mc 318*.

Rhus copallina L. (2:497). Dwarf or Shining Sumac is a common shrub of thickets and waste ground. The leaves turn a brilliant crimson early in fall. Our specimens belong to var. *latifolia* Engler. SP. CIT.: *BS 1108; M 4357*.

Rhus typhina L. (2:497). Staghorn Sumac, often planted for ornament, is native in Illinois in northern counties and in a mesic woodland at Giant City State Park. SP. CIT.: Giant City State Park, September 20, 1953, *M 221*.

Rhus glabra L. (2:497). Smooth Sumac is abundant in woods, along railroads, and in waste ground throughout Illinois. The leaves turn color in fall. SP. CIT.: *M 300; Welch & Fuller 241.*

88. *Staphyleaceae* – Bladdernut Family

Staphylea trifolia L. (2:505). Bladdernut is often the dominant shrub in mesic woodlands. It and *Lindera benzoin* comprise most of the shrubby element in such situations. Flowers appear in April and early May. SP. CIT.: *BS 336; M 657.*

89. *Aceraceae* – Maple Family

Acer L.

1. Leaves 3- to 7-foliolate .   .   .   .   . *Acer negundo*
1. Leaves simple, palmately lobed 2
   2. Leaves white to silver beneath 3
      3. Leaves tomentose beneath at maturity .   .   .   .
        .   .   .   . *Acer drummondii*
      3. Leaves glabrous or nearly so beneath at maturity .   . 4
        4. Leaves with shallow sinuses .   .   . *Acer rubrum*
        4. Leaves with deep narrow sinuses . *Acer saccharinum*
   2. Leaves green or gray beneath 5
      5. Leaves with drooping sides, pubescent beneath   .   .
        .   .   .   . *Acer barbatum*
      5. Leaves flat, usually nearly glabrous beneath .   .   .
        .   .   .   . *Acer saccharum*

Acer negundo L. (2:509). Box Elder sometimes comprises 80 per cent of the woody plants in some of our low swampy woods. It is common in any moist situation. SP. CIT.: *M 49.*

Acer drummondii Hooker & Arnott (2:509) [*A. rubrum* L. var. *drummondii* (Hooker & Arnott) Sarg.]. In Southern Illinois the Drummond's Red Maple is found only in swamps. It often grows with the common Red Maple. SP. CIT.: Lester Swamp, nine miles west of Murphysboro, August 23, 1954, *M 4641.*

Acer rubrum L. (2:509). As with a few other species of plants, Red Maple occurs either in rich woods or on very dry bluffs. SP. CIT.: *Mc 608; M 4094.*

Acer saccharinum L. (2:509). Silver or Soft Maple is a common tree along streams or in moist woods in our area. SP. CIT.: *BS 653; M 2157.*

Acer barbatum Michx. The Southern Sugar Maple has long been listed under the name *Acer nigrum* in Southern Illinois. The presence of fine pubescence on the lower surface of the leaf and on the young ovaries distinguishes this southern species from the more northern Black Maple. Treated by some as a variety of *Acer saccharum*, the Southern Sugar Maple grows in rich beech-maple woods in the most southern counties in Illinois. SP. CIT.: *M 307; Fuller & Welch 174.*

Acer saccharum Marsh. (2:506). Sugar or Hard Maple is common in mesic woodlands where its principal associates include *Liriodendron tulipifera, Acer barbatum, Fagus grandifolia,* and *Nyssa sylvatica.* SP. CIT.: *M 21.*

90. *Hippocastanaceae* – Horse-chestnut Family

Aesculus L.

Aesculus glabra Willd. (2:509). Ohio Buckeye is found in woods locally in Southern Illinois. It is abundant in the Pine Hills area of Union County. The flowers are yellowish. SP. CIT.: Fountain Bluff, French.

Aesculus discolor Pursh (2:511). The beautiful Red Buckeye occurs along edges of woods in five southern counties. Red flowers are produced the last part of April. SP. CIT.: fertile bottoms, *Fuller & Welch 552.*

Horse Chestnut, *Aesculus hippocastanum* L. (2:509), is an attractive tree often cultivated in our area. It is distinguished from *A. glabra* and *A. discolor* by its very sticky buds.

91. *Sapindaceae* – Soapberry Family

Cardiospermum L.

Cardiospermum halicacabum L. (2:511). Balloon Vine occurs along the Mississippi River near Neunert (Jackson County). It is also known in Illinois from St. Clair County. It is a native of tropical America. SP. CIT.: September 5, 1954, *Brewer.*

92. *Rhamnaceae* – Buckthorn Family

1. Leaves with three main veins   .   .   .   .   .*Ceanothus*
1. Leaves pinnately veined .   .   .   .*Rhamnus*

Ceanothus L.

Ceanothus americanus L. (2:515). New Jersey Tea is a rather common small shrub occurring in dry woods or in hill prairies and along railroads. Specimens with leaves pilose above are var. *pitcheri* Torr. & Gray. SP. CIT.: typical: *Welch and Fuller 176; Mc 855; M 255.* Var. *pitcheri: M 3102.*

Rhamnus L.

Rhamnus caroliniana Walt. (2:516). Carolina Buckthorn is a rare species in Illinois, being found only in Monroe, Randolph, Jackson, Gallatin, and Pope counties. Our plants belong to var. *mollis* Fern., the type collected from Grand Tower by George Vasey. SP. CIT.: Devil's Bake Oven, *M 4828.*

93. *Vitaceae* – Grape Family

1. Leaves palmately compound .   .   .   .*Parthenocissus*
1. Leaves pinnately compound or simple 2
   2. Leaves pinnately compound .   .   .   .*Ampelopsis*
   2. Leaves simple 3
      3. Leaves usually lobed, pubescent (at least in vein axils)
      .   .   .   . *Vitis*
      3. Leaves not lobed, glabrous, shaped like leaf of *Populus*
      deltoides .   .   .   .*Ampelopsis*

Parthenocissus PLANCH.

Parthenocissus quinquefolia (L.) Planch. (2:522). Virginia Creeper is one of the most common and handsome of vines. It is very abundant in woodlands throughout Illinois. SP. CIT.: *M 245; M 4084.*

Ampelopsis MICHX.

Ampelopsis cordata Michx. (2:521). Raccoon Grape is found in woods or thickets in most of our counties although it cannot be considered common. It produces violet-colored fruits during late August. SP. CIT.: *French.*

Ampelopsis arborea (L.) Koehne (2:521). This plant, called Pepper Vine, occurs in Illinois only in the southwestern counties where it is usually found in calcareous areas near the Mississippi River. SP. CIT.: *French.*

Vitis L.

1. Leaves gray or reddish pubescent beneath at maturity .　　.　　.　2
   2. Branchlets smooth at maturity .　　.　　.　.*Vitis aestivalis*
   2. Branchlets hairy at maturity　　.　　.　　.　*Vitis cinerea*
1. Leaves not as above although there may be short hairs along veins　3
   3. New branches red in color .　　.　　.　　.　*Vitis palmata*
   3. New branches brown, gray, or green　.　　.　　.　　.　4
      4. Blades strongly lobed　.　　.　　.　　.　*Vitis riparia*
      4. Blades shallowly lobed .　　.　　.　　.　*Vitis vulpina*

Vitis aestivalis Michx. (2:518). Summer Grape is fairly abundant in low woods and along streams in Southern Illinois. SP. CIT.: *M 314; M 2669.*

Vitis cinerea Engelm. (2:518). Winter Grape is common in situations similar to those of Summer Grape. They are often found together. SP. CIT.: *M 230; M 4813; BS 1805.*

Vitis palmata Vahl (2:521). This species is known as Catbird Grape. It grows in rich woods. SP. CIT.: *M 743; BS 1806.*

Vitis riparia Michx. (2:519). Riverbank Grape appears to be the least common species of *Vitis* in our area. We have it only from Jackson, Williamson, Union, Pope, and Pulaski counties. It grows along streams in woods. SP. CIT.: *Mc 710.*

Vitis vulpina L. (2:519). Frost Grape is a common species in woods and along fences throughout Illinois. SP. CIT.: *M 470; M 3507.*

94. *Tiliaceae* – Linden Family

Tilia L.

Tilia americana L. (2:523). Basswood, or American Linden, is a handsome tree of moist woodlands in most of eastern United States. It is not common in Southern Illinois. Leaves are green beneath. SP. CIT.: *M 185; BS 2837.*

White Basswood, *Tilia heterophylla* Vent. (2:523), is rare in Illinois, having been collected only in Hardin, Pope, Massac, and Pulaski counties. Lower leaf surface is white-hairy.

95. *Malvaceae* – Mallow Family

1. Petals yellow, 4 to 6 mm. long; leaves with small spines at base
. . . . *Sida*
1. Petals of various colors, if yellow, then over 1 cm. long; leaves
without small spines at base 2
   2. Leaves deeply 3-parted; petals pale yellow with purple at
base . . . . *Hibiscus*
   2. Leaves not deeply 3-parted (if so, then flowers red), although often lobed 3
      3. Flowers white or pinkish, 6 to 12 mm. broad . . *Malva*
      3. Flowers very large, over 15 mm. broad 4
         4. Leaves and stems harshly pubescent . *Althaea*
         4. Leaves and stems glabrous or velvety . . . 5
            5. Flowers white or red . . . *Hibiscus*
            5. Flowers yellow . . . . *Abutilon*

Sida L.

Sida spinosa L. (2:530). Prickly Sida is a common plant of moist
waste ground. It is a native of tropical America. Flowers appear in
late July. SP. CIT.: *M 444; Mc 412.*

Hibiscus L.

Hibiscus lasiocarpus Cav. (2:534). Velvety Hibiscus is common in
moist ground throughout our area. It is abundant in low pin-oak
woods near Gorham. The plant is velvety, the flowers white or pinkish, and leaves are undivided and unlobed. SP. CIT.: *BS 250; M 4352;
Bell.*

Hibiscus militaris Cav. (2:534). This beautiful plant occurs along
streams and ponds in most southern counties. The plant is glabrous,
the flowers bright red, and leaves usually are three-lobed. SP. CIT.:
Joe Garrison.

Flower-of-an-Hour, *Hibiscus trionum* L. (2:535), a species of roadsides, fields, and waste places, has been collected in our area from
Randolph County. Leaves are deeply three-parted.

Malva L.

Common Mallow, *Malva neglecta* Wallr. (2:526), has been found
in hog lots in Pope and Alexander counties. It is naturalized from
Europe.

Althaea L.

Althaea rosea (L.) Cav. (2:525). Hollyhock is commonly cultivated in Southern Illinois where it freely escapes. It is a native of
China. All colors of petals may be found. SP. CIT.: *French.*

Abutilon MILL.

Abutilon theophrasti Medic. (2:533). Velvet-leaf is abundant in
waste ground, often found along fences. Flowers are produced during August. SP. CIT.: *Mc 410; M 4722.*

96. *Hypericaceae* – St. John's-wort Family

1. Flowers pink or greenish   .   .   .   .   . *Triadenum*
1. Flowers yellow 2
    2. Petals 4   .   .   .   . *Ascyrum*
    2. Petals 5   .   .   .   . *Hypericum*

Triadenum RAF.

Two species of the Marsh St. John's-wort (*Triadenum*) occur in
swampy woods of Southern Illinois, but neither has been recorded
from Jackson County. *Triadenum walteri* (Gmel.) Gleason (2:545)
has been collected in Saline, Union, Johnson, Alexander, Pulaski, and
Massac counties. The minutely punctate leaves are petiolate. *Trade-
num tubulosum* (Walt.) Gleason (2:545) is known only from Alexander, Pulaski, and Massac counties. The non-punctate leaves are sessile. Both species bloom in fall.

Ascyrum L.

Ascyrum multicaule Michx. (2:537) [*A. hypericoides* L. var. *multicaule* (Michx.) Fern.]. This semi-woody species grows in dry oak
woods atop the Shawneetown Ridge bluffs. Flowers appear in mid-
July. SP. CIT.: *M 1369; Mc 917.*

Hypericum L.

1. Stamens over 20   2
    2. Plants shrubby, usually over 3 feet tall 3
        3. Petals over 8 mm. long .   .   . *Hypericum prolificum*
        3. Petals less than 8 mm. long   .   .   .   .   .
             .   .   .   .   .   . *Hypericum lobocarpum* °

2. Plants herbaceous, less than 3 feet tall   4
    4. Petals with black dots, at least around the margins .   .   5
        5. Petals dotted only on margins; branches angled be-
            low each leaf .   .   .   .*Hypericum perforatum*
        5. Petals dotted all over; branches not angled below
            each leaf .   .   .   .   .*Hypericum punctatum*
    4. Petals not black-dotted   6
        6. Petals 4 to 5 mm. long; capsules 3 to 5 mm. long
            .   .   .   .   .   .   *Hypericum denticulatum*
        6. Petals 5 to 12 mm. long; capsules 5 to 7 mm. long
            .   .   .   .   .   .*Hypericum sphaerocarpum*
1. Stamens fewer than 15   7
    7. Leaves ovate, 3- to 5-nerved; plants of moist ground .   .
        .   .   .   . *Hypericum mutilum*
    7. Leaves linear or subulate, 1-nerved; plants of dry ground   8
        8. Leaves 1 to 3 mm. long   .   .*Hypericum gentianoides*
        8. Leaves over 5 mm. long .   .*Hypericum drummondii*

Hypericum prolificum L. (2:539). The beautiful Shrubby St. John's-wort is occassional in woods along streams in most of our counties. It flowers from mid-June to mid-July. SP. CIT.: *M 1160; Mc 924.*

One of the rarest plants in Southern Illinois is *Hypericum lobocarpum* Gatt. (2:539) [*H. densiflorum* Pursh var. *lobocarpum* (Gatt.) Svenson]. It has been found only once in Illinois [*Swayne* (No. *1104*)] near Brookport, Massac County, July 28, 1950.

Hypericum perforatum L. (2:541). Common St. John's-wort is frequent along roads and in fields throughout our area. SP. CIT.: *M 341.*

Hypericum punctatum Lam. (2:541). Spotted St. John's-wort is common in dry woods where it occurs with *Lechea tenuifolia, Linum* spp., and others. SP. CIT.: *M 4392.*

Hypericum denticulatum Walt. (2:542). This species is rare in Illinois, being recorded only from Pope County and Giant City State Park (Jackson County). At the latter station, it occurs with *Hackelia virginiana, Scrophularia marilandica,* and *Impatiens biflora.* SP. CIT.: Giant City State Park, August 9, 1953, *M 515.* Our specimens belong to var. *recognitum* Fern. and Schubert.

Hypericum sphaerocarpum Michx. (2:540). Round-fruited St. John's-wort has a wide range of tolerance as it is found in moist or dry woods or in prairie strips along railroads. SP. CIT.: *M 166; BS 742.*

Hypericum mutilum L. (2:543). Dwarf St. John's-wort occurs in moist soil along streams or in woods. It flowers from August to October. SP. CIT.: *BS 2707; Mc 558.*

Hypericum gentianoides (L.) BSP. (2:544) [*Sarothra gentianoides* L.]. This species is found in dry sandy woods across the Shawneetown Ridge. It sometimes grows with the following species. SP. CIT.: *Welch and Fuller 504.*

Hypericum drummondii (Grev. & Hook.) Torr. & Gray (2:544) [*Sarothra drummondii* Grev. & Hook.]. This species may be found either in dry sandy woods or in fields. It begins to flower about mid-July. SP. CIT.: *M 592; M 4719; McCree & Wilson 1151.*

97. *Cistaceae* – Rockrose Family

Lechea L.

Lechea minor L. (2:551). Pinweed is rare in Southern Illinois, being recorded only from dry sandstone bluffs in Jackson County (see Jones *et al.*, 1955). The leaves are one and one-half to four millimeters broad.

Lechea tenuifolia Michx. (2:551). Narrow-leaved Pinweed is common in dry open woods. Leaves are about one millimeter broad. SP. CIT.: *M 286.*

98. *Violaceae* – Violet Family

1. Flowers greenish-white, petals not spurred .   .   .*Hybanthus*
1. Flowers not greenish-white, lower petal spurred .   .   . *Viola*

Viola L.

1. Plants with leafy stems 2
      2. Stipules leaf-like, deeply pinnatifid; flower lavender or whitish
          .   .   .   .*Viola rafinesquii*
      2. Stipules not deeply pinnatifid; flowers white or yellow .   . 3
          3. Flowers white .   .   .   .   .   . *Viola striata*
          3. Flowers yellow .   .   .   .   .   .*Viola eriocarpa*
1. Plants with the leaves from an underground rhizome .   .   . 4
      4. Leaves (at least some of them) cleft, at least at base .   . 5
          5. Petals without hairs on the inner surface .   .   .   .
              .   .   .   . *Viola pedata*
          5. Petals with hairs on some of them 6

Viola rafinesquii Greene (2:566) [*V. kitaibeliana* R. & S. var. *rafinesquii* (Greene) Fern.]. Johnny-jump-up, or Wild Pansy, is exceedingly abundant in waste ground. Color of petals ranges from white to rose-lavender. SP. CIT.: *M 913; M 1956; M 1889; M 2197; Voigt.*

Viola striata Ait. (2:565). Cream Violet occurs in moist woods. At Giant City State Park, it forms a continuous carpet of flowers along with *Collinsia verna.* SP. CIT.: *BS 60; M 2128; M 2563.*

Viola eriocarpa Schw. (2:564) [*V. pennsylvanica* Michx.]. Smooth Yellow Violet is common in moist woods throughout Illinois. We have recorded it flowering as early as April 10. SP. CIT.: *M 109; M 1944.*

Viola pedata L. (2:555). Bird-foot Violet is found in dry rocky woods in a few of our counties. Petals are either all violet or, in many cases, the upper two petals are deep velvety lavender. The former are segregated as var. *lineariloba* DC. SP. CIT.: Midland Hills, *Voigt.*

Viola falcata Greene (2:559) [*V. triloba* Schw. var. *dilatata* Ell.]. Lobed Violet is found in dry woods. The shape of leaves is variable. SP. CIT.: *M 5; BS 30.*

Viola sagittata Ait. (2:558). Arrow-leaved Violet is not common in Southern Illinois. We have it only from Randolph, Jackson, Union, and Massac counties. SP. CIT.: Lake Murphysboro.

Viola odorata L. (2:560). An occasional escape from cultivation is Sweet Violet. It is introduced from Europe.

Viola sororia Willd. (2:556). Woolly Blue Violet is probably our most common species of Violet. It occurs in moist or dry woodlands. SP. CIT.: *Mc 684; VM 1887.*

Viola cucullata Ait. (2:556). The rare Marsh Blue Violet has been found in our area only in Jackson and Pulaski counties where it grows in deep rich woods. SP. CIT.: Giant City State Park, April 17, 1953, *Stewart 10.*

Viola missouriensis Greene (2:556). This species grows in woods locally in Southern Illinois. We have collected it in flower as early as March 9. SP. CIT.: *M 25.*

Viola papilionacea Pursh (2:555). Butterfly Violet is a common species in Illinois, occurring in woods, waste ground, or other areas. A form with white flowers marked with pale gray or blue lines is often cultivated in our area. It is var. *albiflora* Grover [*V. priceana* Pollard], the Confederate Violet. SP. CIT.: *M 2156; Mc 637; M 57.*

Hybanthus JACQ.

Hybanthus concolor (T. F. Forst) Spreng. (2:567). One must look closely at this small greenish-flowered plant to recognize that it is a member of the violet family. SP. CIT.: three miles south of Murphysboro, *Mc 719; BS 940; BS 2026.*

99. *Thymelaeaceae* – Mezereum Family

Dirca L.

Leatherwood, *Dirca palustris* L. (2:572), a shrub found in dry woods, is rare in Southern Illinois. It is known only from Johnson and Pope counties.

100. *Elaeagnaceae* – Oleaster Family

Elaeagnus L.

Elaeagnus angustifolia L. (2:573). Russian Olive is planted in our area and sometimes escapes. SP. CIT.: Makanda, 1884, *J. Hague* (UI).

101. *Lythraceae* – Loosestrife Family

1. Stem glabrous; leaves sessile 2
    2. Flowers with petals 3
        3. Flowers with 4 petals 4
            4. Leaves auriculate; flowers solitary in the axils . .
                . . . . *Ammannia*
            4. Leaves narrowed at base; flowers usually more than
                one in each axil . . . . . . *Rotala*
        3. Flowers with 5 to 7 petals . . . . . *Lythrum*
    2. Flowers without petals . . . . . . *Peplis* °
1. Stem pubescent; leaves with petioles 5
    5. Mature leaves more than 5 cm. in length, mostly whorled;
        flowers regular; plants shrubby . . . *Decodon* °
    5. Mature leaves less than 4 cm. long, sticky; flowers irregular;
        plants herbaceous . . . . *Cuphea*

Ammannia L.

Ammannia coccinea Rothb. (2:577). This species grows in or near water throughout our area. It often forms dense stands. SP. CIT.: *Mc 1130; M 4340; M 4725.*

Rotala L.

Rotala ramosior (L.) Koehne (2:575). This species is very similar to the preceding and often grows with it. SP. CIT.: *Bailey and Hankla 622; Bell.*

Lythrum L.

Lythrum alatum Pursh (2:579). Common Loosestrife grows in a variety of habitats in Southern Illinois. It is found in moist soil as well as in prairie soil along railroads. Leaves are alternate and flowers are less than one centimeter broad. SP. CIT.: *Mc 932.*

Lythrum salicaria L. (2:579). Purple Loosestrife is known in Southern Illinois only from a moist ditch south of Elkville. Leaves are mostly opposite and flowers are over twelve millimeters broad. SP. CIT.: *Joe Garrison.*

Peplis L.

Water Purslane, *Peplis diandra* Nutt. [*Didiplis diandra* (Nutt.) Wood], grows along edges of ponds or in swamps. We have it from Pope and Union counties.

Decodon   J. F. GMEL.

Water Willow, or Swamp Loosestrife, *Decodon verticillatus* (L.) Ell. (2:577), grows in water or mud very locally in Southern Illinois (Saline, Union, and Pulaski counties). Beautiful flowers are produced the first part of August.

Cuphea   P.BR.

Cuphea petiolata (L.) Koehne (2:577). Waxweed is a sticky plant. While it usually grows in dry soil, it may be found in low woods associated with *Carex muskingumensis, Phlox paniculata,* and others. SP. CIT.: *Welch & McCree 1180; M 4339.*

102. *Passifloraceae* – Passion-flower Family

Passiflora   L.

Passiflora lutea L. (2:268). This species of Passion-flower is fairly common along edges of woods. The flowers are yellow. This species in sterile condition may be mistaken for *Menispermum canadense* but it is distinguished from Moonseed by the presence of tendrils. Our specimens belong to var. *glabriflora* Fern. SP. CIT.: *Mc 914.*

Passiflora incarnata L. (2:568). Maypops occurs along roads and fences occasionally in Southern Illinois. The purple flowers are attractive. Fruits are edible as in the preceding species. SP. CIT.: *M 4352.*

103. *Cactaceae* – Cactus Family

Opuntia   MILL.

Opuntia rafinesquii Engelm. (2:571) [*O. humifusa* Raf.; *O. compressa* (Salisb.) Macbr.]. Prickly Pear is a rather common species of the Shawneetown Ridge (Fig. 53). The handsome flowers are to be found in late May. SP. CIT.: *BS 701; BS 712.*

104. *Loasaceae* – Loasa Family

Mentzelia   L.

In hill prairies of Monroe County has been found rare Stick-leaf, *Mentzelia oligosperma* Nutt. (2:569). It is known in Illinois only from counties bordering the Mississippi River. Evers (1950).

FIG. 53. *Prickly Pear Cactus* (Opuntia rafinesquii) *grows in dry situations on sandstone outcrops.*

105. *Cucurbitaceae* – Gourd Family

1. Stems glabrous . . . .*Melothria* °
1. Stems pubescent 2
    2. Leaves palmately lobed . . . .*Sicyos*
    2. Leaves pinnately divided . . . . . .*Citrullus*

Melothria L.

The Creeping Cucumber, *Melothria pendula* L. (3:313) was collected in rocky soil at the base of limestone bluffs in Alexander County near Thebes by H. M. Franklin in 1949. It should be sought in the Pine Hills area. It flowers during late summer. SP. CIT.: *Stadlebacher*—Union County, 1955.

Sicyos L.

Sicyos angulatus L. (3:313). Bur Cucumber is found occasionally in Southern Illinois where it occurs chiefly in calcareous areas. It is very common along the base of Pine Hills. SP. CIT.: north of Pine Hills, *M 4709.*

Citrullus SCHRAD.

Citrullus vulgaris Schrad. Watermelon often escapes from cultivation but it can hardly be considered established. SP. CIT.: along Gulf, Mobile and Ohio Railroad, Murphysboro, *Mohlenbrock.*

106. *Melastomaceae* – Melastoma Family

Rhexia L.

Rhexia virginica L. (2:581). The handsome Meadow Beauty has been found in moist sandy soil in Jackson, Williamson, Pope, and Massac counties. It is nowhere abundant. The stem is four-winged and leaves are rounded at the base. The brilliant rose-purple flowers appear during August and September. SP. CIT.: Carbondale Reservoir, September 3, 1954, *Brewer*. The even rarer *Rhexia mariana* L. (2:581) has been found in our area (according to Fernald, 1950). We have not seen any specimens.

107. *Onagraceae* – Evening Primrose Family

1. Herbs with opposite leaves   2
    2. Erect plants; leaves dentate; flowers in terminal racemes; flowers white with 2 petals and 2 stamens .   .   .*Circaea*
    2. Procumbent or floating herbs; flowers solitary in axils of leaves; flowers pinkish or petals absent; sepals 4; stamens 4 .   .   .   . *Ludwigia palustris*
1. Herbs with alternate leaves   3
    3. Plants floating on water or growing prostrate on mud flats; leaves petiolate, entire; petals and sepals 5; stamens 10 .   .
        .   .   .   . *Jussiaea diffusa*
    3. Plants erect   4
        4. Stems angled; leaves sessile or subsessile and entire; flowers solitary in axils, yellow; petals 4; stamens 8   .   .
        .   .   .   . *Jussiaea decurrens*
        4. Stems not angled   5
            5. Flowers in terminal spikes or racemes; leaves sessile or subsessile; petals 4; stamens 8; flowers pinkish
            .   .   .   . *Gaura*
            5. Flowers axillary, several or solitary   6
                6. Flowers pink- or rose-colored   .   .   .   7
                    7. Flowers 3 to 6 cm. broad, loosely spicate
                    .   .   .   .   . *Oenothera speciosa*
                    7. Flowers 2 to 3 mm. broad, numerous in axils of leaves   .   .   .   . *Epilobium*
                6. Flowers yellowish-green or yellow .   .   .   8
                    8. Stamens 8 .   .   .   . *Oenothera*
                    8. Stamens 4 .   .   .   . *Ludwigia*

Circaea L.

Circaea latifolia Hill (2:599) [*C. quadrisulcata* (Maxim.) Franch. & Sav. var. *canadensis* (L.) Hara]. This species is known as Enchanter's Nightshade. It grows in moist soil in deep shade. Flowers are produced from mid-June through August. SP. CIT.: *Mc 118; M 482; M 4091.*

Ludwigia L.

Ludwigia palustris (L.) Ell. (2:583). This creeping, prostrate *Ludwigia* with opposite leaves occurs in moist ditches throughout this area. Specimens in the Southern Illinois University herbarium from LaRue Swamp and labelled *Ludwigia natans* Ell. are actually *L. palustris*. Our specimens belong to var. *americana* (DC.) Fern. SP. CIT.: *M 392.*

Ludwigia polycarpa Short & Peter (2:584). This species is confined to wet soil along roads in Jackson, Saline, and Johnson counties. Flowers are sessile and the capsule is about as broad as long. SP. CIT.: (see Jones *et al.,* 1955).

Ludwigia alternifolia L. (2:584). Seedbox is fairly abundant along streams and in other moist situations in Southern Illinois. The calyx lobes turn a brilliant crimson in fall. Flowers are on distinct pedicels. SP. CIT.: *M 1124; V 832.*

A fourth species, *Ludwigia glandulosa* Walt. (2:585), has been collected in low swampy ground in Johnson, Pulaski, and Massac counties. Flowers are sessile and the capsule is about twice as long as broad.

Oenothera L.

1. Petals pink . . . .*Oenothera speciosa*
1. Petals yellow 2
    2. Leaves thread-like; petals 3 to 5 mm. long . . . .
    . . . . *Oenothera linifolia*
    2. Leaves broader; petals larger 3
        3. Leaves pinnatifid . . . .*Oenothera laciniata*
        3. Leaves entire or merely toothed 4
            4. Calyx lobes erect or ascending at maturity . . 5
                5. Leaves and stem with appressed pubescence
            . . . . . . *Oenothera tetragona*

Oenothera speciosa Nutt. (2:593). White or Pink Evening Primrose is an occasional waif along roads where it is .adventive from western United States. SP. CIT.: *BS 2484; M 2368; V 1235.*

Oenothera linifolia Nutt. (2:595). Thread-leaved Sundrops is found locally on standstone bluffs across Southern Illinois. It is usually associated with *Stylosanthes biflora, Lechea tenuifolia,* and *Polygonum tenue.* SP. CIT.: Giant City State Park, May 15, 1953, *M 67.*

Oenothera laciniata Hill (2:592). This species grows in waste ground or in open woods throughout this area. It is abundant along railroads. SP. CIT.: *M 2187; Voigt.*

Oenothera tetragona Roth (2:595). This rare plant grows at edges of woods. We have it from six counties. SP. CIT.: *French.*

Oenothera pilosella Raf. (2:595). Common Sundrops is another plant with a wide range of habitats. It seems equally at home either in the dry prairie soil along railroads or in low swampy woods. SP. CIT.: *Mc 159; Mc 782; Hardy 4.*

Oenothera strigosa (Rydb.) Mack. & Bush (2:591). This Evening Primrose, native in western United States, occurs occasionally in fields or along roads. SP. CIT.: (see Jones *et al.,* 1955).

Oenothera biennis L. (2:591). Common Evening Primrose is abundant in dry waste ground. It is variable in its stature, degree of pubescence, and size of flowers. SP. CIT.: *M 1370; McCree & Wilson 1141.*

Oenothera rhombipetala Nutt. (2:592). This species, which is confined in Illinois chiefly to northern counties, has been collected in sandy soil in Jackson and Pulaski counties. SP. CIT.: in University of Illinois herbarium.

FIG. 54. *Water Evening Primrose* (Jussiaea diffusa) *creeping on mud and floating on water of a shallow pond.*

Jussiaea L.

Jussiaea diffusa Forsk. (2:582) [*J. repens* L.]. Water Evening Primrose often covers small shallow bodies of water (Fig. 54). The stems are not winged and the flowers have five petals. SP. CIT.: *Mc 470; M 4737; Voigt; Bell.*

Jussiaea decurrens (Walt.) DC. (2:582). This is rare in Illinois. It grows in wet soil elsewhere, too. We have it from Jackson, Union, Hardin, Alexander, Pulaski, and Massac counties. The stems are strongly winged and the flowers have four petals. SP. CIT.: one mile north of Pomona, July 8, 1951, *Stewart and Hatcher 108.*

Gaura L.

Gaura biennis L. (2:597). Pink Butterfly Weed is rather common in prairie strips along railroads of Jackson and Randolph counties. The petals are five millimeters or more long and the ovary is sessile. Our specimens belong to var. *pitcheri* Pickering. SP. CIT.: two miles south of Mathews, July 25, 1954, *M 4364.*

Gaura parviflora Dougl. (2:597). This species is known in Southern Illinois only from a station in prairie soil along the Illinois Central Railroad north of Carbondale. The petals are about two millimeters long and the ovary is sessile. SP. CIT.: three miles north of Carbondale, July 20, 1949, *BS 772.*

A third species, *Gaura filipes* Spach (2:597), is known in Illinois only from Hardin County between Cave-in-Rock and Saline Creek. It was found in 1916 by Trelease. The ovary is pedicelled.

Epilobium L.

Epilobium co'oratum Muhl. (2:589). Willow-herb is found in moist ground locally in our area. It flowers in August and September. SP. CIT.: M 475; V 998; Bell.

108. Haloragidaceae – Water Milfoil Family

1. Leaves alternate; stamens 3 . . . . .Proserpinaca
1. Leaves opposite (alternate in one species); stamens 4 . . .
. . . . Myriophyllum

Proserpinaca L.

Proserpinaca palustris L. (2:602). Mermaid-weed has been found in ponds in six of our counties. Our specimens usually are segregated as var. crebra Fern. & Grisc. SP. CIT.: Snyder Lake, October 2, 1952, BS 3103.

Myriophyllum L.

Myriophyllum heterophyllum Michx. (2:601). This species of Water Milfoil is known in our area only from a small pond about one mile north of Murphysboro. The bracts have spinulose teeth. SP. CIT.: June 19, 1951, Hardy 2.

Myriophyllum verticillatum L. (2:601). This species is abundant in the lagoon at Riverside Park, Murphysboro. We also have it from Saline County. The bracts are not spinulose-toothed. SP. CIT.: August 26, 1954, M 4829.

Pinnate Milfoil, Myriophyllum pinnatum (Walt.) BSP. (2:601), is known from shallow water in Randolph County. It is our only Myriophyllum with alternate leaves.

109. Callitrichaceae – Water Starwort Family

Callitriche L.

Callitriche terrestris Raf. (2:491) [C. austini Engelm.]. The terrestrial Water Starwort grows in moist soil throughout our area. It is undoubtedly often overlooked. The fruit is on a short peduncle. SP. CIT.: Giant City State Park, Mohlenbrock.

Callitriche heterophylla Pursh (2:493). This species grows in shallow water where it has been found in Jackson, Union, and Johnson counties. The fruit is sessile and has round margins. SP. CIT.: *BS 1024.*

A third species, *Callitriche palustris* L. (2:491), grows in shallow water. We have it only from Union and Johnson counties. The sessile fruits have acute margins.

110. *Cornaceae* – Dogwood Family

1. Leaves opposite . . . .*Cornus*
1. Leaves alternate 2
   2. Lateral veins of leaf curved forward toward apex .*Cornus*
   2. Lateral veins of leaf running nearly straight toward margin
    . . . . *Nyssa*

Cornus L.

1. Flowers subtended by 4 large white (or pink) bracts . . .
   . . . . *Cornus florida*
1. Flowers not as above 2
   2. Leaves alternate . . . . . .*Cornus alternifolia*
   2. Leaves opposite 3
      3. Upper surface of leaves scabrous . . . . .
      . . . . *Cornus drummondi*
      3. Upper surface of leaves usually pubescent but not scabrous 4
         4. Leaves green on both sides . .*Cornus foemina*
         4. Leaves paler beneath 5
            5. Young branches with appressed hairs or glabrous 6
               6. Mature branches bright red . . . . .
               . . . . . . *Cornus stolonifera*
               6. Mature branches gray or brown . . .
               . . . . . . *Cornus racemosa*
            5. Young branches with dense soft pubescence .
            . . . . . . *Cornus obliqua*

Cornus florida L. (2:643). The beautiful Flowering Dogwood is common in dry woods throughout Southern Illinois. Flowers are produced during late April and May. Unfortunately, the tree is vulnerable to greedy flower-lovers. SP. CIT.: *French 1029.* Occasionally specimens are found with pinkish bracts. These are forma *rubra* (Weston) Palmer & Steyerm. SP. CIT.: Giant City State Park, April 11, 1953, *M 58.*

Cornus alternifolia L. f. (2:644). Alternate-leaved Dogwood is exceedingly rare in Southern Illinois where it occurs in rich woods. We have it from Jackson, Pope, and Alexander counties. SP. CIT.: Illinois State Museum herbarium.

Cornus drummondi C. A. Meyer (2:645). Rough-leaved Dogwood is common in moist ground. It is often found along roads. Flowers are formed during May. SP. CIT.: *M 302.*

Cornus foemina Mill. (2:645) [*C. stricta* Lam.]. This species grows in low swampy pin-oak woods. Other associated shrubs are *Itea virginica* and *Ilex decidua*. It is not common. SP. CIT.: near Howardton, July 11, 1955, *M 5523.*

Cornus stolonifera Michx. (2:645). Red-osier Dogwood is known in Southern Illinois only from along a stream in Giant City State Park. Flowers appear in early May while the white fruits mature during August. SP. CIT.: Giant City State Park, *M 145.*

Cornus racemosa Lam. (2:645). Gray Dogwood is not an uncommon species in woods or along fences. SP. CIT.: *M 471; Mc 1143.*

Cornus obliqua Raf. (2:644). This is the Pale Dogwood. It is found in moist woodlands locally in Southern Illinois. SP. CIT.: Pine Hills, Union County, *M 2436.*

Nyssa L.

Nyssa sylvatica Marsh. (2:646). The handsome Sour Gum or Black Gum is a common tree of mesic forests. Leaves of this species turn a brilliant crimson early in the fall. SP. CIT.: *M 146.*

Tupelo Gum, *Nyssa aquatica* L. (2:646) [*N. uniflora* Wang.], is a Coastal Plain species found in bottomland and swampy woods of Alexander, Pulaski, Massac, Union, Johnson, Pope, and Hardin counties. It has as some of its associates *Taxodium distichum, Carya aquatica, Rosa palustris*, and *Populus heterophylla.*

111. *Araliaceae* – Ginseng Family

1. Leaves in whorls of three (rarely four)   .   .   .   . *Panax*
1. Leaves alternate   .   .   .   . *Aralia*

Panax L.

Panax quinquefolia L. (2:604). Ginseng (Fig. 55) is rare in moist woods. It has become uncommon because of mass collections of its roots during past years. Underground parts are thought to have medicinal value as a demulcent. SP. CIT.: *Hall 1032.*

Aralia L.

Aralia spinosa L. (2:604). Hercules' Club, or Devil's Walking-stick, is a small tree or large shrub found locally in moist woods in the southern counties. The purple berries are eaten by birds, particularly Cedar Waxwings. SP. CIT.: Giant City State Park, June 4, 1953, *M 182.*

Aralia racemosa L. (2:604). American Spikenard is rare in our area where it grows on moist shaded sandstone bluffs. It is herbaceous. SP. CIT.: *BS 514; M 4044.*

112. *Umbelliferae* – Carrot Family

1. Leaves simple 2
   2. Leaves long and narrow with spinulose teeth . .*Eryngium*
   2. Leaves ovate, often cordate . . . . .*Thaspium*
1. Some or all of the leaves compound 3
   3. Fruit and ovary bristly 4
      4. Leaves palmately divided . . . . .*Sanicula*
      4. Leaves pinnately divided 5
         5. Some of the leaf segments over 1 cm. broad . .
            . . . . *Torilis*
         5. Leaf segments much less than 1 cm. broad . *Daucus*
   3. Fruit and ovary not bristly although sometimes pubescent 6
      6. Leaf segments at most 1 mm. broad 7
         7. Flowers yellow . . . . . .*Anethum*

FIG. 55. *A plant carefully looked for by herb collectors is Ginseng* (Panax quinquefolia).

7. Flowers white 8
    8. Inflorescence subtended by many narrowly cleft
       bracts . . . . *Ptilimnium*
    8. Inflorescence with 0 to 3 short bracts . . .
       . . . . *Perideridia* *

6. Leaf segments over 1 mm. broad 9
9. Leaflets three 10
   10. Flowers white . . . . *Cryptotaenia*
   10. Flowers yellow or purple 11
      11. Flowers purple . . . . *Thaspium*
      11. Flowers yellow 12
         12. All flowers on stalks; fruit winged
           . . . . . . *Thaspium*
         12. At least the central flower sessile;
           fruit not winged . . *Zizia*
9. Leaflets more than three 13
   13. Plants 3 to 7 inches tall from a tuber; flowers
      in purple and white clusters appearing in Feb-
      ruary and March . . . . *Erigenia*
   13. Plants not as above 14
      14. Petals yellow 15
         15. Some of the leaflets at least 2 cm.
           broad 16
           16. Leaflets entire . . *Taenidia*
           16. Leaflets serrate . *Pastinaca*
         15. All leaflets at most about 1 cm.
           broad . . . . *Polytaenia*
      14. Petals white 17
         17. Leaves divided into separate leaflets
           which are mostly over 1.5 cm. broad 18
           18. Leaves twice ternately com-
             pound; plants with odor of
             licorice . . . *Osmorhiza*
           18. Leaves 1 to 3 pinnate; no lico-
             rice odor 19
              19. Leaves once-pinnate . 20
                 20. Plants densely pu-
                   bescent . . .
                   . . *Heracleum*
                 20. Plants glabrous . 21
                   21. Inflorescence
                      subtended by
                      bracts . *Sium*
                   21. Inflorescence
                      without bracts
                      . *Oxypolis*

19. Leaves 2- to 3-pinnate .   22
   22. Upper part of stem
        pilose . *Angelica*
   22. Upper part of stem
        glabrous .   .   .
          .   .   . *Cicuta*
17. Leaves much dissected with seg-
    ments less than 1.5 cm. broad .   .   23
   23. Umbel with 1 to 5 rays; plants
        at most 0.7 meters tall .   .
         .   .   . *Chaerophyllum*
   23. Umbel with 5 to many rays;
        plants usually 1 to 3 meters
        tall   .   .   . *Conium*

Eryngium L.

Eryngium yuccifolium Michx. (2:641). This species, called Rattle-snake Master (Fig. 56), occurs in dry woods or prairies along railroads. The leaves resemble those of some monocotyledonous plants while the inflorescence is not a true umbel. SP. CIT.: Giant City State Park, June 17, 1953, *M 1159;* two miles south of Elkville, *BS 525.*

Thaspium NUTT.

Thaspium trifoliatum (L.) Gray (2:633). Meadow Parsnip is a common plant of moist woodlands. Rare specimens occur in which all leaves are simple. Purple flowered specimens also may be found. SP. CIT.: *M 2576; M 4655; M 2412; BS 413.*

FIG. 56. *Rattlesnake Master* (Eryngium yuccifolium), *a plant of unusual appearance, is found growing in prairie soil along railroads.*

Sanicula L.

Sanicula gregaria Bickn. (2:612). Common Snakeroot is abundant in mesic woods. It may be distinguished from other species of *Sanicula* in our area by its styles longer than the bristles and sepals at most five-tenths millimeter long. SP. CIT.: Little Grand Canyon, August, 1954, *M 4035;* Giant City State Park, June 4, 1953, *M 225.*

Sanicula marilandica L. (2:612). Black Snakeroot is found in Southern Illinois only in a few stations. The styles are longer than the bristles and sepals are one to one and one-half millimeters long. SP. CIT.: three miles west of Carbondale, June 27, 1940, *McCree;* Giant City State Park, June 9, 1953, *M 305.*

Sanicula canadensis L. (2:612). Short-styled Snakeroot is common in wet ground in woods or along roads. The styles are shorter than the bristles. SP. CIT.: Giant City State Park, June 11, 1953, *M 295.*

Torilis ADANS.

Torilis japonica (Houtt.) DC. (2:615). Hedge Parsley is rapidly becoming abundant in Southern Illinois where it grows along edges of woods or roads. SP. CIT.: near Howardton, July 9, 1940, *Mc 286.*

Daucus L.

Daucus carota L. (2:617). Queen Anne's Lace or Wild Carrot is abundant in waste ground throughout Illinois. Pink-flowered forms occasionally occur (forma *roseus* Millsp.). The umbel is five to twelve centimeters broad and the ultimate leaf segments are lanceolate. SP. CIT.: typical: *Mc 542; M 202.* Forma *roseus: M 4833.*

Daucus pusillus Michx. (2:617). Small Wild Carrot has been found twice in Illinois. One of the stations is in a woods near Pinckneyville (Perry County) while the other is in a cemetery near Murphysboro. The umbel is two to five centimeters broad and the ultimate leaf segment is linear. SP. CIT.: Zion Cemetery northeast of Murphysboro, June 16, 1954, *M 3003.*

Anethum L.

Anethum graveolens L. (2:633). Dill is sometimes found in waste ground where it has escaped from cultivation. It is a native of Europe.

Ptilimnium RAF.

Ptilimnium nuttallii (DC.) Britt. (2:631). Nuttall's Bishop's-weed is rare in swampy woods of Southern Illinois. We have it only from Jackson, Randolph, and Pulaski counties. The leaf segments are alternate or opposite on the rachis. SP. CIT.: north of Makanda, August 2, 1950, *BS 1114.*

Ptilimnium costatum (Ell.) Raf. (2:631). Ribbed Bishop's-weed also is found in low woods in Southern Illinois where it is uncommon. Near Elkville, where it is rather abundant, it occurs with *Sium suave, Carex muskingumensis, Quercus bicolor, Polygonum ramosissimum,* and others. SP. CIT.: northeast of Howardton, September 13, 1941, *Mc 1197;* six miles northeast of Elkville, *M 1912;* Campbell's Lake (Elkville), July 17, 1941, *Mc 929.*

Perideridia REICHENB.

Two plants of *Perideridia americana* (Nutt.) Reichenb. (2:625) [*Eulophus americanus* Nutt.] were found growing along a road near Blind Hollow in Hardin County. This is the only known Southern Illinois station.

Cryptotaenia DC.

Cryptotaenia canadensis (L.) DC. (2:615). Honewort is common in moist woods and wet roadside ditches. SP. CIT.: *Mc 240; M 340.*

Zizia KOCH

Zizia aurea (L.) Koch (2:620). Golden Alexander is rather uncommon in Southern Illinois. We have it from Jackson, Pope, and Hardin counties. SP. CIT.: Giant City State Park, June 22, 1953, *M 377.*

Erigenia NUTT.

Erigenia bulbosa (Michx.) Nutt. (2:613). Harbinger-of-Spring is one of the first woodland wild flowers to bloom in the spring. We have collected it in flower as early as February 23. SP. CIT.: *Mc 638; M 115.*

Taenidia DRUDE

Taenidia integerrima (L.) Drude (2:620). This species is not common in our area. It usually grows in rather moist soil along roads or edges of woods. We have it listed only from Jackson, Union, and Saline counties. SP. CIT.: *Mohlenbrock.*

Pastinaca L.

Pastinaca sativa L. (2:639). The Parsnip is common in moist waste ground throughout Illinois. SP. CIT.: *BS 731; French.*

Osmorhiza RAF.

Osmorhiza claytoni (Michx.) C. B. Clarke (2:617). Sweet Cicely occurs in moist woodlands. It often grows with the following species. The stipules are ciliate. SP. CIT.: *BS 2311; M 2589.*

Osmorhiza longistylis (Torr.) DC. (2:617). Anise-root is variable in degree of pubescence. Villous forms are var. *villicaulis* Fern. The stipules are velvety pubescent. SP. CIT.: *BS 2312; M 246.*

Heracleum L.

Heracleum lanatum Michx. (2:639) [*H. maximum* Bartr.]. Cow Parsnip is known in Southern Illinois only from moist woods near streams in Pope and Jackson counties. SP. CIT.: northeast of Oraville, May 15, 1955, *M 5261.*

Sium L.

Sium suave Walt. (2:627) [*S. cicutaefolium* Schrank]. Water Parsnip is found locally in swampy pin oak woods. SP. CIT.: near Elkville, July 25, 1954, *M 4749;* Campbell's Lake (Elkville), July 22, 1941, *Mc 931.*

Oxypolis RAF.

Oxypolis rigidior (L.) Raf. (2:636). Cowbane is rare in swampy ground in Illinois. We have it from three southern counties (Jackson, Pope, and Pulaski). SP. CIT.: in Illinois Natural History Survey herbarium.

Angelica L.

Angelica venenosa (Greenway) Fern. (2:635). Angelica is very local in our area where it is found in dry woods. SP. CIT.: *Mc 894; M 434.*

Cicuta L.

Cicuta maculata L. (2:629). Water Hemlock is found in moist ground in woods or along roads. The plant is very poisonous. SP. CIT.: *Mc 1196; M 1172; BS 132.*

Chaerophyllum L.

Chaerophyllum procumbens (L.) Crantz (2:618). Wild Chervil is common in moist soil throughout Illinois. The leaves are glabrous. SP. CIT.: *M 2131; M 1102.*

Chaerophyllum tainturieri Hook. (2:618). This species is spreading rapidly into moist waste areas in Southern Illinois. The leaves are pilose. SP. CIT.: *M 222.*

Conium L.

Conium maculatum L. (2:629). Poison Hemlock is local in Southern Illinois in moist waste ground. We have it only from Jackson and Pulaski counties. The plant is exceedingly poisonous. SP. CIT.: north of Murphysboro, *Mohlenbrock.*

113. *Ericaceae* – Heath Family

1. Plants with no green color . . . . *Monotropa*
1. Plants with green color 2
    2. Leaves with dots . . . . *Gaylussacia* °
    2. Leaves not dotted 3
        3. Leaves softly pubescent beneath; flowers pink, showy . . . . *Rhododendron*
        3. Leaves not softly pubescent beneath; flowers white or greenish, small . . . . *Vaccinium*

Monotropa L.

Monotropa uniflora L. (3:4). Indian Pipe is rare in rich woods in our area. Flowers are solitary. SP. CIT.: Giant City State Park, *R. Sands.*

Monotropa lanuginosa Michx. (3:4) [*M. hypopitys* L.]. The very rare Pinesap has been collected in Southern Illinois only in Jackson County. SP. CIT.: ten miles northwest of Murphysboro, oak-hickory hilltop, July 4, 1951, *BS 1509.*

Gaylussacia HBK.

Black Huckleberry, *Gaylussacia baccata* (Wang.) K. Koch (3:23), has been collected only once in the southern part of Illinois. The plant has been collected in Alexander County.

Rhododendron L.

Rhododendron roseum (Loisel.) Rehd. (3:10). The magnificent Wild Azalea (Fig. 57) grows in acid soil in dry woods in Pine Hills

FIG. 57. *The native Azalea* (Rhododendron roseum) *flowers in May in Pine Hills.*

of Union County and near Little Grand Canyon in Jackson County. Flowers are produced during mid-May. SP. CIT.: *V 1358.*

Vaccinium L.

Vaccinium arboreum Marsh. (3:26). Sparkleberry is perhaps the most common shrub atop sandstone bluffs in Southern Illinois. It grows with *Ulmus alata, Quercus stellata,* and *Quercus marilandica.* The plants usually attain a height of one and one-half to four meters. SP. CIT.: *V 1051; M 790.*

Vaccinium vacillans Torr. (3:29). Hill Blueberry grows along edges of sandstone bluffs in most of the southern counties. Berries are edible. Plants usually are less than one meter tall. SP. CIT.: Midland Hills, April 29, 1948, *BS 384;* Giant City State Park, April 25, 1953, *M 13.*

114. *Primulaceae* – Primrose Family

1. Leaves deeply pinnatifid; plants aquatic; stems inflated . .
   . . . . *Hottonia*
1. Not as above 2
   2. Leaves from base of plant in a rosette 3
   3. Flowers about 2.5 mm. broad . . .*Androsace* °
   3. Flowers much larger with corolla lobes reflexed . .
      . . . . *Dodecatheon*

2. Leaves cauline 4
    4. Leaves alternate; flowers white . . . .*Samolus*
    4. Leaves opposite or whorled; flowers red or yellow . . 5
        5. Flowers bright red . . . . .*Anagallis*
        5. Flowers yellow . . . . . .*Lysimachia*

Hottonia L.

Hottonia inflata Ell. (3:36). American Featherfoil grows in shallow water in three Southern Illinois counties (Jackson, Union, and Johnson). SP. CIT.: Cave Valley, June 16, 1951, *Hardy 72.*

Androsace L.

Tiny *Androsace occidentalis* Pursh (3:36) has been collected from hill prairies in Union and Randolph counties. It flowers during April.

Dodecatheon L.

Dodecatheon meadia L. The attractive Shooting Star is fairly common in dry woods and on sandstone bluffs throughout our area. Flowers are either white or pink. Leaves taper to base. SP. CIT.: *MV 1888; M 795; BS 2301; M 2429; Collins; Greta Intza 35.*

Dodecatheon frenchii (Vasey) Rydb. (3:36). French's Shooting Star (Fig. 58) is Southern Illinois' only endemic species. The plant grows usually in deep shade under overhanging sandstone bluffs. The type specimen has the following notation: "this is thought to be a new species as it differs much from *D. meadia.* Found at Fern Rocks near Carbondale. 1870." The original collector, and the man for whom the species was named, was George Hazen French. We have

FIG. 58. *The rare French's Shooting Star (*Dodecatheon frenchii*) was named for G. H. French, an early professor at Southern Illinois Normal University.*

recorded this species from Jackson, Williamson, Union, Johnson, and Pope counties. The leaves are abruptly narrowed into a distinct petiole. For additional notes concerning this plant, see Voigt and Swayne (1955). SP. CIT.: two miles north of Pomona, July 20, 1951, *V 841;* Giant City State Park, April 3, 1952, *V 1029;* Midland Hills, May 9, 1951, *V 552;* Cedar Ridge south of Hickory Ridge, June 6, 1952, *BS 2482.*

Samolus L.

Samolus parviflorus Raf. (3:36) [*S. floribundus* HBK.]. Brookweed is local in moist shaded soil. Flowers appear during June and July. SP. CIT.: *V 698; M 374; M 4658; Wilson & McCree 1133.*

Anagallis L.

Anagallis arvensis L. (3:43). Pimpernel occurs occasionally in moist roadside ditches where it is naturalized since it came from Europe and Africa. SP. CIT.: four miles east of Ava, June 24, 1949, *Swayne 737.*

Lysimachia L.

1. Corolla lobes entire; leaves usually punctate 2
   2. Stems creeping; corolla punctate . *Lysimachia nummularia*
   2. Stems erect; corolla not punctate . *Lysimachia vulgaris* °
1. Corolla lobes denticulate; leaves not punctate 3
   3. Leaves linear, sessile, lateral veins obscure . . . .
    . . . . *Lysimachia quadriflora* °
   3. Leaves mostly lanceolate to ovate, petiolate (except in *Lysimachia lanceolata*), lateral veins evident 4
     4. Leaves rounded at base 5
       5. Petioles ciliate only at base . . . . .
        . . . . . . *Lysimachia radicans* °
       5. Petioles ciliate to summit . *Lysimachia ciliata*
     4. Leaves tapering to base 6
       6. Plants with basal stolons; upper leaves sessile or nearly so, pale beneath . *Lysimachia lanceolata*
       6. Plants without basal stolons; upper leaves petiolate, green on both sides . . *Lysimachia hybrida*

Lysimachia ciliata L. (3:41). Fringed Loosestrife is usually found in lowlands or wet prairie situations, often in roadside ditches or swampy meadows. Usually associated with it are slough grass (*Spartina pectinata*), bedstraw (*Galium tinctorium*) and fowl manna grass (*Glyceria striata*). SP. CIT.: Fountain Bluff, July 12, 1871, *G. H.*

French; Campbell's Lake, June 24, 1941, *Mc 856;* Giant City State Park, *M 318.*

Lysimachia lanceolata Walt. (3:41). Narrow-leaved Loosestrife is common in moist soil over most of Southern Illinois. It is often found with the preceding species. This is our earliest flowering species of Loosestrife. SP. CIT.: three miles southeast of Elkville, June 17, 1948, *BS 467;* Giant City State Park, April 15, 1954, *M 273.*

Lysimachia nummularia L. (3:39). Moneywort grows on moist ground and is common in this situation locally throughout this area. SP. CIT.: three and one-half miles north of Carbondale, June 5, 1940, *Welch 106.*

Lysimachia hybrida Michx. (3:41). This species, closely allied to the preceding, is known in our area only from a roadside ditch in Jackson County. SP. CIT.: Junction of Illinois highways 3 and 144, August 6, 1954, *M 4333.*

Three additional species of *Lysimachia* occur in Southern Illinois. *Lysimachia quadriflora* Sims (3:39), a rather common plant of central and northern Illinois, has been found in a roadside ditch five miles east of Harrisburg in Saline County and again in Pope County. *Lysimachia radicans* Hook. (3:41), a distinctly southern species, is known only from a single collection (*Winterringer 3209*) in a woods east of Karnak in Pulaski County. The adventive *Lysimachia vulgaris* L. (3:39), an occasional waif in the Chicago area, has one Southern Illinois station in Pope County.

115. *Sapotaceae* – Sapodilla Family

Bumelia   SW.

Woolly Buckthorn, *Bumelia lanuginosa* (Michx.) Pers. (3:44), a rather common species of Missouri limestone glades, occurs in Illinois in Monroe, Hardin, and Pulaski counties. It is readily distinguished by the densely woolly underside of its leaves.

Slightly more common than the preceding, but still rare, is Southern Buckthorn, *Bumelia lycioides* (L.) Gaertn. f. (3:44). This mostly smooth southern shrub has been collected in St. Clair, Alexander, Pulaski, Pope, and Hardin counties.

116. *Ebenaceae* – Ebony Family

Diospyros   L.

Diospyros virginiana L. (3:47). Persimmon is the only temperate region representative of this otherwise tropical family. Persimmon is a pioneer tree with Sassafras in the succession of old fields. These trees are accompanied or preceded by a bramble and weed stage. The wood is marketed for use in making driver heads in golf sets. SP. CIT.: Giant City State Park, May 25, 1948, *BS 420.* Japanese Persimmon, *Diospyros kaki* L. f., with larger fruits, is sometimes cultivated in our area. A large tree occurs in Tower Grove Cemetery west of Murphysboro.

117. *Styracaceae* – Storax Family

1. Petals four . . . . *Halesia* °
1. Petals five . . . . *Styrax* °

Halesia L.

Silver Bell, *Halesia carolina* L. (3:47), is known in Illinois only from along streams in woods in Massac County. Flowers appear in May.

Styrax L.

Storax, *Styrax americana* Lam. (3:47), is a small tree of swamps and moist woods in the extreme southern counties of Illinois. We have it from Johnson, Pope, Pulaski, and Massac counties. A specimen collected by French from Jackson County is from a cultivated plant.

118. *Oleaceae* – Olive Family

1. Leaves simple; fruit a drupe . . . . . *Forestiera*
1. Leaves pinnately compound; fruit a samara . . *Fraxinus*

Forestiera POIR.

Forestiera acuminata (Michx.) Poir. (3:51). Swamp Privet is occasional in swampy woods. It has greenish twigs which are lined as to give them a four-angled appearance. SP. CIT.: Giant City State Park, July 10, 1953, *M 391;* four miles south of Cora, August 6, 1954, *M 4328.*

Fraxinus L.

1. Stems 4-sided . . . . . *Fraxinus quadrangulata*
1. Stems more or less rounded 2

2. Samaras 2 to 3 inches long; leaflet margins entire; swollen based tree of swampy habitats . . . *Fraxinus tomentosa*
2. Samaras 2 inches long or less; leaflet margins entire or serrate 3
   3. Petioles often velvety-pubescent; petiole of leaflets winged nearly to base; leaflets distinctly petiolate . .
     . . . . *Fraxinus pennsylvanica*
   3. Petioles glabrous or nearly so; leaflets nearly sessile . 4
     4. Leaflet margins entire; leaf scars deeply notched at top . . . . . . *Fraxinus americana*
     4. Leaflet margins serrate; leaf scars nearly straight at top . . . . . . *Fraxinus lanceolata*

Fraxinus quadrangulata Michx. (3:51). Blue Ash occurs on almost inaccessible limestone bluffs where it hangs precariously from tiny crevices. Winged twigs make this species easily identified. Water in which a portion of the bark of this small tree is placed turns blue. This species is known in Southern Illinois from Pine Hills of Union County and an adjacent area in Jackson County. SP. CIT.: north of the Pine Hills, August 20, 1954, *Mohlenbrock and Stewart.*

Fraxinus tomentosa Michx. f. (3:49) [*Fraxinus profunda* (Bush) Britt.]. Pumpkin Ash, an inhabitant of swamps in Southern Illinois, has the largest samaras of any of our species of *Fraxinus*, the winged seeds sometimes attaining a length of three inches. Bases of trees which stand in water are usually swollen. In our area, it is recorded from Union, Johnson, Pope, Alexander, Pulaski, and Massac counties.

Fraxinus pennsylvanica Marsh. (3:49). Red Ash, an uncommon tree of river banks and other moist situations, is often confused with Green Ash, but is readily distinguished by velvety petioles. SP. CIT.: Lester Swamp, *M 4643.*

Fraxinus lanceolata Borkh. (3:49) [*Fraxinus pennsylvanica* var. *subintegerrima* (Vahl) Fern.]. Green Ash, sometimes included as a variety of Red Ash, may be distinguished by its glabrous petioles. It is much more common than the preceding. SP. CIT.: Giant City State Park, *M 4643.*

Fraxinus americana L. (3:49). American or White Ash, a common forest tree in Illinois, is important economically for its durable whitish wood. SP. CIT.: Giant City State Park, *M 485.*

Syringa L.

Lilac, *Syringa vulgaris* L., sometimes persists around old dwellings.

119. *Loganiaceae* – Logania Family

Spigelia L.

Spigelia marilandica L. (3:55). Indian Pink, one of the more attractive wildflowers, usually grows in rich, shady woods. In Illinois, it occurs only in the southern quarter of the state. SP. CIT.: west of Crab Orchard Lake Dam, June 1, 1948, *BS 447;* Lake Murphysboro, 1951, *M 1202.*

120. *Gentianaceae* – Gentian Family

1. Plants less than 1 foot tall; stems with scales; sepals 2; corolla
   4-cleft; flowers whitish   .   .   .   . *Obolaria*
1. Plants over 1 foot tall; stems without scales; sepals more than 2;
   flowers colored 2
      2. Plants with verticillate leaves; plants 1 to 2.5 meters high;
         corolla of 4 greenish petals, each with a nectiferous gland
         .   .   .   . *Frasera*
      2. Plants with opposite leaves; plants under 1 meter in height;
         petals without nectiferous glands   3
         3. Flowers pink-rose; stems angled .   .   .   . *Sabatia*
         3. Flowers blue or bluish-white; stems not angled   .   .
         .   .   .   . *Gentiana*

Obolaria L.

Obolaria virginica L. (3:69). Pennywort is one of our rarest species (Fig. 59). Unknown from Missouri, this fleshy little perennial grows in the southernmost counties of Illinois. The pale purplish flowers

FIG. 59. *Pennywort* (Obolaria virginica) *is an inconspicuous plant among fallen leaves in early spring.*

are produced in April and May, while the plant may be found in a healthy vegetative condition the first of January. SP. CIT.: Midland Hills, May 20, 1951, V 658.

Frasera WALT.

Frasera caroliniensis Walt. (3:67) [*Swertia caroliniensis* (Walt.) Ktze.]. American Columbo is one of our largest herbs, often attaining a height of eight feet. Flowers, borne in ample numbers, bear fringed purplish glandular spots on greenish petals. SP. CIT.: one-half mile southeast of De Soto, June 19, 1940, *Mc 215;* Giant City State Park, June 29, 1953, *M 252.*

Sabatia ADANS.

Sabatia angularis (L.) Pursh (3:57). Rose-gentian, or Rose Pink, makes a brilliant sight in moist meadows when flowers appear in July and August. Although found in almost every county, Rose Pink cannot be considered common. SP. CIT.: Giant City State Park, August 8, 1950, *BS 1109;* north of Elkville, July 19, 1948, *BS 523.*

Gentiana L.

Gentiana puberula Michx. (3:63). Downy Gentian or Purple Gentian, a prairie species, is well hidden from view in the fall aspect as it blooms below the level of taller plants associated with it such as Big Bluestem (*Andropogon gerardi*), Little Bluestem (*Andropogon scoparius*), Prairie Dropseed (*Sporobolus heterolepis*), Compass Plant (*Silphium laciniatum*), and Rosinweed (*Silphium integrifolium*). SP. CIT.: one mile south of Elkville, September, 1956, *Joe Garrison.*

Gentiana andrewsii Griseb. (3:63). Closed Gentian is rare in Southern Illinois where it is recorded from Williamson and Pope counties. Collections of *Gentiana saponaria* L. from Pope County in the Southern Illinois University herbarium seem more likely to be *Gentiana andrewsii.* This species ranks high on the list of beautiful wild flowers.

121. *Apocynaceae* – Dogbane Family

2. Plants climbing; corolla funnelform, yellowish . . .
. . . . *Trachelospermum* °
2. Plants not climbing; corolla rotate or campanulate . . . 3
   3. Creeping evergreen plants; corolla rotate; flowers solitary,
      blue . . . . *Vinca*
   3. Erect herbs; corolla campanulate; flowers cymose, whit-
      ish . . . . *Apocynum*

Amsonia WALT.

Amsonia tabernaemontana Walt. (3:71) is a blue flowered plant having alternate leaves and a milky white sap. It is found in swampy ditches or meadows and is usually associated with *Iris shrevei, Ranunculus septentrionalis, Carex squarrosa,* and various other sedges. SP. CIT.: Junction of Illinois highways 3 and 144, September 14, 1940, *Mc 472;* one mile north of Carbondale, June 10, 1953, *M 1350.*

Trachelospermum LEM.

Trachelospermum difforme (Walt.) Gray (3:71) occurs sparingly in the southern counties of Illinois (Franklin, White, Gallatin, Union, Johnson, and Pulaski). It is our only climbing member of its family. The habitat for this species is moist woods. Flowers appear from June to August.

Vinca L.

Vinca minor L. (3:71). Periwinkle or Myrtle has been naturalized from Europe. It is planted frequently in shaded lawns or around foundations; elsewhere its habitat is frequently in cemeteries.

Apocynum L.

1. Corolla white or greenish, 2 to 4 mm. long; seeds 4 to 6 mm.
long; leaves spreading or ascending 2
  2. Leaves mostly petiolate 3
    3. Stems and leaves glabrous . .*Apocynum cannabinum*
    3. Stems and leaves pubescent . . . . .
     . . . . *Apocynum cannabinum* var. *pubescens*
  2. Leaves sessile or subsessile . . . *Apcyonum sibiricum*
1. Corolla pinkish, 4 to 10 mm. long, seeds less than 4 mm. long;
leaves usually drooping or sometimes only spreading . . .
  . . . . *Apocynum androsaemifolium*

Apocynum cannabinum L. (3:73). Indian Hemp is found throughout Illinois in roadsides, fields, or woodlands. The pubescent variety *pubescens* (R. Br.) A. DC. (*Apocynum pubescens* R. Br.), called Velvety Dogbane, is not as common as the typical variety, although the two often grow together. SP. CIT.: typical: Midland Hills, September 23, 1953, *Bell;* Carbondale, May 28, 1941, *Mc 771.* Var. *pubescens:* Giant City State Park, June 5, 1953, *M 205.*

Apocynum sibiricum Jacq. (3:73). This species is likewise common in waste areas, particularly those which are dry and sandy. SP. CIT.: near Murphysboro, July 2, 1954, *M 3101.*

Apocynum androsaemifolium L. (3:73). Spreading Dogbane is a species of upland woods rather than waste areas. SP. CIT.: Elkville, June 15, 1953, *M 3107.*

122. *Asclepiadaceae* – Milkweed Family

1. Plants climbing 2
  2. Corolla campanulate, lobes erect . . . .*Ampelamus*
  2. Corolla rotate, lobes spreading . . . .*Gonolobus*
1. Plants not climbing 3
  3. Leaves opposite or whorled . . . . .*Asclepias*
  3. Leaves alternate. 4
    4. Corolla lobes reflexed . . . . . .*Asclepias*
    4. Corolla lobes erect or spreading; corolla greenish-purple
    . . . .*Asclepiodora*

Ampelamus RAF.

Ampelamus albidus (Nutt.) Britt. (3:83). Climbing Bluevine may be found in nearly all kinds of habitats in our area. SP. CIT.: northeast of Carbondale, August 21, 1941, *Mc and Wilson 1181;* near Jacob, June 22, 1948, *BS 480.*

Gonolobus MICHX.

Although no species of this genus has been found in Jackson County, two species are known to occur elsewhere in Illinois. *Gonolobus gonocarpos* (Walt.) Perry (3:83), with nearly glabrous calyx lobes, has been collected in open woods from Union, Hardin, Alexander, and Pulaski counties. *Gonolobus obliquus* (Jacq.) R. Br. (3:84) with pubescent calyx lobes is known only by a single collection north of Golconda in Pope County.

Asclepiodora A. GRAY

Asclepiodora viridis (Walt.) A. Gray (3:80). The greenish-purple flowered Spider Milkweed is found in prairie soil in the southcentral and southwestern counties of Illinois. SP. CIT.: one mile south of Elkville, *BS 402*.

Asclepias L.

Following Woodson's (1954) monograph of North American species of *Asclepias*, we are including under this genus those species previously assigned to the genus *Acerates* Ell.

1. Flowers orange; no milky sap present; leaves alternate . .
. . . . *Asclepias tuberosa*
1. Flowers not orange; milky sap present; leaves opposite, alternate,
or in whorls 2
  2. Corolla with an incurved horn arising from each hood; leaves
  opposite or in whorls. 3
    3. Leaves narrowly linear, verticillate . . . . .
    . . . . *Asclepias verticillata*
    3. Leaves not as above 4
      4. Leaves, or some of them, in whorls of four . . 5
        5. Flowers pinkish . . . *Asclepias quadrifolia*
        5. Flowers white or suffused with purple at the
        base. . . . . *Asclepias variegata*
      4. All leaves opposite 6
        6. Flowers a vivid purple-red . . . . .
        . . . . . *Asclepias purpurascens*
        6. Flowers not a vivid purple-red 7
          7. Leaves oval to ovate or ovate-oblong . . 8
            8. Flowers lavender or purple . . .
            . . . . *Asclepias syriaca*
            8. Flowers white or occasionally greenish-
            purple. . . *Asclepias exaltata*
          7. Leaves lanceolate or ovate-lanceolate . . 9
            9. Corolla rose . *Asclepias incarnata*
            9. Corolla white . *Asclepias perennis*
  2. Corolla without a horn; leaves often alternate or irregularly
  arranged 10
    10. Leaves oval to oblong . . *Asclepias viridiflora*
    10. Leaves linear to linear-lanceolate . *Asclepias hirtella*

Asclepias tuberosa L. (2:8). Butterfly-weed is brilliant and lacks milky sap. Plants of this area belong to ssp. *interior* Woodson. These plants flower from mid-June through August. SP. CIT.: Giant City State Park, June 14, 1953, *M 481*; Boskydell, August 13, 1941, *Mc 1034.*

Asclepias verticillata L. (2:8). Horsetail Milkweed grows in sandy, open soil, often on top of sandstone bluffs. It flowers in July and August. SP. CIT.: Giant City State Park, August 5, 1953, *M 605*; two miles northeast of Carbondale, August 21, 1941, *McCree & Wilson 1152.*

Asclepias quadrifolia Jacq. (2:8). This species is limited in Illinois to the western counties (Jackson and Union). SP. CIT.: Fountain Bluff, May 19, 1952, *BS 2299*; one mile north of Pomona, May 19, 1951, *Hatcher & Stewart 143.*

Asclepias variegata L. (2:8). White Milkweed is known in Illinois only from dry woodlands in the southern counties. SP. CIT.: Giant City State Park, June 4, 1953, *M 216.*

Asclepias purpurascens L. (2:8). Purple Milkweed is occasional in dry soil, often along railroads. SP. CIT.: seven miles north of Murphysboro, June 4, 1954, *MV 1587*; near Etherton, May 23, 1954, *M 2500*; Giant City State Park, June 4, 1953, *M 217.*

Asclepias syriaca L. (2:8). Common Milkweed becomes weedy throughout this area and elsewhere. Its pods have soft spines. SP. CIT.: Giant City State Park, June 14, 1953, *M 480*; Southern Illinois University campus, June 10, 1940, *Mc 128.*

Asclepias exaltata L. (2:8) [*Asclepias phytolaccoides* Pursh]. Poke Milkweed has leaves resembling those of *Phytolacca*. This species is of rare occurrence. SP. CIT.: Lake Murphysboro, June 12, 1954, *M 2671.*

Asclepias incarnata L. (2:8). Swamp Milkweed is common in low moist areas but usually not in woods. SP. CIT.: Carbondale Reservoir, October, 1953, *Bell*; Giant City State Park, August 6, 1953, *M 548*; Midland Hills, July 23, 1941, *Mc 970.*

Asclepias perennis Walt. (2:8). This plant is found only in the southern counties, sometimes in low woods. It is rather common in swamps of Pulaski County, otherwise rare. SP. CIT.: Turkey Bayou, September 2, 1952, *BS 2879.*

Asclepias viridiflora Raf. (2:8) [*Acerates viridiflora* (Raf.) Eaton] is found occasionally along roads throughout Illinois. On one of our hill prairies is found narrower-leaved var. *linearis* (Gray) Fern. SP. CIT.: typical: (see Jones *et al.*, 1955). Var. *linearis:* Grassy Knob hill prairie, August 28, 1954, *M 4241.*

Asclepias hirtella (Pennell) Woodson (2:8) [*Acerates hirtella* Pennell] grows in dry open areas throughout Illinois except the southern two tiers of counties. SP. CIT.: Elkville, October 3, 1953, *Mohlenbrock.*

123. *Convolvulaceae* – Morning-glory Family

1. Plants parasitic, without chlorophyll, leafless, twining . *Cuscuta*
1. Plants not parasitic, but green with normal leaves 2
    2. Stigma capitate 3
        3. Stamens exserted   .   .   .   .   . .*Quamoclit*
        3. Stamens not exserted .   .   .   .   . .*Ipomoea*
    2. Stigmas two .   .   .   .*Convolvulus*

Cuscuta L.

Flowers are necessary and capsules are desirable for identification. All ten species attributed to Illinois occur in our study area. These species are leafless annuals with yellow or orange stems and scaly appendages in place of leaves. On germination, they become completely parasitic on herbs or shrubs to which they adhere by twining and by means of suckers.

1. Flowers usually in dense, often globose clusters; sepals free nearly to base   2
    2. Bracts and sepals cuspidate; flowers pedicelled .   .   .
       .   .   .   . *Cuscuta cuspidata*
    2. Bracts and sepals blunt at the apex, or with slender recurving tips; flowers sessile 3
        3. Bracts appressed; seeds about 2.5 mm. long .   .   .
           .   .   .   . *Cuscuta compacta* °
        3. Bracts recurved; seeds about 1.5 mm. long .   .   .
           .   .   .   . *Cuscuta glomerata* °
1. Flowers solitary or few together; sepals usually united below 4
    4. Flowers short-pedicelled 5
        5. Flowers mostly 5-merous; scales (appendages at base of stamens) fimbriate 6
            6. Corolla lobes acute   .   . .*Cuscuta indecora* °
            6. Corolla lobes obtuse   .   . .*Cuscuta gronovii*
        5. Flowers mostly 4-merous; scales irregularly fimbriate or absent 7

    7.   Scales none, or minute; corolla lobes acute   .   .
        .   .   .   . *Cuscuta coryli* °
    7.   Scales irregularly fimbriate; corolla lobes obtuse   .
        .   .   .   . *Cuscuta cephalanthi*

4.   Flowers sessile   8
    8.   Flowers mostly 5-merous; scales fimbriate .   .   .   9
       9.   Scales reaching summit of corolla tube .   .   .
          .   .   .   . *Cuscuta campestris*
       9.   Scales reaching only half-way to summit of corolla
          tube .   .   .   .   .   . *Cuscuta pentagona*
    8.   Flowers mostly 4-merous; scales minute or lacking .   .
        .   .   .   . *Cuscuta polygonorum*

Cuscuta cuspidata Engelm. (3:93). Giant City State Park, August 25, 1953, *M 462;* Riverside Park (Murphysboro), August 8, 1954, *M 4740.* Also in Union and Pulaski counties.

Cuscuta compacta Juss. (3:93). Known from swampy areas in Union and Saline counties.

Cuscuta glomerata Choisy (3:93). Known only from Pope County.

Cuscuta indecora Choisy (3:95). Known from Alexander County.

Cuscuta gronovii Willd. (3:95). Fountain Bluff, September 14, 1940, *Mc 448;* Giant City State Park, September 10, 1953, *M 219;* Lake Murphysboro, August 26, 1954, *M 4660.* Also in Union, Alexander, and Pulaski counties.

Cuscuta coryli Engelm. (3:93). Known only from Pulaski County.

Cuscuta cephalanthi Engelm. (3:93). Known from Jackson (see Jones *et al.*, 1955), Gallatin, Pope, and Pulaski counties.

Cuscuta campestris Yuncker. Giant City State Park, July 7, 1953, *M 351.* Very rare in Illinois.

Cuscuta pentagona Engelm. (3:93). Giant City State Park, July 21, 1953, *M 556;* Thompson's Lake, October 2, 1940, *Mc 539.* Also in Williamson, Saline, Union, and Hardin counties.

Cuscuta polygonorum Engelm. (3:93). Giant City State Park, July 12, 1953, *M 622;* Midland Hills, October 29, 1947, *BS 280.*

Convolvulus L.

1.   Bracts 1 to 8 mm. long, inserted 5 to 20 mm. below calyx .   .
    .   .   .   . *Convolvulus arvensis*
1.   Bracts 10 to 20 mm. long, inserted at base of calyx (often mistaken for sepals)   2
    2.   Peduncles glabrous or nearly so, greatly exceeding length of petioles .   .   .   . *Convolvulus sepium*

2. Peduncles pubescent, about equalling length of petioles .
. . . . . *Convolvulus sepium* var. *fraterniflorus*

Convolvulus arvensis L. (3:91). Field Bindweed is a common adventive in fields and waste places. SP. CIT.: Giant City State Park, July 10, 1953, *M 404;* Fountain Bluff, July 24, 1954, *M 4356.*

Convolvulus sepium L. (3:91). The most variable species of bindweed in eastern United States is Hedge Bindweed. The treatment by Tryon (1939) seems to be most satisfactory for this complex. The sagittate-leaved var. *sepium* with long glabrous peduncles is common.

Quamoclit MOENCH

Quamoclit coccinea (L.) Moench (3:89). Red Morning-glory, a native of tropical America, is frequently cultivated in our area. It flowers during the summer. It occurs in Jackson County (see Jones *et al.,* 1955).

Ipomoea L.

1. Stems glabrous; calyx lobes glabrous, obtuse . . . . .
. . . . *Ipomoea pandurata*
1. Stems pubescent; calyx lobes pubescent, acute 2
   2. Flowers white, to 1 inch long . . .*Ipomoea lacunosa*
   2. Flowers usually pink to purple, over 1 inch long . . . 3
      3. Flowers 2 to 3 inches long; leaves ovate, cordate . .
      . . . . *Ipomoea purpurea*
      3. Flowers 1 to 2 inches long; leaves 3-lobed . . .
      . . . . *Ipomoea hederacea*

Ipomoea pandurata (L.) Meyer (3:88). Wild Sweet Potato Vine is our largest flowered species of *Ipomoea.* It is very common, growing abundantly in fields and waste places. SP. CIT.: Giant City State Park, June 14, 1953, *M 483;* two miles south of Mathews, July 25, 1954, *M 4359.*

Ipomoea lacunosa L. (3:88). Small-flowered Morning-glory with bright, white flowers seems to prefer moist disturbed areas. It dominates large areas of low woodlands. SP. CIT.: four miles south of Cora, August 6, 1954, *M 4329.*

Ipomoea purpurea (L.) Roth (3:88). A native of tropical America, this Morning-glory is our least common. It flowers from August to October and is the latest flowering *Ipomoea* in Illinois. SP. CIT.:

Thompson's Lake, October 1, 1941, *Mc 1289;* Fountain Bluff, October 17, 1947, *BS 248.*

Ipomoea hederacea Jacq. (3:88). The three-lobed leaves of Ivy-leaved Morning-glory make this species easily recognized. It is common in waste ground throughout the area. SP. CIT.: Giant City State Park, August 28, 1953, *M 542;* Fountain Bluff, September 14, 1940 *Mc 435.*

124. *Polemoniaceae* – Phlox Family

1. Leaves simple .   .   .   .   *Phlox*
1. Leaves pinnately compound   .   .   .   *Polemonium*

Phlox L.

1. Petals notched; leaves linear   .   .   .   .   *Phlox bifida*
1. Petals entire; leaves lanceolate or broader   2
    2. Calyx lobes not longer than calyx tube; stamens usually somewhat exserted from corolla tube   3
        3. Leaves with obscure lateral veins .   *Phlox glaberrima*
        3. Leaves with conspicuous lateral veins .   .   .
        .   .   .   .   *Phlox paniculata*
    2. Calyx lobes longer than calyx tube; stamens reaching about the middle of corolla tube   4
        4. Stems often with elongate sterile shoots; leaves narrowly ovate.   .   .   .   .   .   *Phlox divaricata*
        4. Stems without elongate sterile shoots; leaves lanceolate
        .   .   .   .   *Phlox pilosa*

Phlox bifida Beck (3:98). Cleft Phlox is a handsome species which blooms early in spring (Fig. 60). At Saltpeter Cave near Etherton,

FIG. 60. *Cleft Phlox* (Phlox bifida) *grows in dense mats on sandy areas, talus cones, and gravelly roadsides.*

the very rare var. *cedaria* Wherry occurs. It is distinguished by its completely glabrous stems. SP. CIT.: typical: Giant City State Park, April 21, 1951, *M 917;* Peter's Cave, *BS 975;* Fountain Bluff, May 3, 1947, *BS 1.* Var. *cedaria:* Saltpeter Cave, April 10, 1954, *M 1933.*

Phlox glaberrima L. (3:100). Smooth Phlox is the common phlox of prairie areas along railroads where it begins to flower about May 18. It occurs also in open dry oak-hickory woods. In this latter habitat, it tends to flower slightly later. SP. CIT.: Campbell Lakes, July 22, 1941, *Mc 897;* railroad prairie seven miles north of Murphysboro, June 4, 1954, *VM 1575.*

Phlox paniculata L. (3:98) is the Garden Phlox. In its native state, it is confined to low, often swampy woodlands. It often is found with Purple Fringeless Orchid (*Habenaria peramoena*) and both these species flower from late July into September. SP. CIT.: north of the Pine Hills, August 6, 1954, *M 4711;* Boskydell, August 13, 1941, *McCree and Wilson 1031;* Thompson's Lake, October 10, 1940, *Mc 560.*

Phlox divaricata L. (3:97). The most common species of phlox in Southern Illinois is Blue Phlox. It is abundant in moist woodlands where it usually begins to flower in early April. Flowers vary in color with white-flowered specimens occurring occasionally. These may be segregated as f. *albiflora* Farw. An anomalous dwarf specimen has been collected at Giant City State Park in which the green corolla was about one-half typical size. SP. CIT.: Midland Hills, May 9, 1951, *V 849;* Giant City State Park, April 30, 1953, *M 40;* Peter's Cave, April 3, 1954, *VM 1878;* Fountain Bluff, May 9, 1947, *BS 23.*

Phlox pilosa L. (3:97). Commonly known as Downy or Pale Phlox, this species occurs commonly in prairie soil along railroads. It, along with *Tradescantia ohiensis,* makes a particularly attractive sight during late May and June. Pale Phlox also may grow along edges of dry open woods. SP. CIT.: one mile south of Elkville, May 19, 1948, *BS 404;* seven miles north of Murphysboro, June 4, 1954, *VM 1578.*

Polemonium L.

Polemonium reptans L. (3:101). Jacob's-ladder is a common species of moist woodlands. It is sometimes incorrectly called Bluebells. The flowering period is from April 25 to about June 10. SP. CIT.: Giant City State Park, *M 36;* one-half mile southeast of De Soto, *Mc 225;* Midland Hills, *V 510.*

125. *Hydrophyllaceae* – Waterleaf Family

1. Leaves entire . . . . *Hydrolea*
1. Leaves pinnately lobed or divided or palmately lobed . . . 2
   2. Flowers solitary; leaves pinnately divided . . . *Ellisia* °
   2. Flowers in cymes or racemes 3
      3. Plants perennial; basal leaves long-petioled . . .
      . . . . *Hydrophyllum*
      3. Plants annual; no conspicuous basal leaves . *Phacelia*

Hydrolea L.

Hydrolea affinis Gray (3:108). This unusual little plant is rare in Illinois. It has been collected only from Jackson, Union, Alexander, Pulaski, and Massac counties (see Jones *et al.*, 1955). It grows in low swampy woods which are usually inundated a portion of each year.

Ellisia L.

Ellisia nyctelea L. (3:107), popularly known as "Aunt Lucy" because of the similarity of the popular name to *Ellisia* (if said rather carelessly), has a peculiar distribution. "Aunt Lucy" is abundant and weedy in the metropolitan area of St. Louis, but stops abruptly along the southern limits of St. Clair County. Only one station is known from Southern Illinois, that being in Union County.

Hydrophyllum L.

1. Principal leaves of the stem palmately lobed 2
   2. Calyx with small appendages in the sinuses; stamens little if
     any exserted . . . *Hydrophyllum appendiculatum*
   2. Calyx without appendages; stamens exserted . . . .
     . . . . *Hydrophyllum canadense*
1. Principle leaves of the stem deeply pinnately lobed . . . 3
   3. Inflorescence with short-appressed hairs . . . . .
     . . . . *Hydrophyllum virginianum*
   3. Inflorescence with stout spreading hairs . . . . .
     . . . . . . *Hydrophyllum macrophyllum* °

Hydrophyllum appendiculatum Michx. (3:107). Like all *Hydrophyllum* in Illinois, this Waterleaf thrives in rich woods. A striking panorama may be seen along the valley floor at Little Grand Canyon during May when *Hydrophyllum appendiculatum* with its rich lavender flowers contrasts with the yellow Celandine Poppy (*Stylopho-*

rum diphyllum). While we have *H. appendiculatum* only from Jackson, Union, and Alexander counties, it probably occurs in all other southern counties. SP. CIT.: Giant City State Park, May 5, 1951, *M 901;* Little Grand Canyon, June 21, 1951, *V 627.*

Hydrophyllum canadense L. (3:107). This Waterleaf is uncommon in Southern Illinois where it has been found in Jackson and Pope counties in moist woods. Our collection dates are in May. SP. CIT.: Giant City State Park, May 3, 1953, *M 176;* Little Grand Canyon, May 30, 1952, *BS 2436.*

Hydrophyllum virginianum L. (3:107). The most common member of the genus in our area is Virginia Waterleaf. It usually is exceedingly abundant where it does occur. SP. CIT.: Giant City State Park, April 15, 1953, *M 659.*

A fourth species, *Hydrophyllum macrophyllum* Nutt. (3:107) is confined in Illinois to four southeastern counties with only Hardin falling into our study area. It has been attributed erroneously to Jackson County.

Phacelia JUSS.

1. Stamens much exserted beyond corolla tube; some leaves twice-pinnate or twice-pinnatifid .   .   .   . *Phacelia bipinnatifida*
1. Stamens equalling or much shorter than the corolla tube; leaves mostly once-pinnate 2
    2. Corolla lobes fringed; filaments villous . *Phacelia purshii*
    2. Corolla lobes entire; filaments glabrous .   .   .   .   .
        .   .   .   . *Phacelia ranunculacea*

Phacelia bipinnatifida Michx. (3:104). This *Phacelia* occurs in mesic habitats, but is rather local. A colony occurs on top of a large isolated boulder at Dixon Springs State Park (Pope County). SP. CIT.: four miles north of Carbondale, April 20, 1951, *V 520;* Giant City State Park, April 17, 1953, *M 650.*

Phacelia purshii Buckley (3:104). Miami Mist is a species of diverse situations. It is most prevalent along edges of moist woods, streams, or roadside ditches. However, occasional specimens are found on the dryest sandstone bluffs, sometimes under Black Jack Oak and Winged Elm. These specimens of dry situations usually are somewhat smaller than the lowland specimens. SP. CIT.: Fountain Bluff, June 18, 1940, *Mc 166;* Worthen Bayou, April 20, 1948, *BS 357.*

Phacelia ranunculacea (Nutt.) Constance (3:104). This plant might be mistaken for *Ellisia nyctelea* or *Nemophila microcalyx;* however, the latter is not known from Illinois. The range for *P. ranunculacea* is interrupted. It is found in Maryland, the District of Columbia, and Virginia, then in Indiana, Illinois, Missouri, and Arkansas. In Illinois, it has been collected in Jackson, Johnson, Union, Wabash, and Washington counties, and is reported from Adams County (Gleason, 1903). SP. CIT.: Little Grand Canyon, *Gregory.*

126. *Boraginaceae* – Borage Family

1. Plants extremely harshly pubescent; mature leaves prominently
   5- to 6-nerved . . . .*Onosmodium*
1. Plants glabrous or merely pubescent; mature leaves with 1 prominent vein 2
   2. Stems and leaves glabrous; flowers 2.5 cm. long or longer, funnelform-campanulate, blue . . . . *Mertensia*
   2. Stems and leaves pubescent; flowers less than 2.5 cm. long   3
      3. Plants with mature lower leaves over 5 cm. wide .   .   4
         4. Sepals 3 to 4 mm. long, blunt; upper leaves sagittate-clasping; flowers small (not over 8 mm. long), white or reddish-purple . . . . *Cynoglossum*
         4. Sepals 6 to 10 mm. long, subulate; upper leaves not sagittate; flowers over 1 cm. long . *Symphytum*
      3. Plants with mature basal leaves less than 5 cm. wide   5
         5. Flowers red or blue 6
            6. Flowers blue, in crowded spikes 2.5 to 10 cm. long; spikes sometimes arched; leaves undulate . . . .*Heliotropium*
            6. Flowers red or blue, in small cymes; leaves up to 4.5 cm. wide, with entire margins . . . .*Hackelia*
         5. Flowers white, orange, or cream 7
            7. Lower leaves spatulate, rounded at tip; flowers small, white; racemes not bracted . *Myosotis*
            7. Lower leaves lanceolate, or ovate-lanceolate, not rounded at tip; racemes bracteate; flowers cream or orange or bright yellow; each flower subtended by a bract . . . .*Lithospermum*

Onosmodium MICHX.

Onosmodium hispidissimum Mack. (3:117). This harshly scabrous species with lustrous white nutlets is rare in Southern Illinois. We have it from Jackson and Johnson counties. In Jackson County, it occurs on the dry and exposed summit of Devil's Bake Oven near Grand Tower. SP. CIT.: Grand Tower, July 12, 1871, *French 2099.* A very rare species, *Onosmodium molle* Michx. (3:117), has its only station in Illinois near Cave-In-Rock in Hardin County.

Mertensia ROTH

Mertensia virginica (L.) Pers. (3:125). Bluebells or Virginia Cow-slip is one of our most attractive spring wild flowers. It is now much less common in Southern Illinois than in previous years. It often may be found in flower in early April. SP. CIT.: south of Carbondale, April 21, 1951, *M 908.*

Cynoglossum L.

Cynoglossum officinale L. (3:121). The common Hound's-tongue, a native of Europe, has been collected in waste ground or along edges of woods in Jackson, Union, and Hardin counties. This softly pubescent plant puts forth reddish-purple flowers during early summer. SP. CIT.: four miles north of Carbondale, June 8, 1940, *Mc 249.*

Cynoglossum virginianum L. (3:121). Wild Comfrey is a fairly common species of woodlands in Southern Illinois. The flowers, appearing in late April or in May, are pale blue or whitish and the plant is harshly pubescent. SP. CIT.: Giant City State Park, May 1, 1953, *M 118;* Midland Hills, May 9, 1951, *V 659;* Fountain Bluff, May 19, 1952, *BS 2316.*

Symphytum L.

Symphytum officinale L. (3:113). The common Comfrey, an escape from cultivation and a native of Europe, was found by George Hazen French near Grand Tower where it apparently was rather abundant before the turn of the century. SP. CIT.: Grand Tower, *French.*

Heliotropium L.

Heliotropium indicum L. (3:111). A native of India, this Heliotrope is found rather commonly in moist ground throughout Southern Illinois. Bluish flowers appear from late May to July. SP. CIT.: Makanda, 1870, *French 2093;* Lake Murphysboro, August 3, 1954, *M 4729.*

A recently discovered plant in Illinois is the small *Heliotropium tenellum* (Nutt.) Torr. (3:111). It was collected along a rock ledge southeast of Fults (Monroe County) in 1950 by R. A. Evers.

Hackelia OPIZ.

Hackelia virginiana (L.) I. M. Johnston (3:123). The globose fruits of this species are of interest. The entire backs of the fruits are covered by short hooked bristles so that they adhere to clothing, animal's fur, etc. Flowering begins the middle of June and continues through September. SP. CIT.: Carbon Lake, July 27, 1954, *M 4389.*

Myosotis L.

Myosotis macrosperma Engelm. (3:119). This species is local in rich woodlands where it blooms during May and June. It is very similar to the next species, but differs in that the calyx tube bears many tiny bristles and the nutlets are larger (about two millimeters broad). SP. CIT.: Giant City State Park, June 19, 1953, *M 86;* Etherton, *M 2333.*

Myosotis verna Nutt. (3:119). Dry woodlands and sandy fields provide the habitat for this rather common species which flowers from April to June. SP. CIT.: Midland Hills, May 24, 1952, *Voigt;* Etherton, *M 2159.*

Lithospermum L.

1. Corolla whitish   2
    2. Leaves without lateral veins .   . *Lithospermum arvense*
    2. Leaves with lateral veins .   .   . *Lithospermum latifolium*
1. Corolla orange .   .   .   .   .   . *Lithospermum canescens*

Lithospermum arvense L. (3:117). Field Gromwell is a very abundant roadside weed, probably occurring in every county. SP. CIT.: Giant City State Park, May 10, 1953, *M 786;* south of Ava, April 22, 1952, *BS 2040.*

Lithospermum latifolium Michx. (3:117). American Gromwell is an occasional plant of dry woodlands. The nutlets are white and glossy. Flowers appear in May and June. SP. CIT.: Midland Hills, May 7, 1952, *BS 2234;* Lake Murphysboro, July 26, 1954, *Mohlenbrock.*

Lithospermum canescens (Michx.) Lehm. (3:117). This species is called Hoary Gromwell. It occurs in sandy soil in most of our counties. Orange colored flowers are produced in April and May. SP. CIT.: Fountain Bluff, May 16, 1947, *BS 40.*

127. *Verbenaceae* – Verbena Family

1. Corolla 5-lobed.　　　　　　　　　　　　2
　　2. Ovary 4-lobed .　　　　　　　. *Verbena*
　　2. Ovary unlobed .　　　　　　　. *Phryma*
1. Corolla 4-lobed .　　　　　　　. *Phyla*

Verbena L.

1. Corolla white .　　　　　　　　　　　2
　　2. Spikes after flowering greatly elongated, mature calyces not overlapping .　　　　　　.*Verbena urticifolia*
　　2. Spikes after flowering not greatly elongated, mature calyces often overlapping .　.*Verbena urticifolia* x *Verbena hastata*
1. Corolla pink, purple, or violet .　　　　　　3
　　3. Corolla 2 to 3 cm. long .　　　.*Verbena canadensis*
　　3. Corolla less than 2 cm. long .　　　　　4
　　　　4. Spikes in panicles at tip of stems .　　　5
　　　　　　5. Leaves glabrous or nearly so beneath .
　　　　　　　　　　　　. *Verbena hastata*
　　　　　　5. Leaves densely hairy beneath .
　　　　　　　　. *Verbena hastata* x *Verbena stricta*
　　　　4. Spikes mostly in 3's or solitary at tips of stems .　6
　　　　　　6. Leaves pinnately lobed or deeply incised .
　　　　　　　　　　　　.*Verbena bracteata*
　　　　　　6. Leaves simply or coarsely toothed .　　　7
　　　　　　　　7. Flowering spikes 7 to 10 mm. thick; stems with long soft hairs .　　.*Verbena stricta*
　　　　　　　　7. Flowering spikes 5 to 6 mm. thick; stems glabrous or densely pubescent, but seldom with long soft hairs .　　　　　8
　　　　　　　　　　8. Leaves linear to lanceolate; stems glabrous or sparsely pubescent .　.*Verbena simplex*
　　　　　　　　　　8. Leaves elliptic; stems densely pubescent .
　　　　　　　　　.　*Verbena simplex* x *Verbena stricta*

Verbena bracteata Lag. and Rodr. (3:134). Bracted Verbena is an inhabitant of waste ground, particularly along roads. Our records for its flowering are from June 15 to July 28. It is the least common of our weedy Vervains. SP. CIT.: *BS 479;* Jacob, June 22, 1948, *BS 479.*

Verbena canadensis (L.) Britt. (3:134). The attractive Wild Verbena or Sweet William, although uncommon in our area, is prevalent in calcareous regions. It is common in limestone talus of Pine Hills in Union County. SP. CIT.: Thompson's Woods, Southern Illinois University campus, M 2578.

Verbena hastata L. (3:130). Blue Vervain is a common species of waste places and dry open woodlands. It usually begins to flower later than other species of Verbena, reaching anthesis from July to late September. SP. CIT.: one mile north of Carbondale, Mc 268; Giant City State Park, M 516.

Verbena simplex Lehm. (3:132). Another of the weedy Verbenas is the Narrow-leaved Vervain. It occurs in all southern counties. SP. CIT.: one and one-half miles north of Etherton, Mc 918.

Verbena stricta Vent. (3:132). Hoary Vervain, a tall, densely pubescent species, is common in waste ground and rocky places. It flowers during summer. SP. CIT.: north of Etherton Mc 915; Fountain Bluff, M 4628.

Verbena urticifolia L. (3:130). This tall plant is the only Vervain with white flowers. Its habitats include both waste places and woodlands. SP. CIT.: one mile south of Carbondale, Mc 295.

Hybridization among the Vervains is easy to observe, and the hybrids are sterile. The following hybrids are known from this area:

Verbena hastata x *stricta* Rydb. Jackson County: one mile north of Pomona, September 28, 1951, V 1002.

Verbena hastata x *urticifolia* Pepoon. Hardin County; August 10, 1950, H. E. Ahles.

Verbena simplex x *stricta* H. E. Ahles. Hardin County; open pastured hillside, near Elizabethtown, June 19, 1931, J. Schopf.

Phryma L.

Phryma leptostachya L. (3:269). This summer flowering species, Lopseed, is often put in its own family, the Phrymaceae. Rarely any oak or hickory woods is without this species. SP. CIT.: Giant City State Park, M 276; Lake Murphysboro, M 4160A.

Phyla LOUR.

Phyla lanceolata (Michx.) Greene (3:136). A common species along shores and in moist roadside ditches is Frog-fruit. Flowering occurs from early June to mid-September. SP. CIT.: Giant City State Park, M 1333; Fountain Bluff, M 4354A.

128. *Labiatae* – Mint Family

1. Leaves lobed (pinnatifid or sharply angled and 3-lobed) . . 2
 2. Leaves pinnatifid or undulate (lyrate); calyx teeth not spine-tipped. 3
  3. Flowers white . . . . *Lycopus*
  3. Flowers purple . . . . *Salvia*
 2. Leaves with lobes sharply angled and 3-lobed; calyx teeth spine-tipped. . . . . *Leonurus*
1. Leaves not lobed, only serrate, dentate, crenate, or entire . . 4
 4. Calyx crested . . . . *Scutellaria*
 4. Calyx not crested 5
  5. Leaves ovate, ovate-lanceolate, or reniform . . . 6
   6. Leaves reniform; stems creeping . . . *Glecoma*
   6. Leaves not reniform; stems erect 7
    7. Flowers in axillary whorls or interrupted spikes or flowers in dense head-like or capitate clusters 8
     8. Flowers in head-like clusters; bracts broad, conspicuous . . . . . *Monarda*
     8. Flowers not in head-like clusters; bracts not broad or conspicuous 9
      9. Leaves distinctly petioled . . . 10
       10. Leaves crenate or dentate . . 11
        11. Calyx with 10 teeth . . . . . . *Marrubium*
        11. Calyx with 5 teeth. . . . . . . *Lamium*
       10. Leaves coarsely serrate, dentate, or only serrate 12
        12. Leaves coarsely serrate or dentate; calyx 13-nerved . . . . . . *Melissa*
        12. Leaves finely serrate, serrate, or entire. 13
         13. Anther-bearing stamens 2 . . . *Blephilia*
         13. Anther-bearing stamens 4 14
          14. Stamens equal in length; plants with aromatic odor . . . *Mentha*
          14. Stamens unequal in length (2 shorter than the other

2); plants not hav-
ing a strong aro-
matic odor . . .
. . . *Stachys*

9. Leaves sessile or subsessile . . . 15

 15. Stems stiffly and corymbosely
branched; leaves serrate or finely
serrate; anthers conspicuously exserted . . . . . *Cunila*

 15. Stems simple; leaves coarsely serrate or dentate; anthers usually not
conspicuously exserted . *Lycopus*

7. Flowers in dense or uninterrupted spikes, or
flowers in cymes, racemes, panicles or few in
axils of leaves 16

 16. Bracts broad and conspicuous . . . 17

  17. Flowers lavender; inflorescence spicate . . . . . *Prunella*

  17. Flowers white, in head-like cymes .
. . . . . *Pycnanthemum*

 16. Bracts not broad and conspicuous . . 18

  18. Leaves somewhat cordate at base;
flowers large (2 to 3 cm. long), white,
solitary in axils of upper leaves . .
. . . . . . *Synandra*

  18. Leaves not cordate at base . . . 19

   19. Flowers in long racemes; upper
lip of corolla wanting . .
. . . . . *Teucrium*

   19. Flowers not in long racemes; upper lip of corolla not wanting 20

    20. Flowers in loose terminal
panicles; leaves large and
ovate, petiolate . . .
. . . . *Collinsonia*

    20. Flowers not in loose termi-
nal panicles; leaves not as
above 21

     21. Stamens conspicuous-
ly exserted beyond
the corolla; stems
simple; plants 2 to 5
feet tall; inflorescence
a spike . *Agastache*

     21. Stamens not conspic-
uously exserted;

plants shorter (1 to 3 feet tall). . . 22
  22. Leaves ovate, red-purplish-green; flowers in elongate spikes . . . *Perilla*
  22. Leaves ovate or ovate-lanceolate, light green, densely canescent; flowers white with purple dots on the corolla lobes . . . *Nepeta*
5. Leaves linear or lanceolate. 23
  23. Leaves sessile or subsessile 24
    24. Leaves entire . . . .*Trichostema* *
    24. Leaves serrate 25
      25. Leaves finely serrate; stamens exserted beyond corolla . . .*Physostegia*
      25. Leaves serrate; stamens not exserted beyond corolla; plants aromatic . *Mentha*
  23. Leaves petiolate; stems corymbosely branched . 26
    26. Calyx tubular, 13-nerved; stamens 2 . . . . *Hedeoma*
    26. Calyx broadly campanulate, 10-nerved; stamens 4. . . . . . .*Isanthus*

Salvia L.

Salvia lyrata L. (3:169). Sage or Cancerweed is known only from the very southern counties of Illinois where it occurs in rocky woods. SP. CIT.: Giant City State Park, *M 94.*

Leonurus L.

Leonurus cardiaca L. (3:160). Motherwort is a native of Europe which has become naturalized in waste places. We have it only from Jackson and Union counties. Spiny calyx teeth make this plant uncomfortable to handle. SP. CIT.: Giant City State Park, *M 199;* Fountain Bluff, *BS 412.*

Scutellaria L.

1. Flowers in axillary or terminal racemes 2
   2. Inflorescence of lateral racemes from axils of cauline leaves
     . . . . *Scutellaria lateriflora*
   2. Inflorescence mostly of terminal racemes. 3
     3. Leaves cordate at base 4
       4. Bracteal leaves longer than calyx; leaves mostly 5 to
         8 cm. long. . . . . . *Scutellaria ovata*
       4. Bracteal leaves shorter than or equalling calyx; leaves
         mostly 8 to 12 cm. long . . . .
         . . . . *Scutellaria ovata* var. *versicolor*
     3. Leaves rounded or tapering at base. 5
       5. Stem with 2 to 5 pairs of leaves below inflorescence
         . . . . *Scutellaria elliptica*
       5. Stem with 7 or more pairs of leaves below inflores-
         cence . . . . . . *Scutellaria incana*
1. Flowers solitary in axils of cauline leaves 6
   6. Leaves conspicuously toothed, some of them 2.5 to 4.5 mm.
     long 7
     7. Corolla blue. . . . . *Scutellaria nervosa*
     7. Corolla white . . . *Scutellaria nervosa* f. *alba*
   6. Leaves entire or only obscurely toothed 8
     8. Leaves mostly glabrous above, with two pairs of lateral
       veins . . . . *Scutellaria leonardi*
     8. Leaves pubescent above, with 3 to 5 pairs of lateral veins 9
       9. Lateral veins uniting near the leaf margin to form a
         submarginal nerve. . . . *Scutellaria australis*
       9. Lateral veins not uniting near the leaf margin . .
         . . . . *Scutellaria parvula*

Scutellaria elliptica Muhl. (3:147). This is another one of the La-
biatae of Illinois which is restricted to the southernmost counties. It
occurs in dry upland woods where it is scarce. Jones *et al.* (1955) report it from Jackson County, but we have seen no specimens.

Scutellaria incana Biehler (3:147). Downy Skullcap is a common
species of prairies and dry oak-hickory woods (Fig. 61). It is a very
attractive species. A form with rose flowers occurs in prairie soil along
a railroad in northern Jackson County just south of Matthews. This is
the type locality of newly proposed f. *rhodantha* Mohlenbrock, *a typo
differt floribus roseis.* SP. CIT.: typical: Clear Creek, *BS 1700;* Midland Hills, *Mc 949;* Giant City State Park, *M 416;* one mile north of

FIG. 61. *Downy Skullcap Mint*
(Scutellaria incana) *is often found on moist roadsides, prairies, or along railroads.*

Etherton, *Mc 916;* Pomona, *V 533;* Lake Murphysboro, *M 4604. F. rhodantha,* two miles south of Mathews, July 25, 1954, *M 356* (type—S.I.U. herb.).

Scutellaria lateriflora L. (3:148). *Scutellaria lateriflora* is found along rivers and streams and in moist woods. Along the Mississippi River near Neunert, specimens with glandular hairs in the inflorescence occur. Flowers appear in August and September. SP. CIT.: Giant City State Park, *M 561;* Midland Hills, *Mc 956;* four miles south of Cora, *M 4752.*

Scutellaria leonardi Epling (3:149) [*S. parvula* Michx. var. *leonardi* (Epling) Fern.]. This small plant is local throughout our area. In Southern Illinois it is known from Jackson, Johnson, Pope, and Hardin counties. SP. CIT.: Giant City State Park, June 15, 1953, *M 795.*

Scutellaria nervosa Epling. (3:150). This is a rare species of moist woodlands. We have it from Johnson, Pope, and Hardin counties. The much smaller and white-flowered f. *alba* Steyerm. occurs in a very low woodlands at Lake Murphysboro. SP. CIT.: f. *alba:* Lake Murphysboro, June 1, 1954, *M 4751.*

Scutellaria ovata Hill. (3:147). Heart-leaved Skullcap is common along sandstone bluffs. It occurs with var. *versicolor* (Nutt.) Fern. SP. CIT.: typical: Giant City State Park, *M 900;* one mile southeast of De Soto, *Mc 220;* Pomona, *V 697;* Grand Tower, *French 2053.* Var.

versicolor: Boskydell, *Mc and Wilson 1030;* Little Grand Canyon, *M 4027.*

Scutellaria australis (Fassett) Epling (3:150) [*S. parvula* var. *australis* Fassett]. This species is abundant in prairie soil along the railroad near Finney. SP. CIT.: seven miles north of Murphysboro, *Mohlenbrock.*

Scutellaria parvula Michx. (3:149). This skullcap has been collected only a few times. It may be mistaken for S. *australis*. SP. CIT.: Thompson's Woods, May 17, 1952, *V 1028.*

Glecoma L.

Glecoma hederacea L. var. *micrantha* Moricand (3:153) [*G. heterophylla* Waldst. & Kit.]. Ground Ivy is a frequent plant of waste places, although it occasionally may be found along edges of woods. It is native of Europe. It flowers from mid-April through May. SP. CIT.: Campbell Hill Brick Plant, *M 2588;* Giant City State Park, *French.*

Monarda L.

Monarda bradburiana Beck (3:171). Bergamot Mint or Bee Balm is common in dry open woods. It is absent from northern Illinois. The leaves are sessile and the stems glabrous or nearly so. Fernald (1944) gives reason for substituting the name *M. russeliana* Nutt. for *M. bradburiana.* We have recorded it in flower from May 20 to June 18. SP. CIT.: Giant City State Park, *M 970;* Little Grand Canyon, *M 4067.*

Monarda fistulosa L. (3:171). The showy Wild Bergamot occurs along the edges of dry woodlands. It flowers in June and July. The leaves are petiolate and the stems are usually pubescent. SP. CIT.: Giant City State Park, *M 379;* Etherton, *Mc 920.*

Marrubium L.

Marrubium vulgare L. (3:150). Horehound is an infrequent escape from cultivation. It is naturalized from Eurasia. SP. CIT.: near Carbondale, in 1870, *French.*

Lamium L.

Lamium amplexicaule L. (3:158). One of the most common weeds of lawns and waste places is Henbit, a native of Europe. The earliest flowers are cleistogamous. The upper leaves are sessile and clasping. SP. CIT.: *M 35; M 2132; Mc 48.*

Lamium purpureum L. (3:158). This native of Eurasia, Purple Dead Nettle, is much less common than the preceding. The upper leaves are petiolate. SP. CIT.: *M 2562.*

Melissa L.

Melissa officinalis L. (3:174). A native of Europe, the Balm has been collected infrequently. It is a casual escape from cultivation. SP. CIT.: Grand Tower, July 11, 1871, *French.*

Blephilia RAF.

Blephilia ciliata (L.) Benth. (3:173). Wood Mint, very similar to the following, is common in dry oak-hickory woodlands. It differs from *B. hirsuta* in that the nearly odorous leaves are on petioles two to ten mm. long. SP. CIT. Giant City State Park, *M 200;* Pomona *V 676.*

Blephilia hirsuta (Pursh) Benth. (3:173). Among the distinguishing characteristics of this species are fragrant and long-petioled leaves. It often grows with the preceding. Flowering dates for this species are from late May to mid-September. SP. CIT.: Beaucoup Creek, *Mc 883;* Fountain Bluff, *Mc 454.*

Mentha L.

1. Flowers in subglobose axillary clusters .   . *Mentha arvensis*
1. Flowers mostly in terminal spikes 2
    2. Leaves sessile or nearly so; spikes crowded . *Mentha spicata*
    2. Leaves sessile; spikes interrupted 3
        3. Leaves with wavy or crisped margins . *Mentha crispa*
        3. Leaves merely toothed .   .   .   .*Mentha piperita*

Mentha arvensis L. var. *villosa* (Benth.) S. R. Stewart (3:187) [*M. canadensis* L.]. Wild Mint does not seem to be as common in Southern Illinois as in the northern section of the state. It is an inhabitant of moist soil. SP. CIT.: Makanda, *BS 864.*

Mentha spicata L. (3:188). Spearmint is naturalized from Europe. It is cultivated occasionally in our area. SP. CIT.: southern Murphysboro, *Mohlenbrock.*

Mentha crispa L. (3:187). This Eurasian species of *Mentha* is known in Illinois only from Cook, Lake, Champaign, and Jackson counties. SP. CIT.: Carbondale, July 16, 1950, *H. H. Stone.*

Mentha piperita L. (3:188). Peppermint, like Spearmint, is a cultigen which sometimes escapes. It flowers during August and September. SP. CIT.: Makanda, *BS 864;* Murphysboro, *M 4628.*

Stachys L.

1. Sides of upper part of stems pubescent . *Stachys arenicola* °
1. Sides of upper parts of stems glabrous or nearly so, but sometimes hirsute on the angles 2
    2. Leaves glabrous above 3
        3. Leaves sessile or nearly so. . . . *Stachys aspera*
        3. Leaves on distinct petioles . . . *Stachys tenuifolia*
    2. Leaves pubescent above 4
        4. Calyx lobes nearly as long as calyx tube . . . . .
        . . . .*Stachys hispida* °
        4. Calyx lobes about one-half as long as calyx tube . .
        . . . . *Stachys clingmanii* °

Stachys aspera Michx. (3:163) [*S. hyssopifolia* Michx. var. *ambigua* Gray]. This species, while fairly common in central Illinois, is known in the extreme north only from Cook County and from the south only from Jackson County where it was collected along the edge of a swamp. SP. CIT.: Lester Swamp, August 23, 1954, *M 4640.*

Stachys tenuifolia Willd. (3:163). Common Hedgenettle occurs in moist ground throughout the state. Common associates of it are *Impatiens biflora, Solidago flexuosa, Mimulus alatus,* and others. It flowers from early July to mid-September. SP. CIT.: four miles west of De Soto, *Mc 887;* Giant City State Park, *M 393;* Carbon Lake, *M 4396.*

Three other species of *Stachys* have been recorded from Southern Illinois. *Stachys arenicola* Britt. (3:163), of wet ground, was found in Randolph County on Menard Penitentiary grounds near the Mississippi River in the summer of 1955 by Mohlenbrock. *Stachys hispida* Pursh (3:164) was unknown south of Fayette County until 1955 when it likewise was found in moist soil in Randolph County by Mohlenbrock. *Stachys clingmanii* Small (3:165), confined to southeastern United States, has been collected in Illinois from Alexander, Hardin, and Massac counties.

Cunila L.

Cunila origanoides (L.) Britt. (3:183). Stone Mint is a characteristic plant of dry bluff tops along the Shawneetown Ridge. It flowers from August to late October and has a very characteristic minty odor. SP. CIT.: Giant City State Park, August 30, 1954, *M 4682;* Little Grand Canyon, August, 1954, *M 4062;* Pomona Natural Bridge, October 2, 1947, *BS 200 A.*

Lycopus L.

1.   Calyx lobes less than 1 mm. long, mostly obtuse .   .   .   .
     .   .   .   .   *Lycopus virginicus*
1.   Calyx lobes 1 to 2 mm. long, acuminate.   2
    2.   Leaves deeply incised to pinnatifid . *Lycopus americanus*
    2.   Leaves shallowly toothed   3
        3.   Leaves elliptic, 1 to 4 cm. broad   4
            4.   Stems glabrous.   .   .   .   . *Lycopus rubellus*
            4.   Stems pubescent . *Lycopus rubellus var. arkansanus*
        3.   Leaves lanceolate, 0.5 to 2.0 cm. broad .   .   .   .
       .   .   .   .   . *Lycopus rubellus* var. *lanceolatus*

Lycopus virginicus L. (3:184). An occasional plant of moist situations throughout Illinois is Water Horehound. It occurs frequently with *Diodia teres, Spermacoce glabra,* and other amphibious species. SP. CIT.: Giant City State Park, July 28, 1953, *M 585.*

Lycopus americanus Muhl. (3:184). This is our most abundant Water Horehound. It is often found growing with other species of *Lycopus.* SP. CIT.: Giant City State Park, August 2, 1953, *M 559;* Lester Swamp, August 23, 1954, *M 4644.*

Lycopus rubellus Moench (3:184). This species is variable with respect to leaf morphology. The typical variety is our most common one, although var. *lanceolatus* Benner and var. *arkansanus* (Fresn.) Benner have single stations in Jackson County. These last two are known in Illinois only from their Jackson County stations. SP. CIT.: typical: Giant City State Park, August 2, 1953, *M 587.* Var. *lanceolatus:* one mile north of Pomona, September 28, 1951, *V 1009.* Var. *arkansanus:* low ground, Finney railroad prairie, August 28, 1954, *M 4663A.*

Prunella L.

Prunella vulgaris L. (3:155). Selfheal or Heal-all may be found in almost any wasteland and occasionally in woods. It flowers from August to October. SP. CIT.: Midland Hills, *Bell; Mc and Wilson 1028*.

Pycnanthemum MICHX.

1. Leaves sessile 2
   2. Stems glabrous; bracts small, appressed . . . . .
       . . . .*Pycnanthemum flexuosum*
   2. Stems pubescent on the angles; bracts foliaceous . . .
       . . . . . . *Pycnanthemum virginianum* *
1. Leaves petiolate 3
   3. Upper leaves whitish . *Pycnanthemum pycnanthemoides*
   3. Upper leaves green 4
      4. Leaves lanceolate . . .*Pycnanthemum pilosum*
      4. Leaves ovate . . . .*Pycnanthemum incanum*

Pycnanthemum flexuosum (Walt.) BSP. (3:179) [*P. tenuifolium* Schrad.]. This is our most common species of Mountain Mint, occurring in dry woodlands and fields. We have noted it in flower in June and July. SP. CIT.: *Stewart; Welch 281; Mc 882; M 479.*

Pycnanthemum incanum (L.) Michx. (3:183). This species is limited in Illinois to only a few southern counties where it is found in oak-hickory woods, particularly in limestone areas. SP. CIT.: Giant City State Park, August 5, 1953, *M 596;* five miles north of Carbondale, August 14, 1940, *Mc 368.*

Pycnanthemum pilosum Nutt. (3:180). Jackson County is the only county in our area in which this plant has been found, and there not since 1871. It is much more common to the north. SP. CIT.: Carbondale, July 18, 1871, *French.*

Pycnanthemum pycnanthemoides (Leavenw.) Fern. (3:183). This mint is common in Southern Illinois, but it has not been recorded north of Randolph County. It flowers in July and August. The pale upper leaves make it easily recognized. SP. CIT.: *M 601;* Boskydell, *BS 158.*

A fifth species, *Pycnanthemum virginianum* (L.) Dur. and Jacks. (3:180), is rather common in northern and central Illinois but is known in our area only from Perry and Saline counties.

Synandra NUTT.

Synandra hispidula (Michx.) Britt. (3:157). This is one of the rarer plants in Illinois (Fig. 62). It is known only from Giant City

State Park and the Pomona Natural Bridge area, both in Jackson County. It grows in deep rich woods where it produces attractive white flowers in May. Associated with this species are *Aplectrum hyemale, Stylophorum diphyllum, Collinsia verna,* and *Valeriana radiata.* SP. CIT.: Giant City State Park, *Mc 768; V 661; Stewart and Voigt 491; M 682.*

Teucrium L.

Teucrium canadense L. (3:143). Wood-sage is common on moist ground. It is particularly plentiful in prairie soil along a railroad near Finney. Its flowering period is June and July. SP. CIT.: Giant City State Park, *M 345;* near Carbondale, *French 1975;* south of De Soto, *Mc 221.*

Collinsonia L.

Collinsonia canadensis L. (3:189). Richweed is rare and local in Illinois. Besides being recorded from two counties in east-central Illinois we have it from Jackson, Union, Hardin, Alexander, and Pulaski counties. SP. CIT.: Clear Springs area, *BS 2971.*

Agastache CLAYTON

Agastache nepetoides (L.) Kuntze (3:152). Giant Hyssop is a rather infrequent inhabitant of sandy or gravelly soils. It seems to be most common and vigorous in soil along railroads. SP. CIT.: Fountain Bluff, August 15, 1954, *Mohlenbrock.*

Perilla L.

Perilla frutescens (L.) Britt. (3:190). Beefsteak plant is becoming an aggressive weed along streams. It receives its name from the fact that the lower surface of the leaves is similar in color to that of a piece of raw beefsteak. It flowers from mid-August until the first of October. SP. CIT.: Midland Hills, *Bell;* Fountain Bluff, *Mc 437;* Giant City State Park, *Voigt;* Pomona Natural Bridge, *BS 202;* Little Grand Canyon, *M 4037.*

Nepeta L.

Nepeta cataria L. (3:153). Catnip, a native of Europe, is a local escape into waste ground in our area. SP. CIT.: along railroad at base of Fountain Bluff, *M 4357.*

FIG. 62. *This large-flowered White Mint* (Synandra hispidula) *is known only in Illinois from Jackson County.*

Physostegia BENTH.

Physostegia virginiana (L.) Benth. (3:156). False Dragonhead, which occurs sparingly in moist habitats, is a striking late summer wildflower. SP. CIT.: two miles south of the junction of Illinois highways 144 and 3, *Mc 480;* Finney railroad prairie, *M 4662.*

Hedeoma PERS.

Hedeoma hispida Pursh (3:173). Rough Pennyroyal, while common in the central and northern parts of Illinois, has been found only once in Southern Illinois. It occurs in a dry woodland at Giant City State Park. The plant is strongly scented. The linear sessile leaves have entire margins. SP. CIT.: Giant City State Park, August 25, 1953, *M 784.*

Hedeoma pulegioides (L.) Pers. (3:173). American Pennyroyal is common in dry fields or dry woodlands. It is strongly scented. The elliptic and petiolate leaves have some serrations on their margins. It blooms from mid-July until the first of October. SP. CIT.: Giant City State Park, *M 560;* Thompson's Lake, *Mc 409.*

Isanthus MICHX.

Isanthus brachiatus (L.) BSP. (3:143). False Pennyroyal occurs in a variety of dry habitats, particularly along edges of hill prairies.

We have it only once from Jackson County, although it has been found in Randolph, Gallatin, Union, Pope, Johnson, and Hardin counties. SP. CIT.: one mile north of Pine Hills, September 9, 1954, *M 4757*.

Trichostema L.

Bluecurls, *Trichostema dichotomum* L. (3:143), has been collected in a few northern counties and in Union, Johnson, Pope, Hardin, and Massac counties in Southern Illinois, but it is unknown from Jackson County. It is most often found in sandy soil.

129. *Solanaceae* – Nightshade Family

1. Shrubby plants, climbing or trailing; fruit a berry . *Lycium*
1. Herbs, usually erect.   2
   2. Fruit enclosed by the calyx making an inflated saccate structure   3
      3. Flowers yellowish or cream-colored; calyx not split to its base.   .   .   .   .*Physalis*
      3. Flowers purple or blue; calyx split to its base .   .   .
      .   .   .   . *Nicandra* °
   2. Fruit not enclosed by an inflated calyx   4
      4. Fruit a berry; corolla about 2 cm. long or less . *Solanum*
      4. Fruit a rather prickly capsule; corolla 5 to 15 cm. long or sometimes longer.   .   .   .   .   .   .*Datura*

Lycium L.

Lycium halimifolium Mill. (3:202). Matrimony Vine occurs in waste areas in a few of our counties. Lavender flowers appear from May to early August and are followed by bright crimson berries. It was introduced from the Old World.

Lycopersicum MILL.

Lycopersicum esculentum Mill. (3:203). Tomato occasionally escapes from cultivation but rarely persists for more than one season.

Physalis L.

1. Stems and leaves glabrous or glabrate   2
   2. Plants annual; corolla yellow, without dark spots .   .   .
   .   .   .   . *Physalis pendula* °

2. Plants perennial; corolla yellow, with a purple center . .
. . . . *Physalis subglabrata*
1. Stems and leaves villous and often viscid 3
3. Leaves tapering to the base; pubescence sparse . . .
. . . . *Physalis virginiana*
3. Leaves rounded or cordate at the base; plants densely villous
and often viscid. 4
4. Plants with some viscid hairs 5
5. Plants annual; anthers purple . *Physalis pruinosa*
5. Plants perennial; anthers yellow. . . . .
. . . . *Physalis heterophylla*
4. Plants without viscid hairs 6
6. Leaves pale green, weak, entire or undulate . .
. . . . *Physalis pubescens*
6. Leaves dark green, firm, with sharp teeth . .
. . . . *Physalis barbadensis*

Physalis subglabrata Mack. and Bush (3:197). Smooth Ground-cherry is a common species of cultivated ground and waste land. It flowers throughout summer.

Physalis virginiana Mill. (3:198). Virginia Ground-cherry is similar to the preceding species in characteristics and in habitat requirements but usually begins to flower about one month earlier. SP. CIT.: near Murphysboro, July 21, 1954, *Mohlenbrock.*

Physalis pruinosa L. (3:195). This rare species is known in our area mostly from limestone talus in Randolph and Jackson counties. SP. CIT.: Giant City State Park, *M 334;* Lake Murphysboro, *M 4730.*

Physalis heterophylla Nees var. *ambigua* (Gray) Rydb. (3:197). This plant is common throughout Illinois in cultivated ground or along roads. It flowers during summer. SP. CIT.: Giant City State Park, *M 599;* north of Murphysboro, *VM 1579.*

Physalis pubescens L. (3:195). This diffusely branched Ground-cherry is abundant in most southern counties. SP. CIT.: Grand Tower; Giant City State Park, *M 414.*

Physalis barbadensis Jacq. (3:196). The only record of this species for Illinois is a George Vasey specimen from Carbondale, Jackson County. It is without date or number.

A rare species, *Physalis pendula* Rydb. (3:196), with a completely yellow corolla, is known in Illinois only from Union County where it was collected in 1950 along a road at the base of the Pine Hills bluffs by Harry E. Ahles (*2864*).

Solanum L.

1. Plants glabrous or glabrate 2
   2. Leaves thin; berries shiny black . *Solanum americanum* *
   2. Leaves firm; berries dull black . . . *Solanum nigrum*
1. Plants prickly 3
   3. Plants with numerous bright yellow prickles; flowers yellow
   . . . . *Solanum rostratum*
   3. Plants with pale prickles; flowers white or lavender . .
   . . . . *Solanum carolinense*

Solanum americanum Mill. American Nightshade, similar to the introduced Black Nightshade (S. *nigrum*), is rare in Illinois. We have it only along a limestone ledge in the Pine Hills (Union County). At this station, the flowers were lavender and were produced in September and October.

Solanum nigrum L. (3:198). Black Nightshade occurs in waste ground throughout Illinois. It sometimes attains a height of nearly three feet. The dull black berries are poisonous. SP. CIT.: Giant City State Park, M *345;* Midland Hills, *Mc 955.*

Solanum rostratum Dunal (3:200). Buffalo-bur, a native of western United States, has escaped into a few localities in Southern Illinois. The entire plant is densely beset with yellow prickles. SP. CIT.: Jacob, June 22, 1948, *BS 478.*

Solanum carolinense L. (3:201). Horse-nettle is probably the commonest *Solanum* in Southern Illinois. It may be found in almost any locality and habitat. SP. CIT.: Giant City State Park, M *292;* Thompson's Lake, *Mc 1216;* Murphysboro, M *904.*

Nicandra ADANS.

Apple of Peru, *Nicandra physalodes* (L.) Gaertn. (3:193), is adventive in our area only in Johnson County. It flowers during the late summer.

Datura L.

Datura stramonium L. (3:205). Jimson-weed is abundant in waste ground, particularly around farms. Flowers are two to four inches long, and leaves are coarsely toothed. The plant is glabrous. SP. CIT.: M *399; BS 3217.*

Datura innoxia Mill. (3:205). This large introduction from tropical

America is rarely collected in Illinois and is known only from Jackson among the southern counties where it was collected in the Grand Tower park. SP. CIT.: Grand Tower, July 25, 1955, *M 4249.*

130. *Scrophulariaceae* – Figwort Family

1. Flowers spurred 2
    2. Leaves linear; flowers in spikes or spike-like racemes, yellow; plants erect . . . . *Linaria*
    2. Leaves triangular, hastate; flowers solitary, axillary, yellowish above, purplish beneath; plants creeping . . . *Kickxia*
1. Flowers not spurred 3
    3. Anther bearing stamens 5; corolla nearly regular, rotate; leaves alternate . . . . *Verbascum*
    3. Anther bearing stamens 2 or 4; corolla irregular; leaves usually opposite, rarely whorled, or upper ones alternate . . 4
        4. Anther bearing stamens 2 5
            5. Leaves whorled; flowers bluish or white in dense peduncled spike-like racemes, terminal or in upper axils . . . . *Veronicastrum*
            5. Leaves opposite or some of upper leaves rarely alternate 6
                6. Flowers with 4 sepals and 4-lobed corollas; flowers blue or white; usually small plants . . . . *Veronica*
                6. Flowers with 5 sepals and 5-lobed corollas; plants of moist ground or wet shores . . . 7
                    7. Corolla purplish . . . . *Lindernia*
                    7. Corolla whitish or yellowish . . *Gratiola*
        4. Anther bearing stamens 4 8
            8. Leaves pinnately compound or pinnatifid . . . 9
                9. Flowers borne in a crowded terminal spike or spike-like raceme; flowers yellow, 2-lipped . . . . *Pedicularis*
                9. Flowers solitary or 2 in axils of leaves . . 10
                    10. Flowers 2 in axils of leaves, greenish-white, 2-lipped, 3 to 5 mm. long . *Leucospora*
                    10. Flowers solitary in axils of leaves, flowers yellow, the corolla spreading, with slightly unequal lobes, somewhat campanulate . 11
                        11. Leaves pinnatifid, the lower ones long-petioled; flowers 2.5 to 7.5 cm. long . . . . . *Seymeria*
                        11. Leaves entire or sinuate-margined or

some of the lower pinnatifid; lower
leaves short-petioled; flowers 3.5 to
5.0 cm. long . . . . *Aureolaria*
8. Leaves not lobed 12
    12. Corolla spreading, slightly unequally 5-lobed,
    somewhat campanulate or rotate . . . 13
        13. Plants with oval, sessile leaves; creeping
        plants of mud flats; corolla white . .
        . . . . *Bacopa*
        13. Plants with linear or lanceolate leaves;
        erect plants; corolla purple to rose . .
        . . . . *Gerardia*
    12. Corolla rather distinctly 2-lipped . . . 14
        14. Leaves with entire margins . *Penstemon*
        14. Leaves with serrate or dentate margins . 15
            15. Flowers in dense spikes, flowers
            whitish; plants of swampy or marshy
            places . . . . . *Chelone*
            15. Flowers not in dense spikes . . 16
                16. Flowers solitary in axils of
                leaves 17
                    17. Flowers pink, purple, or
                    rose . . . *Mimulus*
                    17. Flowers bicolored, lower
                    lip blue, upper lip white
                    . . . . *Collinsia*
                16. Flowers in terminal panicles . 18
                    18. Tall plants mostly over 3
                    feet tall and up to 8 feet;
                    flowers small (1 to 1.5
                    cm. long), greenish-ma-
                    roon, or purplish . .
                    . . . *Scrophularia*
                    18. Shorter plants, mostly less
                    than 3 feet tall; flowers
                    white, purplish or violet;
                    larger flowers (1.5 to 2.5
                    cm. long) . *Penstemon*

Linaria MILL.

Linaria vulgaris Mill. (3:229). The familiar Butter-and-Eggs, native of Europe, flowers from late spring throughout the summer. The
flowers are yellow. SP. CIT.: *M 4808.*

Linaria canadensis (L.) Dumort. (3:229). Blue Toadflax occurs
in sandy soil in Alexander and Johnson counties. The spur is usually

two to five millimeters long. Plants with spurs six to nine millimeters long have been collected in Alexander County. These are segregated as var. *texana* (Scheele) Pennell [*L. texana* Scheele].

Kickxia DUMORT.

Kickxia elatine (L.) Dumort. (3:229). Canker-root, native of Eurasia, is known only from a few Illinois counties with Jackson being the sole county in our area. The species occurs in moderate abundance along the Gulf, Mobile and Ohio Railroad in Murphysboro. SP. CIT.: along Gulf, Mobile and Ohio Railroad, Murphysboro, June 21, 1955, *Mohlenbrock.*

Verbascum L.

Verbascum blattaria L. (3:219). Moth Mullein, a predominantly glabrous species, occurs sparingly in waste ground in Southern Illinois. Flowers are usually yellow, although a white-flowered form (*f. albiflora* (Don) House) does occur. SP. CIT.: Giant City State Park, *M 472;* De Soto, *Mc 816.*

Verbascum thapsus L. (3:218). Woolly Mullein is a common waste ground species which has been naturalized from Europe. Leaves are densely covered by wool on both surfaces. This plant sometimes attains a height of twelve to fifteen feet. SP. CIT.: Giant City State Park, *M 293;* Carbondale, *BS 128.*

Veronicastrum FABR.

Veronicastrum virginicum (L.) Farwell (3:233). Culver-root, or more appropriately Candelabra Plant, is found in open woods. Flowers appear in July and August. SP. CIT.: Giant City State Park, *M 438;* north of Carbondale, *V 890;* south of Elkville, *BS 1512.*

Veronica L.

1. Flowers in racemes   .   .   .   .   . *Veronica serpyllifolia*
1. Flowers solitary in axils of upper leaves   2
    2. Corolla white   .   .   .   .   .   . *Veronica peregrina*
    2. Corolla blue   3
        3. Corolla 8 to 10 mm. long; pedicels of fruit at least 1 cm. long   .   .   .   .   .   . *Veronica persica*
        3. Corolla 2 to 4 mm. long; pedicels of fruit minute or up to 4 mm. long   4

    4. Upper leaves mostly smaller than lower ones .   .
       .   .   .   .*Veronica arvensis*
    4. Upper leaves like lower ones .   . *Veronica polita*

Veronica serpyllifolia L. (3:233). The collection by G. H. French from Grand Tower in 1871 is the only known record in Southern Illinois. It is naturalized from Eurasia. SP. CIT.: Grand Tower, July 12, 1871, *French*.

Veronica peregrina L. (3:234). The small White Speedwell is common in waste places, particularly moist areas. It flowers from mid-April to mid-July. SP. CIT.: *Mc 28; M 104; M 2181*.

Veronica persica Poir. (3:235). A collection of this species from the Southern Illinois University campus in 1954 is the first from the southern counties. SP. CIT.: *Voigt*.

Veronica arvensis L. (3:234). The tiny Blue Speedwell is a common inhabitant of lawns. It begins to flower in mid-April. SP. CIT.: *Mc 23; M 2133; M 2404*.

Veronica polita Fries [*V. didyma* Tenore]. Unknown in Illinois until its collection from Lawrence County in 1949, this small Speedwell is the first *Veronica* to flower. It is found in lawns where it bears some resemblance to *Veronica arvensis*. SP. CIT.: *Mohlenbrock*.

Lindernia ALL.

Lindernia anagallidea (Michx.) Pennell (3:216). A rather common species of stream banks and lake shores is this summer-flowering annual. The pedicels are much longer than subtending leaves. SP. CIT.: Carbondale Reservoir, October, 1953, *Bell;* Grand Tower slough, August 20, 1954, *M 4636*.

Lindernia dubia (L.) Pennell (3:216). False Pimpernel seems to be somewhat more common than the preceding although they are sometimes found together. The pedicels are shorter than the subtending leaves. SP. CIT.: three miles southeast of Elkville, July 16, 1940, *Mc 289*.

Gratiola L.

Gratiola neglecta Torr. (3:212). Common Hedge-hyssop is one of the characteristic species of stream banks, lake shores, and moist ditches. It flowers during May. Plants are somewhat glandular-puberulent. SP. CIT.: Giant City State Park, May 24, 1953, *M 192;* Saltpeter Cave, May 13, 1954, *M 2921*.

Gratiola virginiana L. (3:212). This species is much rarer than the preceding. It is usually found in moist roadside ditches. Along Illinois highway 3 near Gorham, it occurs with *Isoetes melanopoda*. This *Gratiola* is glabrous. SP. CIT.: along Illinois highway 3, May 31, 1948, *BS 410*.

Pedicularis L.

Pedicularis canadensis L. (3:251). Common Lousewort is a most attractive flowering plant. The spike of bright yellow, two-lipped flowers overtopping the basal fern-like leaves is a brilliant sight at the end of April and first part of May. The name Lousewort, commonly applied to any species of *Pedicularis*, actually belongs to *Pedicularis palustris*. Early Europeans thought that this latter species, if eaten by cattle, would cause the cattle to become covered with lice. *Pedicularis canadensis* is found in rather dry open woods. It is rare in this area. SP. CIT.: Giant City State Park, April 25, 1954, *M 2565*.

Leucospora NUTT.

Leucospora multifida (Michx.) Nutt. (3:214) [*Conobea multifida* (Michx.) Benth.]. This is another of the small annual or biennial *Scrophulariaceae* which are found along banks or shores. This group includes *Lindernia dubia*, *Lindernia anagallidea*, *Bacopa rotundifolia*, *Gratiola neglecta*, *Gratiola virginiana*, and *Leucospora multifida*. SP. CIT.: Giant City State Park, July, 1871, *French*; Lake Murphysboro, August 3, 1954, *M 4728*; near Murphysboro, August 14, 1954, *M 4623*.

Seymeria PURSH

Seymeria macrophylla Nutt. (3:239) [*Dasistoma macrophylla* (Nutt.) Raf.]. Mullein Foxglove is a species of moist woods. The woolly yellow flowers are to be found during July and August. A common associate is *Solidago latifolia*. SP. CIT.: Beaucoup Creek, July 1, 1941, *Mc 892*; Carbon Lake, July 27, 1954, *M 4819*; north of Pine Hills, August 8, 1954, *M 4743*.

Aureolaria RAF.

Aureolaria flava (L.) Farwell (3:241) [*Gerardia flava* L.]. This species, somewhat similar in appearance to *Seymeria macrophylla*, occurs chiefly in dry open woods. SP. CIT.: Giant City State Park, *M 565;* Carbondale, September 7, 1878, *French.*

Bacopa   AUBL.

Bacopa rotundifolia (Michx.) Wettst. (3:209). Water Hyssop occurs infrequently throughout this area. It grows in mud of pond or lake shores. SP. CIT.: Carbondale Reservoir, October, 1953, *Bell;* Turkey Bayou, September 2, 1952, *BS 2878B;* Thompson's Lake, September 16, 1948, *Bailey and Hankla 623.*

Gerardia   L.

1. Calyx conspicuously veiny   2
   2. Stems and leaves harshly scabrous . *Gerardia skinneriana* °
   2. Stems and leaves glabrous or nearly so . *Gerardia gattingeri*
1. Calyx with few or no conspicuous veins   3
   3. Stems and leaves harshly scabrous .   .   .*Gerardia aspera*
   3. Stems and leaves glabrous or nearly so   4
      4. Pedicels 10 to 25 mm. long; corolla mostly 10 to 15 mm. long   .   .   .   . *Gerardia tenuifolia*
      4. Pedicels 2 to 5 mm. long; corolla mostly 15 to 35 mm. long   5
         5. Flowers 15 to 20 mm. long . *Gerardia paupercula*
         5. Flowers 20 to 35 mm. long .   . *Gerardia purpurea*

Gerardia skinneriana Wood (3:245). A single collection from Saline County represents the only record of this species from Southern Illinois.

Gerardia gattingeri Small (3:245). This plant occurs on wooded slopes and ridges in a few places in Southern Illinois. We have recorded it from Jackson (see Jones *et al.,* 1955), Williamson, Johnson, and Pope counties.

Gerardia aspera Dougl. (3:245). This harshly scabrous plant of gravelly or sandy soil is rare in Southern Illinois where it is known only from Jackson (see Jones *et al.,* 1955) and Pope counties.

Gerardia tenuifolia Vahl (3:245). Two variants of this species occur in our area. The abundant var. *parviflora* Nutt. with small, somewhat scabrous leaves occurs in low moist areas throughout Illinois. The larger and smooth-leaved var. *macrophylla* Benth. has been collected along a bluff overlooking Cedar Creek in southwestern Jackson County. SP. CIT.: var. *parviflora:* Grand Tower, August 20,

1954, *M 4632;* one mile north of Pomona, September 28, 1951, *V 1008;* Campbell's Lake, September 12, 1941, *Welch and McCree 1182;* Giant City State Park, September 6, 1953, *M 527;* Midland Hills, October 15, 1947, *BS 237.* Var. *macrophylla:* Cedar Creek Bluff, August 6, 1954, *M 4722.*

Gerardia purpurea L. (3:243). This rare plant of moist soil is known in our area only from Jackson (see Jones *et al.,* 1955) and Pulaski counties. It flowers from August to October.

Gerardia paupercula (Gray) Britt. (3:243). This is our rarest *Gerardia.* It is known in Illinois from Lake, McHenry, Cook, Lee, Lawrence, Richland, and Jackson counties. It grows in moist areas. SP. CIT.: Thompson's Woods, September 21, 1948, *Bailey and Hankla 625.*

Penstemon MITCH.

1. Leaves and stems glabrous (stems often puberulent in *P. arkansanus*) 2
   2. Flowers usually 2 to 3 cm. long; sepals 5 to 10 mm. long at flowering 3
      3. Capsule 9 to 12 mm. long; anthers hairy . . . .
         . . . . *Penstemon digitalis*
      3. Capsule 5 to 8 mm. long; anthers glabrous; flowers purplish . . . . *Penstemon calycosus*
   2. Flowers up to 2 cm. long; sepals 2 to 5 mm. long at flowering 4
      4. Corolla glabrous within . . *Penstemon arkansanus*
      4. Corolla minutely glandular within . . . . .
         . . . . *Penstemon tubaeflorus*
1. Leaves and stems pubescent 5
   5. Leaves entire or with sparse low serrations 6
      6. Corolla mostly less than 2 cm. long; plants minutely pubescent . . . . . *Penstemon deamii* °
      6. Corolla usually 2 to 2.5 cm. long, white, purple-lined inside; plants usually densely soft pilose . . . .
         . . . . *Penstemon pallidus*
   5. Leaves usually with sharp serrations 7
      7. Sterile filament densely hairy; corolla with a purplish throat and white lobes . . . *Penstemon hirsutus*
      7. Sterile filament only slightly hairy; corolla whitish or violet tinged 8
         8. Anthers glabrous; capsule 7 to 8 mm. long; plants generally of dry habitats . . *Penstemon calycosus*
         8. Anthers hairy; capsule 8 to 9 mm. long; plants of low swampy woods . . . *Penstemon alluviorum* °

Penstemon digitalis Nutt. (3:223). Foxglove Penstemon, a handsome large-flowered species of open woodlands, is common in almost all sections of Illinois. It flowers during May and June. SP. CIT.: two miles north of Carbondale, May 25, 1951, *V 864;* Giant City State Park, June 4, 1953, *M 189.*

Penstemon calycosus Small (3:224). This species is variable in degree of pubescence. Most specimens are puberulent, although a few are glabrous. It is one of our commonest Foxgloves. SP. CIT.: Giant City State Park, June 9, 1953, *M 304;* Etherton, May 23, 1954, *M 2507.*

Penstemon arkansanus Pennell. This Ozarkian species, recently discovered in Southern Illinois, occurs in rocky woods in Jackson and Randolph counties. It is similar to *Penstemon pallidus* except that it is nearly glabrous throughout. Pale flowers are borne in May and June. SP. CIT.: north of Murphysboro, May 15, 1956, *M 5354.*

Penstemon tubaeflorus Nutt. (3:222). This native midwestern species is rare in Illinois. In our area it is recorded from Jackson and Union counties where it grows along railroads. SP. CIT.: south of Elkville, June 12, 1951, *BS 1400.*

Penstemon deamii Pennell (3:224). Named for the famous Indiana field botanist, this species is native only to southern Indiana and Southern Illinois. It grows in dry, often sandy, woodland. We have it only from Union, Pope, and Wabash counties.

Penstemon pallidus Small (3:224). This species is known from all parts of Illinois. It is a conspicuous member of the Shawneetown Ridge bluff-tops flora where it has been observed to thrive under dry conditions. SP. CIT.: Saltpeter Cave, May 13, 1954, *M 2424;* Giant City State Park, April 30, 1954, *M 2569;* Midland Hills, April 28, 1948, *BS 377;* Etherton, April 23, 1954, *M 2147;* seven miles north of Murphysboro, June 4, 1954, *M 1585.*

Penstemon hirsutus (L.) Willd. (3:223). Hirsute Foxglove is known from Cook and Will counties in the northeast and Jackson, Union, and Hardin counties in Southern Illinois. It occurs mostly in limestone areas. The white corolla lobes contrast with the lavender corolla tube. SP. CIT.: two miles south of Carbondale, May 17, 1940, *Mc 3524.*

Penstemon alluviorum Pennell (3:223). A collection of this species from a swampy woods in the LaRue Swamp of Union County marks the only record for it in our area. It is considered by some authors as insufficiently distinct from *Penstemon digitalis.*

Chelone L.

Chelone glabra L. (3:220). The beautiful White Turtlehead is a species of swampy areas. Numerous variations have been named, but our specimens all seem to belong to var. *linifolia* Coleman. At the "marsh" in Murphysboro, it is found with *Dryopteris thelypteris*, *Viburnum recognitum*, *Rosa palustris*, and other swamp species. SP. CIT.: Murphysboro "marsh," August 14, 1954, *M 4815*.

Chelone obliqua L. var. *speciosa* Pennell & Wherry (3:220). The attractive Purple or Showy Turtlehead, a plant of low woodlands, is known in our area from Jackson, Gallatin, Johnson, Alexander, and Pulaski counties. It has not been found in Jackson County since 1878. SP. CIT.: Carbondale, September 24, 1878, *French 1855*.

Mimulus L.

Mimulus alatus Ait. (3:214). Common Monkey-flower grows in wet areas in all southern counties. The leaves are petiolate. SP. CIT.: Thompson's Lake, July 16, 1941, *Mc 923*; Giant City State Park, July 28, 1953, *M 447*.

Mimulus ringens L. (3:214). Sessile-leaved Monkey-flower, rarer than the preceding, grows along streams in sandy soil. We have it from Jackson and Union counties. SP. CIT.: Grand Tower, *Gleason*.

Collinsia NUTT.

Collinsia verna Nutt. (3:227). Blue-eyed Mary is one of our most attractive spring wild flowers. In low woods at Giant City State Park, hundreds of plants cover a small area, giving it a bluish tone during April and May. Flowers have some blue petals and some white. It is recorded from Jackson, Union, and Johnson counties. SP. CIT.: Giant City State Park, April 25, 1950, *M 2582*.

Scrophularia L.

Scrophularia marilandica L. (3:226). This species sometimes attains a height of five to six feet. The unequally-lobed double-lipped flowers which are produced during summer are inconspicuous. SP. CIT.: Giant City State Park, August 26, 1953, *M 1321*; Clear Springs, September 11, 1952, *BS 2976*.

131. *Acanthaceae* – Acanthus Family

1. Flowers mostly solitary; stamens 4 . . . . . . *Ruellia*
1. Flowers in spikes or heads or borne in paniculate clusters; stamens 2 2
    2. Upper lip of corolla notched; flowers without conspicuous bracts . . . . *Justicia*
    2. Upper lip of corolla entire or at most emarginate; flowers with conspicuous bracts . .. . . . *Dicliptera* °

Ruellia L.

1. Calyx lobes 2 to 4 mm. broad . . . . *Ruellia strepens*
1. Calyx lobes 0.5 to 1.5 mm. broad 2
    2. Leaves sessile or nearly so . . . . *Ruellia humilis*
    2. Leaves petiolate 3
        3. Plants generally villous or hirsute; flowers in glomerules on short peduncles . . . . *Ruellia caroliniensis*
        3. Plants merely puberulent; flowers solitary on long peduncles . . . . *Ruellia pedunculata*

Ruellia strepens L. (3:264). Smooth Ruellia is a rather common species in Southern Illinois where it is found in rich moist soil. It flowers during June and early July. SP. CIT.: four miles north of Carbondale, *Mc 132*.

Ruellia humilis Nutt. (3:264). Hairy Ruellia is abundant along roads and in dry oak-hickory woods throughout our area. It is our most common species of Wild Petunia. It flowers from June to mid-August. SP. CIT.: Giant City State Park, *M 365;* Cedar Creek bluff, *M 4723;* seven miles north of Murphysboro (along railroad), *VM 1582.*

Ruellia pedunculata Torr. (3:264). This is the Stalked Ruellia, so-called because of long peduncles on which solitary flowers are borne. It is known in Illinois only from Randolph County southward. It begins to flower the last week in May and continues until the end of July. SP. CIT.: Pomona, *V 158;* Giant City State Park, *M 215;* ten miles northwest of Murphysboro, *BS 1508;* Etherton, *M 2507.*

The most recent species of *Ruellia* to be added to the Illinois flora, *R. caroliniensis* (Walt.) Steud. (3:264), was collected in 1955 (Mohlenbrock, 1956) in Hardin County where it is fairly common at the edge of a dry oak-hickory woods. It may be found in most of our counties since it is not uncommon in either Indiana or Missouri.

Justicia L.

Justicia americana (L.) Vahl [*Dianthera americana* L.] (3:267). Water-willow is a handsome species which grows in our shallow streams. It is particularly abundant in Pope County along Lusk Creek below Indian Kitchen. The lanceolate to linear leaves are sessile or nearly so. This species usually occurs with several amphibious *Scrophulariaceae*. SP. CIT.: Midland Hills, September 23, 1953, *Bell*.

The rarer Swamp Water-willow, *Justicia ovata* (Walt.) Lindau (3:267) [*Dianthera lanceolata* (Chapm.) Small], is restricted to Alexander and Pulaski counties where it has been found in swampy river bottoms. It flowers from mid-July to mid-August. The elliptic-ovate leaves are short-petioled.

Dicliptera JUSS.

One of the rarer species of the Middle West, *Dicliptera brachiata* (Pursh) Spreng. (3:267) has been found in Southern Illinois in Massac County (Evers, 1956). In addition to this station, it is known from a few stations in adjacent Indiana and Missouri.

132. *Lentibulariaceae* – Bladderwort Family

Utricularia L.

Utricularia gibba L. (3:263). This Bladderwort is known from Cook, Lake, and McHenry counties in northeastern Illinois, and from five counties along the western border of Southern Illinois. Immersed, finely-dissected leaves bear scattered traps or "bladders" which enable the plant to float at flowering time. The slender flowering scape bears one to six flowers that are about one centimeter long. SP. CIT.: Campbell Lake, October 2, 1952, *BS 3107*.

Common Bladderwort, *Utricularia vulgaris* L. (3:263), found scattered throughout Illinois, is recorded in our area only from LaRue Swamp in Union County. The stout scape bears six to twenty flowers that are about one and one-half centimeters long.

133. *Orobanchaceae* – Broom-rape Family

1. Lower flowers cleistogamous and fertile, the upper flowers complete but usually sterile . . . . *Epifagus*

1. All flowers complete and perfect, flowers solitary, creamy white to
    lilac   .   .   .   . *Orobanche*

Epifagus NUTT.

Epifagus virginiana (L.) Barton (3:258). Beech-drops, a parasitic
species on roots of beech trees, is uncommon in Illinois where it is
known from only eight southern and southeastern counties. It is rare
in adjacent Missouri. The slender, brown, much branched stems bear
numerous small scattered scales. Racemes of dull white and brown
flowers appear from August to October. SP. CIT.: Fountain Bluff,
October 2, 1948, *Evers 15417;* Pomona Natural Bridge, October 2,
1947, *BS 201;* Little Grand Canyon, August, 1954, *M 4033.*

Orobanche L.

Orobanche uniflora L. (3:257). Broom-rape, a parasite on the roots
of various herbs, is not common in most of Illinois. In Southern Illinois, it has been collected in Jackson, Union, and Alexander counties,
and is nowhere abundant. Delicate cream or lilac flowers are borne
singly on a long naked stalk during May. SP. CIT.: Little Grand
Canyon, May 2, 1950, *BS 977;* Fountain Bluff, May 16, 1947, *BS 43.*

134. *Bignoniaceae* – Trumpet Creeper Family

1. Climbing or trailing shrubs with compound leaves   .   .   .   2
    2. Leaflets 7 to 11, the margins serrate .   .   .   . *Campsis*
    2. Leaflets 2, the margins entire   .   .   .   . *Bignonia* °
1. Trees having simple ovate leaves, cordate at base   3
    3. Flowers white; pods elongate; leaves whorled .   . *Catalpa*
    3. Flowers purple; pods ovoid; leaves opposite .   . *Paulownia*

Campsis LOUR.

Campsis radicans (L.) Seem. (3:257). Trumpet Creeper, a colorful red and yellow flowered vine, is common in this area. A species
of southeastern and east-central United States, this woody perennial
is known only from the southern two-thirds of Illinois. SP. CIT.:
Giant City State Park, May 29, 1953, *M 248;* Carbondale, June 9,
1940, *Mc 282.*

Bignonia L.

Cross-vine, *Bignonia capreolata* L. (3:254), so named because

FIG. 63. *Cross-vine* (Bignonia capreolata) *is a woody climber in some upland rocky woods.*

transverse sections of the woody stem show a cross, usually grows in swampy situations in the extreme southern counties of Illinois. Paradoxically, it has been found growing abundantly on very dry, exposed bluffs along Lusk Creek in Pope County. It is not common (Fig. 63).

Catalpa SCOP.

Catalpa bignonioides Walt. (3:254). Southern Catalpa, a native tree of southeastern United States, is commonly planted in our area. The long slender pods are often called "Lady Cigars." The beautiful, densely purple-spotted flowers make this one of our most attractive ornamental trees. SP. CIT.: near Carbondale, June 14, 1951, *BS 1421.*

Catalpa speciosa Warder (3:254). Northern Catalpa is found in low, rich woods of Southern Illinois, southern Indiana, and adjacent sections of Missouri, Arkansas, and Tennessee. It is often cultivated elsewhere in Illinois. We have recorded it as a native species from the seven southernmost counties. The large flowers are sparsely spotted with purple.

Paulownia SIEB. AND ZUCC.

Paulownia tomentosa (Thunb.) Steud. (3:209). This handsomely flowering tree, a common ornamental in our area, is frequently placed among the *Scrophulariaceae* because of capsular structure and presence of endosperm. It is native of China and is called the Empress

Tree, being named for Anna Paulowna, Queen of Netherlands during the first part of the nineteenth century. SP. CIT.: roadside, Grand Tower, *Winterringer 1814*.

135. *Martyniaceae* – Martynia Family

Proboscidea   KELLER

Proboscidea louisianica (Mill.) Thell. (3:263) [*Martynia louisianica* Mill.; *Proboscidea jussieui* Van Eseltine]. The most interesting feature of Unicorn-plant is its fruit. The five- or six-inch long capsule attenuates at one end into an upwardly curved horn. The entire capsule splits into two valves at maturity. Flowers of this plant are whitish and spotted with purple. Although reported as native in Illinois, we have recorded it only as an escape from cultivation. SP. CIT.: Grand Tower, July 11, 1871, *French*.

136. *Plantaginaceae* – Plantain Family

Plantago   L.

1. Leaves linear; stamens 2 or 4   2
    2. Bracts much longer than the flowers, with an awn 1 to 5 cm. long; stamens 4   .   .   .   .   .   *Plantago aristata*
    2. Bracts at most twice as long as the flowers, awnless; stamens 2   3
        3. Capsule 4-seeded   .   .   .   .   *Plantago pusilla*
        3. Capsule 7- to 30-seeded   .   *Plantago heterophylla* °
1. Leaves ovate, oval, or lanceolate; stamens 4   4
    4. Veins of leaves branching from the mid-vein .   .   .   .   *Plantago cordata*
    4. Veins of leaves prominent and parallel, not branching conspicuously from the mid-vein   5
        5. Spikes dense, ovoid   .   .   .   *Plantago lanceolata*
        5. Spikes slenderly cylindric   6
           6. Leaves thick, rough on both surfaces .   .   .   .   *Plantago major*
           6. Leaves thin, mostly smooth .   .   *Plantago rugelii*

Plantago aristata Michx. (3:269). Bracted Plantain is abundant in fields and open woods. Large scarious bracts give this plant a silvery appearance. SP. CIT.: Giant City State Park, June 11, 1953, *M 298*.

Plantago pusilla Nutt. (3:271). This small plant occurs in two distinct habitats in our area. It is common in waste ground, but also grows in moist depressions of sandstone bluffs. At these latter stations, it is found with other small, narrow-leaved species such as *Isoetes butleri, Cyperus aristatus,* and *Bulbostylis capillaris.* SP. CIT.: Giant City State Park, May 15, 1953, *M 70.*

A similar small plant, *Plantago heterophylla* Nutt. (3:272), is known in Illinois from a single collection in Union County near Anna in 1923 (*Benke 3959*).

Plantago cordata L. (3:270). Heart-leaved Plaintain leaves sometimes attain a size greater than one foot long. A single station at Lake Murphysboro is known in this area for this plant. At this location, scores of these plants grow in the bed of a narrow stream and are often completely submerged. SP. CIT.: Lake Murphysboro, August 27, 1954, *M 4656.*

The remaining three species of *Plantago* in Southern Illinois are rank weeds. The common *Plantago lanceolata* L. (3:270) and the rarer *Plantago major* L. (3:271) are natives of Europe. The latter is known from Jackson and Union counties. The third of these weed species is the native and very common *Plantago rugelii* Dec. (3:269). SP. CIT.: *P. lanceolata:* Giant City State Park, May 29, 1953, *M 253. P. major:* Giant City State Park, July 7, 1953, *M 354. P. rugelii:* Midland Hills, September 23, 1953, *Bell.*

137. *Rubiaceae* – Madder Family

1. Shrubs of wet habitat; flowers in dense globose heads . . .
   . . . . *Cephalanthus*
1. Herbs; flowers not in dense globose heads 2
   2. Leaves in whorls of 4 to 8; stems square . . . *Galium*
   2. Leaves opposite 3
      3. Flowers sessile or nearly so, axillary 4
         4. Plants pubescent . . . . . . *Diodia*
         4. Plants glabrous . . . . . *Spermacoce*
      3. Flowers pedicellate, solitary or cymose 5
         5. Plants low-growing, creeping; fruit-globose, red, about 1 cm. in diameter . . . . *Mitchella*
         5. Plants erect; flowers in cymes or solitary; fruit a capsule . . . . *Houstonia*

Cephalanthus L.

Cephalanthus occidentalis L. (3:279). Buttonbush is an aquatic shrub which is common throughout Southern Illinois along shores of streams, ponds, lakes, and in swamps. It is often associated with *Fraxinus, Salix, Alisma,* and other amphibious or emergent forms. Specimens with pubescent twigs and leaves are known as var. *pubescens* Raf. SP. CIT.: typical: Giant City State Park, June 5, 1953, *M 254;* Carbondale Reservoir, October, 1953, *Bell.* Var. *pubescens:* Murphysboro "marsh," August 16, 1954, *M 4140.*

Galium L.

1. Leaves four in a whorl 2
   2. Leaves with 3 prominent veins . . . *Galium circaezans*
   2. Leaves with one nerve, or with 3 obscure nerves near the base 3
      3. Flowers purplish . . . . . *Galium pilosum*
      3. Flowers white 4
         4. Flowers sessile; leaves 4 to 10 mm. long . . . . *Galium virgatum* °
         4. Flowers stalked; leaves 15 to 30 mm. long . . . . *Galium obtusum*
1. Leaves (on most of them) 6 to 8 in a whorl 5
   5. Stems with retrorse bristles . . . . *Galium aparine*
   5. Stems smooth or nearly so 6
      6. Leaves oval . . . . . . *Galium triflorum*
      6. Leaves linear 7
         7. Leaves with a cusp; stems erect; plants of dry woodlands . . . . . . *Galium concinnum*
         7. Leaves blunt; stems mostly prostrate and matted; plants of swamps and ponds . . .*Galium tinctorium*

Galium circaezans Michx. (3:283). This is one of the characteristic herbaceous species of dry open woodlands. It is common in all of our counties where it flowers from May to August. SP. CIT.: Giant City State Park, May 27, 1952, *V 1244;* Fountain Bluff, August 5, 1952, *BS 2706.*

Galium pilosum Ait. (3:284). Purple Bedstraw is local over the entire state, although it seems to be most common in the southern counties. It occurs in dry woodlands, and blooms during summer. SP. CIT.: Giant City State Park, June 19, 1953, *M 370.*

Galium obtusum Bigel. (3:288). This blunt-leaved species often becomes matted, choking out weaker vegetation. It flowers in May and June. SP. CIT.: Pomona, June 21, 1951, *V 629;* two miles north of Fountain Bluff, May 21, 1948, *BS 407.*

Galium aparine L. (3:284). Goosegrass is a very familiar plant to those who explore in nature. Almost any field botanist has had to pull from his trousers retrorsely barbed stems of this species. This is the first species of *Galium* to bloom in this area. SP. CIT.: Giant City State Park, April 24, 1953, *M 908;* Lake Murphysboro, May 16, 1951, *M 793;* Etherton, April 19, 1954, *M 2178.*

Galium triflorum Michx. (3:285). Sweet-scented Bedstraw is common in woods throughout Illinois. The flowering period begins in May and extends into September. SP. CIT.: Giant City State Park, May 27, 1952, *V 1242;* Fountain Bluff, September 14, 1940, *Mc 452.*

Galium concinnum Torr. & Gray (3:287). This very narrow-leaved, smooth-stemmed species is abundant in dry woodlands, occurring frequently with species of *Blephilia* and *Hypericum.* It flowers during June and early July. SP. CIT.: Giant City State Park, June 5, 1953, *M 213;* Little Grand Canyon, June 21, 1951, *V 876.*

The slender *Galium virgatum* Nutt. (3:283), a plant of Ozark glades, has been collected on a rock ledge near Fults (Monroe County) by R. A. Evers. It should also be looked for on borders of hill prairies in Randolph, Jackson, and Union counties.

Galium tinctorium L. (3:289). Swamp Bedstraw is one of the rarest species in our flora. Although reported previously only from the extreme northeastern counties of Illinois, it has recently been collected in two counties of Southern Illinois. It occurs in water near a spring at the base of a bluff in the LaRue Swamp (Union County) and in a pond north of Makanda (Jackson County). SP. CIT.: north of Makanda, July 8, 1948, *BS 512.*

Diodia L.

Diodia teres Walt. (3:279). Rough Buttonweed is very common in dry or sandy situations throughout our area. Leaves of this species are about five millimeters broad while flowers are about five millimeters long. It flowers throughout the summer. SP. CIT.: Giant City State Park, July 23, 1948, *BS 534;* south of De Soto, June 19, 1940, *Mc 218.*

Diodia virginiana L. (3:279). Virginia Buttonweed is somewhat larger than the preceding species, and less common. Of southern

FIG. 64. *Partridge-berry* (Mitchella repens), *a northern relict, may occasionally be found on moist north-facing slopes. It has pea-sized red berries.*

origin, this plant often occurs with *Spermacoce glabra, Phyllanthus caroliniensis,* and *Ammannia coccinea.* Leaves are five to fifteen millimeters broad while flowers are about ten millimeters long. SP. CIT.: one mile north of Pomona, July 5, 1951, *BS 1517.*

Spermacoce L.

Spermacoce glabra Michx. (3:280). Smooth Buttonweed is characteristic of wet roadside ditches and low woods in the southern half of Illinois. It can be found in almost any moist situation. SP. CIT.: Campbell Lake, June 10, 1941, *Mc 218;* Fountain Bluff, October 18, 1947, *BS 271.*

Mitchella L.

Mitchella repens L. (3:279). Botanists in Illinois are thrilled when they find new stations for Partridge-berry, for not only is it a pretty plant, but it is rare since it is known only from four widely scattered counties in northeastern Illinois and from six southern counties. The brilliant red berries, almost hidden among attractive and persistent dark green foliage, make these plants a welcome sight when most other plants are dormant for the winter. Although seldom seen, the flowers are also attractive. This species is found in deep moist ravines (Fig. 64). SP. CIT.: Midland Hills, April 21, 1941, *Mc 614;* Saltpeter Cave, March 8, 1954, *M 4785.*

Houstonia L.

1. Corolla salverform; stamens included 2
    2. Corolla lobes 3 to 5 mm. broad; plants perennial from rhi-
        zomes . . . .*Houstonia caerulea*
    2. Corolla lobes 1 to 3 mm. broad; annuals 3
        3. Corolla tube about twice as long as the calyx lobes . .
            . . . . *Houstonia pusilla*
        3. Corolla tube and calyx lobes about equal in length . .
            . . . . *Houstonia minima*
1. Corolla funnelform; stamens exserted 4
    4. Upper stipules bristle-tipped . . . *Houstonia nigricans*
    4. Upper stipules rounded or triangular, not bristle-tipped . 5
        5. Leaves tapering to base 6
            6. Flowers on pedicels 5 to 15 mm. long; capsules
                broader than high . . . *Houstonia tenuifolia*
            6. Flowers on pedicels 1 to 4 mm. long; capsules higher
                than broad . . . . .*Houstonia longifolia*
        5. Leaves rounded at base 7
            7. Leaves 10 to 30 mm. broad; sepals at flowering time
                1.5 to 4.0 mm. long . . . *Houstonia purpurea*
            7. Leaves 5 to 10 mm. broad; sepals at flowering time 3
                to 7 mm. long . . . . *Houstonia lanceolata*

Houstonia caerulea L. (3:277). Yellow-centered Bluet is local in its distribution in Illinois. We have it only from Jackson, Johnson, and Massac counties in the southern part of the state. It occurs principally in fields. SP. CIT.: Giant City State Park, March 22, 1953, *M 740.*

Houstonia pusilla Schoepf (3:277) [*Houstonia patens* Ell.]. This species and the following are both small southern annuals which bloom early in the spring. SP. CIT.: Giant City State Park, April 25, 1950, *M 2566;* Midland Hills, May 9, 1951, *V 653.*

Houstonia minima Beck (3:277). This small inhabitant of dry situations is local in Illinois. It is recorded in Southern Illinois from Jackson (see Jones *et al.,* 1955), Saline, and Johnson counties.

Houstonia nigricans (Lam.) Fern. (3:277). Although seldom attaining a height of more than ten inches, this hill prairie species becomes less herbaceous near the base. It is known in Illinois from seven counties bordering the Mississippi River where it occurs in hill prairies or on limestone ledges that border these prairies. SP. CIT.: north of the Pine Hills, September 9, 1954, *M 4753.*

Houstonia tenuifolia Nutt. (3:277). Limited in Illinois to rocky woods in the southern counties, this relatively uncommon species is one of the first perennial Bluets to flower. We have collected it in flower as early as May 7. SP. CIT.: Giant City State Park, May 8, 1953, *M 88;* Midland Hills, May 7, 1952, *BS 2233.*

Houstonia longifolia Gaertn. (3:277). While this species is not listed from Southern Illinois, recent collections show it to occur in Randolph, Jackson, and Johnson counties. SP. CIT.: Fountain Bluff, *Welch and Fuller 267.*

Houstonia purpurea L. (3:277). This species is the largest of Illinois Bluets. It grows in dry habitats and occurs occasionally in prairie soil along railroads. SP. CIT.: five miles north of Carbondale, June 10, 1940, *Mc 3610.* The pubescent forma *pubescens* (Britt.) Fern. has been collected in Williamson County.

Houstonia lanceolata (Poir.) Britt. (3:277) [*Houstonia purpurea* var. *calycosa* Gray]. Characteristics which are used by most authors to distinguish this species from the preceding overlap to some extent, although most specimens from Southern Illinois are easily referred to one species or the other. SP. CIT.: Giant City State Park, May 19, 1953, *M 92;* one mile south of Elkville, May 19, 1948, *BS 401.*

138. *Caprifoliaceae* – Honeysuckle Family

1. Shrubs or small trees 2
    2. Weak shrubs with pinnately compound leaves . *Sambucus*
    2. Shrubs or small trees with simple leaves 3
        3. Leaves entire, or with undulating margins . . . 4
            4. Leaves glabrous . . . . *Symphoricarpos*
            4. Leaves pubescent, at least below . . . *Lonicera*
        3. Leaves serrate, dentate, or shallowly lobed . *Viburnum*
1. Erect herbs or climbing or trailing vines 5
    5. Erect herbs . . . . *Triosteum*
    5. Climbing or trailing vines . . . . . . *Lonicera*

Sambucus L.

Sambucus canadensis L. (3:296). Elderberry is a common, weak shrub which sometimes attains a height of ten feet. According to Tehon (1951), "the bark is used as a cathartic, the berries are used in cooling drinks, the flowers as a mild astringent in eye lotion." SP. CIT.: Giant City State Park, May 29, 1953, *M 262;* Little Grand Canyon, August, 1954, *M 4040.*

Symphoricarpos DUHAM.

Symphoricarpos orbiculatus Moench (3:302). Coralberry, or Buck-brush, is a common species in dry thickets. It is unknown from the northern one-third of Illinois. Flowers appear during the end of June and into July. SP. CIT.: Giant City State Park, *M 1096*.

Lonicera L.

Lonicera japonica Thunb. (3:298). The cream-flowered, sweet-scented Japanese Honeysuckle is becoming an obnoxious weed in some parts of our area. Margins of some young leaves occasionally are lobed. SP. CIT.: Midland Hills, May 24, 1952, *Voigt;* Giant City State Park, May 24, 1953, *M 178*.

Lonicera sempervirens L. (3:300). Trumpet Honeysuckle or Fire-cracker Vine produces bright red and yellow flowers during summer months. SP. CIT.: Scenic View, July 27, 1954, *M 4340*.

The shrubby *Lonicera tatarica* L. (3:298), a native of Asia and frequently planted in our area, sometimes is an escape in Illinois.

Viburnum L.

1. Lateral veins of leaves extending to marginal teeth . . .
. . . . *Viburnum recognitum*
1. Lateral veins uniting before they reach marginal teeth . . . 2
  2. Petioles with reddish tomentum . . *Viburnum rufidulum*
  2. Petioles glabrous or nearly so 3
    3. Leaves acuminate at apex . . . *Viburnum lentago*
    3. Leaves at most acute at apex . *Viburnum prunifolium*

Viburnum recognitum Fern. (3:295) [*Viburnum dentatum* var. *lucidum* Ait.; *V. dentatum* var. *deamii* (Rehder) Fern.; *V. dentatum* var. *indianense* (Rehder) Gleason]. This Arrow-wood usually occurs in moist situations. It is frequently found growing with *Ilex decidua*. SP. CIT.: Murphysboro "marsh," June 5, 1954, *M 4753;* Giant City State Park, August 8, 1953, *M 520*.

Viburnum rufidulum Raf. (3:293). Southern Black Haw is com-mon in dry rocky woods throughout the southern counties. Bark of this and the following two species contains chemicals which have been used as astringents, tonics, and nerve medicines. SP. CIT.: Lake Murphysboro, April 12, 1954, *M 1915;* Little Grand Canyon, May 9, 1950, *BS 1005*.

Viburnum lentago L. (3:293). Nannyberry is rare in Southern Illinois. It occurs in a rich mesic woods at Giant City State Park. Grow-ing nearby is a small colony of Staghorn Sumac, the only station for this species in Southern Illinois. SP. CIT.: Giant City State Park, April 30, 1953, *M 641*.

Viburnum prunifolium L. (3:294). Black Haw is apparently much less common in Southern Illinois than Southern Black Haw. It grows in moist woods. SP. CIT.: Giant City State Park, June 16, 1953, *M 501*.

Triosteum L.

1. Leaves perfoliate   .   .   .   .   . *Triosteum perfoliatum*
1. Leaves tapering to base 2
  2. Flowers reddish   .   .   .   .   . *Triosteum illinoense* °
  2. Flowers yellowish .   .   .   . *Triosteum angustifolium*

Triosteum perfoliatum L. (3:303). The perfoliate leaves of this Horse Gentian make it easily identified. It is known in our area in woods from Jackson and Massac counties. SP. CIT.: Giant City State Park, April 25, 1953, *M 110*.

Triosteum illinoense (Wieg.) Rydb. [*Triosteum aurantiacum* var. *illinoense* (Wieg.) Palmer and Steyerm.]. A station in Union County is the only one known for this species in our area although several collections have been made of it in Monroe County.

Triosteum angustifolium L. (3:303). Our commonest species of Horse Gentian is this yellow-flowered perennial. It occurs in moist thickets. SP. CIT.: Midland Hills, May 19, 1953, *V 1388*; Giant City State Park, May 1953, *M 138*.

139. *Campanulaceae* – Bell-flower Family

1. Leaves sessile and mostly clasping, cordate at base; flowers ses-sile, solitary in the axils of the leaves .   .   .   . *Triodanis*
1. Leaves petiolate or subsessile, not cordate at base; flowers in a terminal inflorescence .   .   .   . *Campanula*

Triodanis RAF.

Triodanis biflora (Ruiz & Pavon) Greene (3:317) [*Specularia bi-flora* (Ruiz & Pavon) Fisch. and Mey.]. Attractive Venus' Looking-glass is an occasional inhabitant of dry open places. None of the leaves clasp the stem. SP. CIT.: Midland Hills, May 24, 1952, *Voigt;*

Giant City State Park, June 5, 1953, *M 206;* Etherton, May 11, 1954, *M 2410.*

Triodanis perfoliata (L.) Nieuwl. (3:317) [*Specularia perfoliata* (L.) A.DC.]. This Venus' Looking-glass is common in dry sandy soil throughout Southern Illinois. It often becomes weedlike. At least some of the leaves clasp the stem. SP. CIT.: Giant City State Park, May 24, 1953, *M 79;* Little Grand Canyon, August, 1954, *M 4063.* *Triodanus perfoliata* (L.) Nieuwl., has bluish-purple flowers. A white color form has been observed. This white-flowered form was observed July 5, 1952, along a roadside between Simpson and Jackson Hollow in Pope County. For future reference this form may be known as *Triodanus perfoliata* f. *alba* Voigt, f. *nov., a typo differt floribus albis.* SP. CIT.: *Voigt, J. W., No. 1105,* type in herbarium of Illinois State Museum, Springfield, Illinois.

Campanula L.

Campanula americana L. (3:315). Bellflower is one of our most attractive summer wild flowers. The spike-like racemes of blue rotate flowers make moist woodlands colorful. While the ovate cauline leaves of most specimens taper to the base, those of others are abruptly contracted to the petiole. These latter may be segregated as var. *illinoensis* (Fresn.) Farw. SP. CIT.: typical: Giant City State Park, July 15, 1953, *M 407;* Fountain Bluff, July 24, 1954, *M 4350.* Var. *illinoensis:* Brownsville, August 9, 1954, *M 4721.*

Campanula intercedens Witasek (3:315) [*Campanula rotundifolia* L. var. *intercedens* (Witasek) Farw.]. This rare species has a disjunct range in Illinois. It grows in rich sandstone ravines in a few of our northern counties and also occurs in similar habitats at Fountain Bluff in Jackson County. The specific epithet *rotundifolia* pertains to early deciduous basal leaves. Cauline leaves are linear to linear-lanceolate. SP. CIT.: Fountain Bluff, June 8, 1940, *Welch and Fuller 148.*

140. *Lobeliaceae* – Lobelia Family

Lobelia L.

1. Flowers 3 cm. or more long; plants of wet places; flowers red (white forms occur) . . . . *Lobelia cardinalis*
1. Flowers less than 3 cm. long; plants of wet or dry places; flowers blue or bluish white 2

2. Flowers 1 to 2.5 cm. long   3
   3. Calyx with large auricles; frequent plants of wet ground
    . . . . *Lobelia siphilitica*
   3. Calyx with small auricles; rare plants of woods . .
    . . . . *Lobelia puberula*
2. Flowers up to 1 cm. long   4
   4. Flowers in spike-like racemes; stems simple . . .   5
     5. Calyx lobes not appendaged . . *Lobelia spicata*
     5. Calyx lobes each with 2 reflexed slender appendages
      equaling the calyx tube in length . . . .
      . . . . *Lobelia spicata* var. *leptostachys*
   4. Flowers loosely racemose; stems paniculate-branched;
   pods inflated . . . . . . *Lobelia inflata*

Lobelia cardinalis L. (3:319). The brilliant scarlet-flowered Cardinal-flower is a species of low wet ground. It is most frequent in marshy areas of Southern Illinois. SP. CIT.: Cave Valley, September 28, 1951, *V 997;* Thompson's Lake, October 10, 1940, *Mc 556.*

Lobelia siphilitica L. (3:319). Great Blue Cardinal-flower, more common than the preceding species, likewise occurs in wet situations and it probably grows along almost every stream in Southern Illinois. Occasional specimens may be found in flower in November. SP. CIT.: Lake Murphysboro, September 4, 1954, *M 4696.*

Lobelia puberula Michx. (3:320). Downy Lobelia, a species of southern United States, is rare and local in counties of Southern Illinois. Our specimens belong to var. *simulans* Fern. SP. CIT.: Giant City State Park, September 13, 1949, *BS 851;* Midland Hills, September 23, 1953, *Bell.*

Lobelia spicata Lam. (3:322). Spiked Lobelia is an inhabitant of dry sandy soil. Apparently more common in our area is var. *leptostachys* (A.DC.) Mack. [*Lobelia leptostachys* A.DC.]. SP. CIT.: var. *leptostcahys:* Giant City State Park, July 19, 1949, *BS 765.*

Lobelia inflata L. (3:322). "Indian Tobacco contains several alkaloids which are used as antispasmodics in laryngitis and . . . [as] expectorant[s] and emetic[s]," remarks Tehon (1951). SP. CIT.: Fountain Bluff, September 14, 1940, *Mc 445;* Giant City State Park, July 23, 1953, *M 444.*

141. *Valerianaceae* – Valerian Family

1. Plants annual; leaves entire or dentate   .   .   . *Valerianella*
1. Plants perennial; leaves pinnately divided .   .   . *Valeriana*

Valerianella MILL.

Valerianella radiata (L.) Dufr. (3:307). Lamb's Lettuce is a frequent species in moist waste areas. It apparently is becoming more common. SP. CIT.: Midland Hills, *Voigt;* two miles north of Carbondale, *V 873;* Giant City State Park, *M 2568.*

Valeriana L.

Valeriana pauciflora Michx. (3:305). The beautiful delicately pink-flowered Valerian is an inhabitant of rich woods in only a few of our counties. At Giant City State Park it occurs with *Collinsia verna, Polemonium reptans,* and *Aplectrum hyemale.* SP. CIT.: Giant City State Park, *M 75.*

142. *Dipsacaceae* – Teasel Family

Dipsacus L.

Teasel, *Dipsacus sylvestris* Huds. (3:308), an adventive from Europe, has been collected along a road in Jackson County for its only Southern Illinois station. SP. CIT.: near Elkville, *Joe Garrison.*

143. *Compositae* – Composite Family

Southern Illinois is rich in number of composites. Fifty-five genera and one hundred and eighty-four species are recorded from this area while Jackson County alone has fifty genera and one hundred fifty-five species. The greater number of species are late summer or fall flowering plants, although some species (*Antennaria* spp., *Chrysanthemum leucanthemum, Ghaphalium purpureum,* etc.) flower as early as April and May.

Some Illinois genera of *Compositae* have numerous species in our area. There are nineteen species of Goldenrod (*Solidago*) in Southern Illinois (seventeen in Jackson County), seventeen species of Aster (*Aster*) (sixteen in Jackson County), thirteen species of Sunflower (*Helianthus*) (twelve in Jackson County), ten species of Beggar-ticks (*Bidens*) (nine in Jackson County), and nine species of Snakeroot or Joe Pye-weed (*Eupatorium*) (six in Jackson County). Composites are predominantly yellow or blue-flowered.

1. Heads of flowers with ligulate corollas; sap usually milky; leaves alternate or basal . . . . *Liguliflorae*
1. Heads of both ray and disk flowers or of disk flowers only; sap not milky . . . . *Tubuliflorae*

SUBFAMILY LIGULIFLORAE

1. Flowers yellow or orange (sometimes blue in *Lactuca*) . . 2
  2. Plants without leafy stems or stems with 1 leaf . . . 3
    3. Scapes bearing 1 yellow head . . . . *Taraxacum*
    3. Scapes bearing 1 to several yellow-orange heads; leaves glaucous . . . . *Krigia*
  2. Plants with leafy stems 4
    4. Short plants, usually less than 2 dm. high . . *Krigia*
    4. Plants over 2 dm. high 5
      5. Stem and leaves copiously hairy . . *Hieracium*
      5. Stem and leaves not copiously hairy 6
        6. Heads less than 2.5 cm. broad 7
          7. Heads fewer than 50-flowered . *Lactuca*
          7. Heads more than 50-flowered; leaves auriculate, prickly . . . . . *Sonchus*
        6. Heads over 2.5 cm. broad 8
          8. Leaves entire, grass-like, clasping; involucre simple . . . . . *Tragopogon* °
          8. Leaves oblong-lanceolate, entire, cut or pinnatifid; stem leaves partly clasping; involucre double . . . . *Pyrrhopappus*
1. Flowers usually blue, sometimes yellow or white 9
  9. Plants of roadsides and waste places; flowers usually blue or yellow 10
    10. Flowers 2.5 to 4 cm. in diameter (blue, purple, or pink) . . . . *Cichorium*
    10. Flowers less than 2 cm. in diameter (blue or yellow) . . . . *Lactuca*
  9. Plants of rich woods; leaves broad, triangular, sinuate, 3- to 5-cleft; flowers white, cream, or pink . . . *Prenanthes*

SUBFAMILY TUBULIFLORAE

1. Flowers purple, pink, or blue (cream in *Kuhnia*) . . . 2
  2. Bracts in 1 or 2 series 3

    3. Heads in glomerules . . . . . *Elephantopus*

    3. Heads borne singly or in cymose-corymbose or paniculate inflorescences 4

        4. Ray flowers pinkish or often white; inflorescence paniculate . . . . *Erigeron*

        4. Flowers lavender, inflorescence cymose-corymbose . . . . *Eupatorium*

2. Bracts imbricated in more than 2 series 5

    5. Involucral bracts hooked at the tip or spiny (toothed at the tips in *Centaurea*) 6

        6. Involucral bracts hooked; leaves not spiny . . 7

            7. Corolla and pappus present . . . *Arctium*

            7. Corolla and pappus absent; monoecious annuals . . . . *Xanthium*

        6. Involucral bracts spiny or notched or toothed at the apex, but not hooked 8

        8. Leaves very spiny 9

            9. Leaves sparingly tomentose above . . . . *Carduus*

            9. Leaves not tomentose above . . *Cirsium*

        8. Leaves not spiny; involucral bracts toothed at the tips . . . . . *Centaurea*

    5. Involucral bracts not hooked or spiny 10

        10. Leaves entire 11

            11. Inflorescence spicate or racemose; flowers usually purple . . . . . . *Liatris*

            11. Inflorescence paniculate or corymbose; flowers mostly blue . . . . . . *Aster*

        10. Leaves dentate, dissected, or pinnatifid . . . 12

            12. Leaves sessile 13

                13. Leaves not punctate . . . *Aster*

                13. Leaves punctate, resinous dotted . . . . *Kuhnia*

            12. Leaves petiolate 14

                14. Leaves punctate . . . . *Kuhnia*

                14. Leaves not punctate 15

                    15. Heads wholly of tubular flowers; pappus capillary 16

                        16. Plants not odorous . *Vernonia*

                        16. Plants odorous . . *Pluchea*

                    15. Heads with pinkish, long ligulate flowers . . . . *Echinacea*

1. Flowers yellow, white, greenish, or flesh-colored 17

    17. Leaves always opposite or the lower sometimes alternate . 18

        18. Plants climbing . . . . . . *Mikania* °

32. Heads yellow 33
    33. Rays conspicuously notched or, if entire, the leaves whorled; leaves not edgewise to a north-south direction . . *Coreopsis*
    33. Rays not notched; leaves edgewise to a north-south direction, scabrous . . . *Silphium*
32. Heads greenish-yellow, inconspicuous; staminate flowers in separate heads . . . . *Ambrosia*
31. Leaves ovate-lanceolate, perfoliate or basal 34
    34. Leaves perfoliate or basal, scabrous . . . . *Silphium*
    34. Leaves ovate, lanceolate, usually verticillate in 3's or 6's or opposite . . . . *Eupatorium*
29. Leaves always alternate or alternate with upper leaves rarely opposite 35
  35. Leaves dentate, dissected, or pinnatifid . . 36
    36. Bracts in 1 or 2 series 37
      37. Leaves decurrent along the stem . 38
        38. Stems glabrous; rays yellow . . . . . . *Helenium*
        38. Stems pubescent, winged; rays white or yellow . . *Verbesina*
      37. Leaves not decurrent along stem . 39
        39. Leaves not pinnatifid or dissected 40
          40. Leaves not scabrous . . 41
            41. Heads whitish, rayless, or apparently so . . . *Erechtites*
            41. Heads radiate, bright yellow . . *Senecio*
          40. Leaves scabrous . . . . . *Parthenium*
        39. Leaves pinnatifid . . . 42
          42. Rays white, few . . . 43
            43. Inflorescence corymbose . . *Achillea*
            43. Inflorescence paniculate or heads solitary; disk flowers yellow; rank odor . *Anthemis*

42. Rays absent; plants with
odor . . . *Matricaria*
36. Bracts in more than 2 series . *Tanacetum*
35. Leaves not wholly dentate, pinnatifid, or dissected 44
44. Bracts in 1 or 2 series 45
45. Rays cream or white, or flowers small
and greenish 46
46. Leaves scabrous, rays few (about
5) . . . . *Parthenium*
46. Leaves not scabrous . . . 47
47. Leaves large, hastate .
. . . . *Cacalia*
47. Leaves whitish canescent .
. . . . *Artemisia*
45. Rays yellow and disk brown . .
. . . . . *Helenium*
44. Bracts in more than 2 series . . . 48
48. Plants white-woolly and low growing 49
49. Leaves broad; plants dioecious
. . . . *Antennaria*
49. Leaves not broad; plants not dioecious 50
50. Pappus single, capillary .
. . . *Gnaphalium*
50. Pappus none . . .
. . . . *Artemisia*
48. Plants not white-woolly and not low
growing 51
51. Heads conspicuous, over 4 mm.
broad 52
52. Heads all yellow or rays
yellow 53
53. Heads all yellow . 54
54. Heads small, 1
cm. or less
broad; leaves
linear or ovate-
lanceolate .
. . *Solidago*
54. Heads larger
than 1 cm.
broad, bright
yellow . . 55

55. Leaves ses-
sile and en-
tire .   .
.   *Chrys-
opsis* °
55. Leaves
dentate .
*Hetero-
theca* °
53. Heads with yellow
rays and brown disk   56
56. Rays 1 to 3
inches   long,
m a r k e d l y
drooping,   the
center   elon-
gated;   leaves
pinnately 3- to
7-divided .   .
.   . *Ratibida*
56. Rays not droop-
ing,   shorter;
center or disk
usually   not
higher   than
broad   .   .
. *Rudbeckia*
52. Heads with all white or
cream flowers or white rays
and yellow disks or no rays
with yellow disks   .   .   57
57. Tall plants of wood-
land; large hastate
and pubescent leaves
with pineapple odor
when crushed .   .
.   .   . *Polymnia*
57. Plants 1 to 4 feet tall,
of   meadows,   road-
sides or waste places   58
58. Solitary   heads
on long pedun-
cles   .   .   .
Chrysanthemum
58. Paniculate inflo-
rescence .   .   59

> 59. Leaves de-
>   current on
>   stem;
>   stems
>   lined   .
>   .   *Boltonia*
> 59. Leaves not
>   decurrent;
>   stems not
>   lined   .
>   .   *Aster*
> 51. Heads not conspicuous, about
>   4 mm. or less in diameter .   .   60
>   60. Flowers small and greenish
>   .   .   .   .   *Artemisia*
>   60. Flowers (rays) white   .
>   .   .   .   .   *Erigeron*

Taraxacum   WEBER

Taraxacum officinale Weber (3:533). Common Dandelion is a naturalized weed from Europe. It is particularly common on lawns, on roadside shoulders, and in over-grazed bluegrass pastures. In our territory a few plants have been observed to bloom during the warmer days of December, January, or February. Blooming is profuse during April and May and throughout the summer months. SP. CIT.: *Mc* 3569.

Taraxacum erythrospermum Andrz. (3:533). Red-seeded Dandelion differs from the preceding species in having less than one hundred and sixty flowers in the head (usually seventy-five to one hundred and twenty-five), flowers are sulphur yellow, and leaves are more deeply pinnatifid. This species is also found in waste places though less frequently than *T. officinale*. We have it from Gallatin, Johnson, and Pulaski counties.

Krigia   SCHREB.

> 1. Plants with leafless scapes   2
>   2. Plants with a small tuber   .   .   .   .*Krigia dandelion*
>   2. Plants without a small tuber .   .   .   *Krigia virginica*
> 1. Plants with 1 to several stem leaves   3
>   3. Leaves 1 to 3, clasping   .   .   .   .   *Krigia biflora*
>   3. Leaves several, not clasping   .   .   *Krigia oppositifolia*

Krigia dandelion (L.) Nutt. (3:541). A common plant of open woods, this species usually occurs wherever the rare orchid, *Spiranthes gracilis,* occurs. SP. CIT.: Midland Hills, May 24, 1952, *Voigt;* Giant City State Park, May 3, 1953, *M 1337;* Saltpeter Cave, May 13, 1954, *M 2432.*

Krigia virginica (L.) Willd. (3:541). This uncommon species grows in dry soil usually in open areas. It is local in our area. SP. CIT.: Giant City State Park, May 24, 1953, *M 680.*

Krigia biflora (Walt.) Blake (3:541). The attractive False Dandelion, our most common species of *Krigia,* is found on wooded slopes where it begins to flower the first part of May. SP. CIT.: Giant City State Park, May 3, 1951, *V 652:* Saltpeter Cave, May 13, 1954, *M 2428.*

Krigia oppositifolia Raf. (3:541) [*Serinia oppositifolia* (Raf.) Kuntze]. This southern species is fairly common in moist sandy soil of Southern Illinois. SP. CIT.: Midland Hills, May 24, 1952, *Voigt;* three miles south of Boskydell, June 20, 1952, *BS 2559.*

Hieracium L.

1. Leaves and stems copiously long-hirsute; whitish hairs about 1 to 2 cm. long . . . .*Hieracium longipilum* °
1. Leaves and stems with hairs shorter (less than 1 cm.), not as copious as above 2
    2. Bracts leafy . . . . . .*Hieracium scabrum*
    2. Bracts small, not leafy . . . .*Hieracium gronovii*

Hieracium longipilum Torr. (3:525) possesses the longest pubescence of any species in our area, the hairs sometimes reaching a length of about two centimeters.

Hieracium scabrum Michx. (3:526). Rough Hawkweed has been found in dry, open areas in Jackson and Johnson counties. SP. CIT.: (see Jones *et al.,* 1955).

Hieracium gronovii L. (3:526). Although this species is our commonest Hawkweed, it is rare in the central and northern sections of Illinois. It occurs in dry oak-hickory woods. SP. CIT.: Giant City State Park, July 30, 1953, *M 433;* Midland Hills, July 23, 1941, *Mc 966;* Fountain Bluff, August 10, 1948, *BS 546.*

Lactuca L.

1. Flowers blue, lavender, or whitish 2
    2. Leaves lyrate (broad, pinnatifid, terminal lobe large); outer achenes thick-beaked . . . . .*Lactuca floridana*
    2. Leaves ovate-lanceolate, uncleft, long-pointed; achenes beak-less . . . . . .*Lactuca floridana* var. *villosa*
1. Flowers yellow 3
    3. Leaves spinulose-toothed . . . .*Lactuca scariola*
    3. Leaves not spinulose-toothed 4
        4. Leaves narrow, often entire . . .*Lactuca saligna* °
        4. Leaves toothed or lobed 5
            5. Main stem leaves usually denticulate; leaf bases sagittate . . .*Lactuca canadensis* var. *obovata*
            5. Main stem leaves broadly falcately lobed . . .
                . . . . .*Lactuca canadensis* var. *latifolia*

Lactuca floridana (L.) Gaertn. (3:537). Blue Lettuce is an occasional woodland species throughout Illinois. The typical variety is much more common than var. *villosa* (Jacq.) Cronq. SP. CIT.: typical: Clear Springs, September 11, 1952, *BS 2981.*

Lactuca scariola L. (3:537). Prickly Lettuce is often confused with species of *Sonchus* because of prickly leaves. It is a native of Europe and is frequently found in Illinois. SP. CIT.: Carbondale, August 21, 1941, *Mc 1127;* Giant City State Park, July 7, 1953, *M 352.*

Willow Lettuce, *Lactuca saligna* L. (3:537), named for its narrow leaves, is a rapidly spreading adventive in Illinois. We have it from all our counties except Randolph and Jackson.

Lactuca canadensis L. (3:536). Common Wild Lettuce in Southern Illinois may be divided into two varieties based on characteristics of leaves. Merely denticulate leaves are known as var. *obovata* Wieg. while lobed-leaved specimens are var. *latifolia* Kuntze. SP. CIT.: var. *obovata:* Giant City State Park, August 8, 1953, *M 534.* Var. *latifolia:* Carbondale, August 12, 1949, *BS 798;* Giant City State Park, July 2, 1953, *M 169.*

Sonchus L.

Sonchus asper (L.) Hill (3:534). Spiny Sow-Thistle, a common annual weed from Europe, is a rather small-flowered species. Flowers seldom attain a width of over one inch. SP. CIT.: Carbondale, July 16, 1878, *French 1632.*

Sonchus oleraceus L. (3:534). An occasional adventive in our area is the perennial Common Sow-Thistle. Its flowers are over one inch wide. SP. CIT.: Giant City State Park, July 2, 1953, *M 167.*

Tragopogon L.

Two species of Goat's-beard, or Oyster Plant, occur commonly in central and northern Illinois but are nearly lacking in the southern part of the state. The purple-flowered *Tragopogon porrifolius* L. (3:545) and the yellow-flowered *T. dubius* Scop. (3:545) have both been collected in Randolph County. The large fuzzy heads make these species easy to spot in autumn (Fig. 65).

Pyrrhopappus DC.

Pyrrhopappus carolinianus (Walt.) DC. (3:530). False Dandelion is a common species along roads and in fields of Southern Illinois. It is entirely lacking from the northern counties. Flowers often bloom early in the morning and are closed by nine o'clock. SP. CIT.: Fountain Bluff, June 6, 1952 *BS 2491;* two miles south of Elkville, July 9, 1951, *BS 1524.*

Cichorium L.

Cichorium intybus L. (3:539). The blue or sometimes white flowered Chicory, an abundant species in waste areas of central and northern Illinois, is an occasional roadside waif in our area. According to Tehon (1951), roots are collected and used as a coffee substitute and as an appetite stimulant. SP. CIT.: near Murphysboro, *Mohlenbrock.*

FIG. 65. *Large globose fruiting heads of Goat's-beard* (Tragopogon dubius) *are infrequently seen in Southern Illinois. It is common in the central part of the state.*

Prenanthes L.

1. Lower leaves palmately lobed or hastate-dentate 2
   2. Leaves palmately lobed; flowers yellow-white   .   .   .
   .   .   .   . *Prenanthes altissima*
   2. Leaves not palmately lobed, but usually hastate; flowers
   cream colored   .   .   .   .   . *Prenanthes crepidinea* °
1. Lower leaves entire, dentate-denticulate, or pinnatifid .   .   . 3
   3. Stems and leaves glabrous; flowers purplish .   .   .   .
   .   .   .   . *Prananthes racemosa* °
   3. Stems and leaves scabrous; flowers light yellow .   .   .
   .   .   .   . *Prananthes aspera*

Prenanthes altissima L. (3:521). This glaucous-leaved species is found in woodlands in Southern Illinois where it flowers during September. Plants sometimes attain a height of three feet. SP. CIT.: Giant City State Park, September 20, 1953, *M 769;* Clear Springs, September 11, 1952, *BS 2980.*

Prenanthes aspera Michx. (3:518). A single collection from Jackson County is the only record we have for this species in Southern Illinois. It was found in prairie soil along the railroad near Elkville. SP. CIT.: one mile south of Elkville, September 11, 1949, *BS 850.*

Two additional rare species of *Prenanthes* have been found at single stations in Southern Illinois. *Prenanthes crepidinea* Michx. (3:519) is known from a moist woods in Pulaski County while *Prenanthes racemosa* Michx. (3:518) occurs in a moist area in Alexander County.

Elephantopus L.

Elephantopus carolinianus Willd. (3:505). To the casual observer, Elephant's-foot does not recall *Compositae* to mind. Closer examination of the flower heads, however, reveals clearly the characteristic composite structure. This mostly southern species occurs in dry woods in the southern one-third of Illinois. SP. CIT.: Giant City State Park, August 28, 1953, *M 531;* Ava Cave, August 26, 1954, *M 4660.*

Erigeron L.

1. Heads over 1 cm. broad 2
   2. Leaves clasping 3
      3. Rays pinkish-lavender, less than 100 in number   .   .
      .   .   .   . *Erigeron pulchellus*

> 3. Rays usually white, 150 to 200 in number . . .
> . . . . *Erigeron philadelphicus*
> 2. Leaves not clasping 4
> 4. Basal leaves ovate, stem leaves ciliate and serrate .
> . . . . *Erigeron annuus*
> 4. Basal leaves spatulate, stem leaves not ciliate or serrate
> . . . . *Erigeron strigosus*
> 1. Heads not over 1 cm. broad 5
> 5. Stems much branched; short plants with linear leaves . .
> . . . . *Erigeron divaricatus*
> 5. Stem strict, up to 1.5 meters high; leaves lanceolate . .
> . . . . *Erigeron canadensis*

Erigeron pulchellus Michx. (3:471). Beautiful Robin's Fleabane is an occasional species of open woods. It flowers from mid-April to mid-May. SP. CIT.: Lake Murphysboro, April 30, 1950, *M 2567;* Midland Hills, April 28, 1948, *BS 376.*

Erigeron philadelphicus L. (3:471). One of our most common native weeds is the Daisy Fleabane. Some pink flowered forms often resemble Robin's Fleabane. SP. CIT.: near Murphysboro, May 5, 1941, *Mc 708.*

Erigeron annuus (L.) Pers. (3:473). Common Fleabane is abundant along roads, in fields, and at borders of woods. SP. CIT.: Giant City State Park, May 24, 1953, *M 196.*

Erigeron strigosus Muhl. (3:473). Very similar to the preceding species is this equally common plant. It is sometimes referred to as Daisy Fleabane. SP. CIT.: Midland Hills, May 24, 1952, *Voigt.*

Erigeron divaricatus Michx. (3:475). This small, often prostrate plant occurs in dry open areas in most of our counties, although it is not commonly found. SP. CIT.: Grand Tower, August 24, 1954, *Bailey 586.*

Erigeron canadensis L. (3:475). A native of Europe, the Muletail is an aggressive weed throughout Illinois. SP. CIT.: Carbondale, September 18, 1947, *BS 179.*

Eupatorium L.

> 1. Lower leaves opposite 2
> 2. Flowers pink or purple 3
> 3. Receptacle conical; flowers more than 30 . . . .
> . . . . *Eupatorium coelestinum*
> 3. Receptacle flat; flowers less than 30 . . . . .
> . . . . *Eupatorium incarnatum* °

Eupatorium coelestinum L. (3:493). Mist-flower occurs in moist ground in the southern half of Illinois. It is often found with *Lobelia siphilitica, Mimulus alatus,* and other wet meadow species. SP. CIT.: Giant City State Park, August 2, 1953, *M 590;* one mile north of Pomona, September 28, 1951, *V 1003.*

Eupatorium serotinum Michx. (3:492). Late Boneset is our most common species in this genus. The plant is poisonous to cattle. SP. CIT.: Giant City State Park, July 29, 1953, *M 431.*

Eupatorium rugosum Houtt. (3:492). Like the preceding species, this White Snakeroot is poisonous to cattle. It is probably not quite as common as the Late Boneset. SP. CIT.: Giant City State Park, July 20, 1953, *M 696;* Fountain Bluff, September 14, 1940, *Mc 439.*

Eupatorium altissimum L. (3:490). Narrow-leaved Snakeroot, or Honeyweed, is an occasional species of stream banks, although it sometimes is found in moist areas of prairie soil along railroads. SP. CIT.: Fountain Bluff, September 14, 1940, *Mc 467;* Finney railroad prairie, August 28, 1954, *M 4665.*

Eupatorium perfoliatum L. (3:491). Boneset is our most easily recognized species of the genus. The connate-perfoliate leaves are densely tomentose. At the "marsh" north of Murphysboro, it occurs with *Solidago patula, Scirpus cyperinus, Galium obtusum,* and others.

Eupatorium purpureum L. (3:487). Joe Pye-weed attains a height of five to eight feet in our area. The purple flowers are borne in huge inflorescences. This species is often confused with *E. maculatum* and *E. fistulosum,* but the former is unknown from Southern Illinois (although erroneously attributed to Jackson and Massac counties) and the latter is known only from Alexander County. SP. CIT.: Giant City State Park, July 12, 1953, *M 398;* Carbon Lake, July 27, 1954, *M 4390.*

Three species of *Eupatorium* occur in Southern Illinois but have not as yet been discovered in Jackson County. *Eupatorium incarnatum* Walt. (3:493) is a lowland species which has been collected in Illinois only from Alexander, Pulaski, and Union counties. *Eupatorium sessilifolium* L., a local species of Illinois oak-hickory woods, is known in our area from Union County. *Eupatorium fistulosum* Barratt has its only Illinois station in low ground near Olive Branch in Alexander County (*J. R. Swayne 1152*).

Arctium L.

Arctium minus (Hill) Bernh. (3:505). Common Burdock is an occasional waif in moist waste ground. It is naturalized from Europe. It flowers during July and August. A white-flowered form (forma *pallidum* Farw.) has been found once in Southern Illinois. SP. CIT.: typical: Fountain Bluff, *BS 549;* Southern Illinois University campus, *Mc 921.* F. *pallidum:* Saltpeter Cave, August 2, 1954, *M 4816.*

Xanthium L.

1. Plants bearing 3-parted axillary, yellow spines; leaves lanceolate
   . . . . *Xanthium spinosum* *
1. Plants without axillary spines; leaves broad or ovate . . . 2
   2. Prickles numerous, beaks hooked . .*Xanthium commune*
   2. Prickles few, beaks nearly straight . .*Xanthium chinense*

Xanthium commune Britt. (3:375) [*Xanthium italicum* Moretti]. This and the following species of Cocklebur are about equally abundant in waste places in Southern Illinois. SP. CIT.: Carbondale Reservoir, October, 1953, *Bell;* Giant City State Park, August 31, 1953, *M 463.*

Xanthium chinense Mill. (3:375) [*Xanthium pennsylvanicum* Wallr.]. SP. CIT.: Southern Illinois University farm, October 3, 1947, *BS 207.*

A rare adventive from tropical America, *Xanthium spinosum* L. (3:375), has been found three times in Illinois by George Vasey. It is known from Alexander, Pulaski, and Cook counties.

Carduus L.

Musk Thistle (*Carduus nutans* L.) (3:507), an introduced species from Europe, has been collected in Southern Illinois from Alexander and Union counties.

Cirsium MILL.

1. Leaves whitish or brownish tomentose beneath; involucral bracts
    some or all prickly-pointed 2
      2. All involucral bracts prickly-pointed . .*Cirsium vulgare*
      2. Only the outer involucral bracts prickly-pointed . . . 3
         3. Leaves shallowly lobed or only toothed, densely white
            tomentose beneath . . . .*Cirsium altissimum*
         3. Leaves deeply pinnately lobed, segments narrow, whitish
            tomentose beneath . . . . .*Cirsium discolor*
1. Leaves green on both sides, somewhat pubescent on underneath
    side . . . .*Cirsium arvense* °

Cirsium vulgare (Savi) Airy-Shaw (3:508) [*Cirsium lanceolatum* (L.) Scop.]. An occasional weed of fields and other waste areas is Bull Thistle. This native of Europe is our most prickly species of *Cirsium*. SP. CIT.: two miles west of Carbondale, July 20, 1948, *BS 524.*

Cirsium altissimum (L.) Spreng. (3:509). Tall Thistle is a native of rather dry woods throughout Illinois. It flowers in August. SP. CIT.: Hickory Ridge, August 24, 1949, *BS 814;* Giant City State Park, August 6, 1953, *M 600.*

Cirsium discolor (Muhl.) Spreng. (3:509). Our most common species of *Cirsium* is Field Thistle. It usually occurs in rich soil in woods, fields, or even waste places. SP. CIT.: Giant City State Park, September 20, 1953, *M 458.*

Canada Thistle, *Cirsium arvense* (L.) Scop. (3:511), a native of Europe, has been collected in our area only from Pulaski County.

Centaurea L.

Centaurea cyanus Lam. (3:515). Bachelor's Button is a common escape from cultivation. It is variable in flower color.

Liatris SCHREB.

1. Inflorescence usually densely spicate 2
      2. Bracts recurved, acuminate . . .*Liatris pycnostachya*

    2. Bracts ascending, acute .   .   .   .   *.Liatris spicata* °
1. Inflorescence racemiform, or heads occasionally solitary   .   .   3
    3. Bracts recurved and spreading .   .   *.Liatris squarrosa*
    3. Bracts ascending and mostly appressed   4
        4. Bracts hirtellous .   .   .   .   .   *.Liatris scabra*
        4. Bracts glabrous or nearly so   5
           5. Bracts mucronate .   .   .   *.Liatris cylindracea*
           5. Bracts obtuse   6
               6. Stems pubescent throughout .   *.Liatris aspera*
               6. Stems glabrous throughout or pubescent only
                 near the summit .   .   .   .
               .   .   .   .   *Liatris aspera* var. *intermedia*

Liatris pycnostachya Michx. (3:498). This beautiful Blazing Star, or Kansas Gay Feather (Fig. 66), presents a brilliant purple color to prairies during August and September. It occurs with species of *Silphium, Gaura biennis, Asclepias tuberosa,* and numerous prairie grasses. SP. CIT.: two miles south of Elkville, July 14, 1948, *BS 521.*

Liatris squarrosa (L.) Michx. (3:499). Occurring with the preceding species, but not as common, is this attractive Blazing Star. SP. CIT.: two miles south of Elkville, July 14, 1948, *BS 520.*

Liatris scabra (Greene) K. Schum. (3:496). This uncommon species is distinguished by its pedunculate flower-heads. The bracts (phyllaries) are not scarious margined. The type locality for this species is in the Pine Hills of Union County where it was collected by F. S. Earle on September 23, 1890.

Liatris cylindracea Michx. (3:499). This Blazing Star, rare in Southern Illinois, is perhaps our most brightly colored species of

FIG. 66. *Tall Blazing Star* (Liatris pycnostachya), *a showy plant of the fall, has purple-rose flowers.*

Liatris. In this area, it has been found only on a hilltop prairie in the Grassy Knob area of Jackson County. SP. CIT.: Grassy Knob hill prairie, September 9, 1954, *M 4758.*

Liatris aspera Michx. (3:497). This plant is uncommon in open woods in Illinois. We have it only from Jackson, Union, and Williamson counties. The nearly smooth-stemmed var. *intermedia* Lunell has been found once in our area. SP. CIT.: typical: southeast of Carbondale, October 22, 1940, *Mc 579.* Var. *intermedia:* Cedar Creek bluff, August 6, 1954, *M 4724.*

Liatris spicata (L.) Willd. (3:498) has been collected in dry open areas in only a few of our counties (Pope, Saline, Gallatin, Pulaski).

Aster L.

1. At least the basal leaves cordate and petiolate   .   .   .   2
    2. Bracts recurved; heads with 25 or more flowers .   .   .
        .   .   .   . *Aster anomalus*
    2. Bracts usually appressed and ascending; heads usually less than 25-flowered   3
        3. Leaves entire or nearly so   .   .   . *Aster shortii*
        3. Leaves toothed   4
            4. Leaves and stems usually hirtellous throughout   .
                .   .   .   .   .   . *Aster drummondii*
            4. Leaves and stems glabrous or nearly so .   .   .
                .   .   .   .   .   . *Aster sagittifolius*
1. None of the leaves cordate or, if cordate, then sessile and often clasping   5
    5. Cauline leaves clasping   6
        6. Stems glabrous   .   .   .   .   . *Aster laevis*
        6. Stems hairy   7
            7. Heads 5 to 8 mm. high   8
                8. Leaves strongly cordate-clasping . *Aster patens*
                8. Leaves scarcely cordate-clasping .   .   .   .
                    .   .   .   .   . *Aster oblongifolius* °
            7. Heads 8 to 10 mm. high .   . *Aster novae-angliae*
    5. Cauline leaves not clasping   9
        9. Bracts with twisted, subulate, green tips . *Aster pilosus*
        9. Bracts not as above   10
            10. Heads 7 to 12 mm. high   . *Aster turbinellus*
            10. Heads less than 7 mm. high   11
                11. Bracts mucronate, squarrose . *Aster exiguus*
                11. Bracts not mucronate, mostly appressed .   . 12
                    12. Leaves puberulent over the surface   .
                        .   .   .   .   . *Aster ontarionis*

12. Leaves glabrous or puberulent only
along the midrib beneath . . . 13
    13. Heads 15 to 25 mm. across . .
         . . . . *Aster praealtus*
    13. Heads less than 15 mm. across . 14
        14. Rays less than 15 . . .
           . . . *Aster lateriflorus*
        14. Rays 15 to 25 . . . 15
          15. Leaves of the branchlets
            mostly less than 15 mm.
            long . . . .
             . *Aster vimineus* var.
                   subdumosus
          15. Leaves of the branchlets
            mostly over 15 mm. long 16
            16. Heads over 4 mm.
              high . . .
                 Aster simplex
            16. Heads 3 to 4 mm.
              high . . .
                 Aster interior

Aster anomalus Engelm. (3:447). A handsome species of rocky woods and bluffs; variable with rays often "fringed"; Jackson, Union, Alexander, and Williamson counties; rare. SP. CIT.: seven miles west of Murphysboro, *Mohlenbrock.*

Aster shortii Lindl. (3:447). Along streams and in woods; common. SP. CIT.: Giant City State Park, September 20, 1953, *M 457.*

Aster drummondii Lindl. (3:449). Dry woods; not common; one of our last species to flower. SP. CIT.: Midland Hills, *BS 240.*

Aster sagittifolius Wedemeyer (3:451). Dry woodlands; common. SP. CIT.: Giant City State Park, *M 402.*

Aster laevis L. (3:455). Smooth Aster; sandy woods, usually near streams; rare; only Jackson, Union, and Williamson counties (see Jones *et al.,* 1955).

Aster patens Ait. (3:451). Dry woods; common; limited in Illinois to the southern counties. SP. CIT.: Midland Hills, *BS 225;* Giant City State Park, *M 464.*

Aster oblongifolius Nutt. (3:453). Wooded bluffs; known in our area only from the Pine Hills of Union County.

Aster novae-angliae L. (3:453). New England Aster; known only in Southern Illinois from prairie soil along railroads in Jackson

County. SP. CIT.: two miles south of Elkville, *Bailey and Hankla 637;* four miles north of Murphysboro, *BS 3172.*

Aster pilosus Willd. (3:461). All open habitats; very common. SP. CIT.: east of Howardton, *Mc 1198;* Giant City State Park, *M 526.*

Aster turbinellus Lindl. (3:463). Very pretty species of rocky woodlands; occasional; flowers from late September through October. SP. CIT.: Giant City State Park, *M 117;* Midland Hills, *BS 226.*

Aster exiguus (Fern.) Rydb. (3:464) [*Aster ericoides* L. f. *exiguus* Fern.]. Very common, particularly around strip mines and in fields. SP. CIT.: Giant City State Park, *M 455.*

Aster ontarionis Wieg. (3:465). Moist areas; infrequent; Jackson, Alexander, and Pulaski counties. SP. CIT.: Murphysboro, *Mohlenbrock.*

Aster praealtus Poir. (3:466) [*Aster salicifolius* Ait.]. Moist soil, especially roadside ditches. SP. CIT.: Giant City State Park, October 9, 1953, *M 326.*

Aster lateriflorus (L.) Britt. (3:465). One of our more common species of woods and roadsides. SP. CIT.: Giant City State Park, September 20, 1953, *M 460.*

Aster vimineus Lam. var. *subdumosus* Wieg. (3:465). Moist, open areas; occasional. SP. CIT.: northeast of Grand Tower, *BS 891.*

Aster simplex Willd. (3:466). An occasional species of moist ground; Jackson, Saline, and Union counties. SP. CIT.: Giant City State Park, October 9, 1953, *M 324.*

Aster interior Wieg. [*Aster simplex* Willd. var. *interior* (Wieg.) Cronq.; *Aster tradescanti* of auth., non L.] Moist soil; not common; only in Jackson County. SP. CIT.: Carbondale Reservoir, October, 1953, *Bell;* Cave Valley (near Pomona), September 28, 1951, *Hardy 120.*

Kuhnia L.

Kuhnia eupatorioides L. (3:501). False Boneset is a species along railroads and in hill prairies. On Grassy Knob hill prairie, it is associated with *Liatris cylindracea, Houstonia nigricans,* and other prairie species. We have this plant only from Jackson, Union, Saline, and Hardin counties. SP. CIT.: Grassy Knob hill prairie, September 9, 1954, *M 4755.*

Vernonia SCHREB.

1. Heads 30- to 60-flowered .   .   .   . .*Vernonia missurica*
1. Heads less than 30-flowered   2
  2. Leaves glabrous on both surfaces, or glabrous above and
     puberulent beneath   3
     3. Leaves ½ inch to 1½ inches wide .   .*Vernonia altissima*
     3. Leaves less than one-half inch wide .   .   .   .   .
        .   .   .   .   *Vernonia fasciculata*
  2. Leaves scabrous above, densely tomentose beneath   .   .
        .   .   .   .   *Vernonia baldwini*

Vernonia missurica Raf. (3:502). This is our commonest species of Ironweed. It occurs along roads, in fields, and sometimes at the edges of woods. SP. CIT.: Giant City State Park, August 8, 1953, *M 513*.

Vernonia altissima Nutt. (3:502). This narrow-leaved Ironweed occurs in open fields. SP. CIT.: Ava Cave, August 26, 1954, *M 4660*.

Vernonia fasciculata Michx. (3:503). This plant is local in Southern Illinois. It is abundant around Lake Murphysboro. SP. CIT.: Lake Murphysboro, September 4, 1954, *M 4698;* three miles northeast of Elkville, July 24, 1941, *Mc 983*.

Vernonia baldwini Torr. (3:503). This is less common than other Ironweeds. It is known only from a few south-central counties, but extends southward into Randolph and Jackson counties. SP. CIT.: Giant City State Park, August 8, 1953, *M 512*.

Pluchea CASS.

Pluchea camphorata (L.) DC. (3:477). Marsh Fleabane, limited in Illinois mostly to the southern half of the state, occurs in wet areas in most of our counties, although it is nowhere common. SP. CIT.: Fountain Bluff, *Cranwill*.

Echinacea MOENCH

Echinacea purpurea (L.) Moench (3:351). Purple Cone-flower is a beautiful species of rather damp woodlands. The brilliant purple rays are produced in August. We have this species only from Jackson and Union counties. SP. CIT.: Fountain Bluff, August 5, 1952, *BS 2705;* Ava Cave, August 26, 1954, *M 4650*.

Prairie Cone-flower, *Echinacea pallida* Nutt. (3:351), occurs on a few hill prairies in Union and Randolph counties. It is also listed for Johnson County. The pink rays are produced in June and July.

Mikania WILLD.

Climbing Hempweed, *Mikania scandens* (L.) Willd. (3:493), our only climbing species of *Compositae*, has been found in swampy situations in only a few southern counties.

Bidens L.

1. Rays small (1 to 5 mm. long) or absent   2
    2. Heads on long slender peduncles   3
        3. Stems purplish, round in cross-section . *Bidens frondosa*
        3. Stems not purplish, quadrangular in cross-section .   .
            .   .   .   . *Bidens bipinnata*
    2. Heads on stout peduncles   4
        4. Leaves petiolate   5
            5. Awns of the achenes upwardly barbed, about one-fourth as long as the achenes .   . *Bidens discoidea*
            5. Awns of the achenes downwardly barbed, about one-half as long as the achene .   .   . *Bidens vulgata*
        4. Leaves sessile or subsessile   6
            6. Heads globose; leaves somewhat perfoliate .   .
                .   .   .   . *Bidens cernua*
            6. Heads not globose; leaves not perfoliate .   .   .   7
                7. Outer bracts leafy and elongate . *Bidens comosa*
                7. Outer bracts not elongate .   . *Bidens connata* °
1. Rays conspicuous (at least 10 mm. long)   8
    8. Heads on stout peduncles   9
        9. Leaves petiolate   10
            10. Awns upwardly barbed .   .   . *Bidens polylepis*
            10. Awns downwardly barbed .   .   .   .   .
                .   .   .   .   . *Bidens polylepis* var. *retrorsa*
        9. Leaves sessile or nearly so .   .   .   . *Bidens cernua*
    8. Heads on long slender peduncles   11
        11. Awns less than one-half as long as the achene .   .
            .   .   .   . *Bidens coronata*
        11. Awns one-half to as long as the achene .   .   .   .
            .   .   .   . *Bidens aristosa*

Bidens frondosa L. (3:356). Very common in woodlands and fields. SP. CIT.: Giant City State Park, August 30, 1953, *M 557;* one mile north of Pomona, September 28, 1951, *V 1006.*

Bidens bipinnata L. (3:356). Spanish Needles; common in fields and along roads; the achenes stick in clothing; flowers from late August through September. SP. CIT.: Giant City State Park, *M 525;* Fountain Bluff, *Mc 455.*

Bidens discoidea (Torr. and Gray) Britt. (3:355). Swampy ground; rare; Jackson, Union, Johnson, and Alexander counties. SP. CIT.: Snyder's Lake, October 2, 1952, *BS 3102.*

Bidens vulgata Greene (3:356). Tall Beggar-ticks; moist ground; rare in Southern Illinois; Jackson and Union counties. SP. CIT.: Jackson County Recreation Lake, September 24, 1952, *BS 3048.*

Bidens cernua L. (3:354). Wet ground; known in Southern Illinois from Jackson and Union counties; rays may be present or absent. SP. CIT.: Grand Tower slough, August 20, 1954, *M 4633.*

Bidens comosa (Gray) Wieg. (3:355) [*Bidens tripartita* L., in part]. Moist ground, rarely in rich woods; known only from Jackson and Union counties, but a little more common than *Bidens cernua.* SP. CIT.: Giant City State Park, August 28, 1953, *M 532;* Grand Tower slough, August 20, 1954, *M 4635;* Lake Murphysboro, September 4, 1954, *M 4695.*

Bidens connata Muhl. (3:355) [*Bidens tripartita* L., in part]. Known in Southern Illinois from the LaRue Swamp of Union County.

Bidens polylepis Blake (3:357). Swampy ground; not common; Jackson, Johnson, and Hardin counties. SP. CIT.: Giant City State Park, October 9, 1953, *M 323;* Grand Tower slough, August 20, 1954, *M 4634.* Even rarer is var. *retrorsa* Sherff. SP. CIT.: Ava Cave, August 26, 1954, *M 4655.*

Bidens coronata (L.) Britt. (3:356). Moist ground; local. SP. CIT.: Lake Murphysboro, August 27, 1954, *M 4662.*

Bidens aristosa (Michx.) Britt. (3:357). Very common in moist ground; flowers in August and September. SP. CIT.: Giant City State Park, *M 545;* west of Carbondale, *BS 169.*

Iva L.

Iva ciliata Willd. (3:372). Marsh-elder occurs in moist waste ground locally throughout our area. Flowers are inconspicuous. SP. CIT.: along Cedar Creek, August 6, 1954, *M 4714.*

Polymnia L.

Polymnia canadensis L. (3:366). Leafcup is found in moist woodlands locally throughout Illinois. It is usually found in limestone areas. The rays are white. SP. CIT.: Grand Tower, July 11, 1871, *French;* north of the Pine Hills, August 6, 1954, *M 4708.*

Polymnia uvedalia L. (3:366). Bearsfoot is limited in Illinois to six southern counties with Jackson being the farthest north. It occurs at the base of a limestone bluff in Jackson County with a rare sedge, *Carex oligocarpa.* SP. CIT.: north of the Pine Hills, August 6, 1954, *Mohlenbrock.*

Galinsoga   RUIZ & PAVON

Galinsoga ciliata (Raf.) Blake (3:364). Peruvian Daisy or Quickweed is an infrequent adventive into waste ground in our area. We have it only from Jackson County. The pappus of the rays is well developed. SP. CIT.: Southern Illinois University campus, May 23, 1952, *Stephens.*

Galinsoga parviflora Cav. (3:364). Collection of this rare adventive in Jackson County marks the second station for this species in Illinois. It had been known from Champaign County. The pappus of the rays is nearly lacking. SP. CIT.: Carbondale, along the Illinois Central Railroad, October 16, 1952, *BS 3212.*

Eclipta   L.

Eclipta alba (L.) Hassk. (3:342). This inconspicuous plant is common in moist situations throughout Southern Illinois. SP. CIT.: Giant City State Park, August 5, 1953, *M 602;* one-half mile south of Carbondale, August 21, 1941, *McCree and Wilson 1134.*

Verbesina   L.

1. Rays yellow   2
    2. Heads 1 to 10; bracts pubescent . *Verbesina helianthoides*
    2. Heads 10 to numerous; bracts glabrous or nearly so .   .
        .   .   .   . *Verbesina alternifolia*
1. Rays white   .   .   .   .   .   . *Verbesina virginica* °

Verbesina helianthoides Michx. (3:341). Yellow Crownbeard has winged stems. It occurs in dry oak-hickory woods in all of our counties. It begins to flower the first of June. SP. CIT.: Fountain Bluff, June 18, 1940, *Welch and Fuller 177;* Giant City State Park, June 4, 1953, *M 227.*

Verbesina alternifolia (L.) Britt. (3:343) [*Actinomeris alternifolia*

(L.) DC.]. Yellow Ironweed is found locally in woodlands. Stems are sometimes winged in this species. It usually flowers later than the preceding species. SP. CIT.: Giant City State Park, August 18, 1954, *M 4812;* Fountain Bluff, September 14, 1940, *Mc 447.*

White Tickweed, *Verbesina virginica* L. (3:341), is presently known from only the following four Illinois counties: Saline, Gallatin, Johnson, and Pope. Flowers are produced in August and September. The plant is usually found in rocky situations.

Silphium L.

1. Stems leafless or nearly leafless; basal leaves ovate, cordate, scabrous . . . . . . *Silphium terebinthinaceum*
1. Stems leafy; leaves opposite or alternate 2
    2. Leaves alternate, large, scabrous, pinnatifid . . . .
    . . . . *Silphium laciniatum*
    2. Leaves opposite 3
        3. Stems not square; leaves sessile, the upper ones sometimes alternate . . . *Silphium integrifolium*
        3. Stems square; leaves perfoliate . *Silphium perfoliatum*

Silphium terebinthinaceum Jacq. (3:369). The basal leaves of Prairie-dock are large. Arising among these leaves is a six to nine foot scape on which sunflower-like heads form. Specimens in which the upper leaf surface is scabrous only around the edges have been named var. *lucy-brauniae* Steyerm. SP. CIT.: typical: four miles south of Elkville railroad prairie, August 11, 1948, *BS 554;* Carbondale Reservoir, September 9, 1954, *BS 1820.* Var. *lucy-brauniae:* Perry County, five miles south of Pinckneyville, *Voigt.*

Silphium laciniatum L. (3:369). Compass-plant, so named because the leaves supposedly line themselves up edgewise in a north-south direction, is a frequent species of prairie soil along railroads. SP. CIT.: seven miles north of Murphysboro, August 21, 1954, *M 4357.*

Silphium integrifolium Michx. (3:368). Rosinweed is a common species of prairie soil along railroads and open woodlands. SP. CIT.: three miles south of Elkville, August 4, 1948, *BS 312;* Giant City State Park, September 6, 1953, *M 537.*

Silphium perfoliatum L. (3:368). Cup-plant, named for the connate-perfoliate leaves in which water collects, is local in Illinois in rather moist situations. SP. CIT.: south of Murphysboro, July 27, 1954, *M 4017.*

Dyssodia CAV.

Dyssodia papposa (Vent.) Hitchc. (3:381). Dogweed, a native species, is presently known from only Johnson and Jackson counties. SP. CIT.: Grand Tower, July 11, 1871, *French.*

Heliopsis PERS.

Heliopsis helianthoides (L.) Sweet (3:345). Ox-eye Sunflower is an occasional species of oak-hickory woodlands. Somewhat less common is var. *scabra* (Dunal) Fern. with harshly scabrous leaves. SP. CIT.: typical: Giant City State Park, July 12, 1953, *M 401;* Lake Murphysboro, September 4, 1954, *M 4702.* Var. *scabra:* Fountain Bluff, October 22, 1947, *BS 270.*

Helianthus L.

1. Disk red, brown, or purple 2
    2. Leaves chiefly alternate 3
        3. Plants 3 to 9 feet tall; leaves toothed . . . .
            . . . . *Helianthus annuus*
        3. Plants 1 to 3 feet tall; leaves entire . . . . .
            . . . . *Helianthus petiolaris*
    2. Leaves chiefly opposite 4
        4. Bracts acute; leaves lanceolate . *Helianthus rigidus*
        4. Bracts obtuse; leaves ovate . *Helianthus silphioides* °
1. Disk yellow 5
    5. Stems glabrous or nearly so below the inflorescence . . 6
        6. Leaves sessile or on petioles 1 to 5 mm. long . . .
            . . . . *Helianthus divaricatus*
        6. Leaves on petioles over 5 mm. long 7
            7. Disk 5 to 10 mm. broad; rays 5 to 8 . . . .
                . . . . . . *Helianthus microcephalus*
            7. Disk over 10 mm. broad; rays 8 to 20 . . . 8
                8. Leaves broadly lanceolate 9
                    9. Leaves entire or nearly so . . . . .
                        . . . . . *Helianthus strumosus*
                    9. Leaves serrate . *Helianthus decapetalus*
                8. Leaves narrowly lanceolate . . . . .
                    . . . . . *Helianthus grosseserratus*
    5. Stems pubescent below the inflorescence 10
        10. Leaves cordate-clasping . . . *Helianthus mollis*
        10. Leaves not cordate-clasping 11
            11. Leaves tapering all the way to the base, thus

being sessile or subsessile . . . . .
. . . . . . *Helianthus maximiliani*
11. Leaves on distinct petioles 12
12. All the leaves opposite . *Helianthus hirsutus*
12. At least the upper leaves alternate . .
. . . . . *Helianthus tomentosus*

Helianthus annuus L. (3:333). Garden Sunflower; escape from cultivation; native to western United States. SP. CIT.: Howardton cutoff, September 11, 1954, *M 4777.*

Helianthus petiolaris Nutt. (3:333). Adventive from western United States; Randolph and Jackson counties. SP. CIT.: Illinois Central Railroad in Carbondale, August 13, 1954, *M 4619.*

Helianthus silphioides Nutt. (3:335) [*Helianthus atrorubens* L. var. *pubescens* Kuntze]. Known only from Cairo, Alexander County, where it was collected by Otto Kuntze in 1874.

Helianthus divaricatus L. (3:335). Open woodlands; common; begins to flower the middle of July. SP. CIT.: Giant City State Park, September 6, 1953, *M 403;* Little Grand Canyon, August, 1954, *M 4054.*

Helianthus microcephalus Torr. and Gray (3:336). Dry, open woodlands; known only from the southern counties of Illinois where it is not common. SP. CIT.: Giant City State Park, August 9, 1953, *M 514.*

Helianthus strumosus L. (3:336). Along roads and in open woodlands; common. SP. CIT.: Giant City State Park, August 2, 1953, *M 582;* Lake Murphysboro, August 27, 1954, *M 4659.*

Helianthus decapetalus L. (3:336). A species of dry, sandy woods. SP. CIT.: Giant City State Park, August 2, 1953, *M 584.*

Helianthus grosseserratus Martens (3:338). Mostly open areas; leaves may be entire or serrate. SP. CIT.: Giant City State Park, August 5, 1953, *M 603;* Howardton cutoff, September 11, 1954, *M 781.*

Helianthus mollis Lam. (3:335). Prairie areas; not common. SP. CIT.: seven miles north of Murphysboro, June 21, 1952, *M 3137.*

Helianthus maximiliani Schrad. (3:338). Maximilian Sunflower; brilliant flowers; adventive from western United States; only Southern Illinois record listed by Jones *et al.* (1955) is from Williamson County although this specimen is probably from Jackson County. SP.

CIT.: south of Crab Orchard Lake spillway, September 19, 1947, *BS 191*.

Helianthus hirsutus Raf. (3:336). A species of fields and other open habitats; specimens we have seen belong to var. *trachyphyllus* Torr. and Gray. SP. CIT.: Giant City State Park, September 4, 1953, *M 496*.

Helianthus tomentosus Michx. (3:338) [*Helianthus tuberosus* L.]. Moist soil; occasional. SP. CIT.: Giant City State Park, September 10, 1953, *M 113;* west of Carbondale, September 24, 1947, *BS 194*.

Coreopsis L.

1. Leaves palmately lobed or divided or the uppermost entire .   .   2
   2. Lower leaves petiolate, 3-parted   3
      3. Stems glabrous; plants 3 to 9 feet tall   .   .   .   .
         .   .   .   . *Coreopsis tripteris*
      3. Stems pubescent; plants 1 to 3 feet tall .   .   .   .
         .   .   .   . *Coreopsis pubescens*
   2. Lower leaves sessile, with 3 linear lobes; plants 1 to 3 feet tall .   .   .   . *Coreopsis palmata*
1. Leaves simple, entire, or pinnately-parted   4
   4. Leaves entire   .   .   .   .   . *Coreopsis lanceolata*
   4. Leaves pinnately-parted   .   .   . *Coreopsis tinctoria*

Coreopsis tripteris L. (3:360). Tall Tickseed is fairly abundant in open woods and along roads where it begins to flower the last of August. SP. CIT.: Midland Hills, October, 1953, *Bell*.

Coreopsis pubescens Ell. (3:358). This softly pubescent southern species is known in Illinois from only four counties—St. Clair, Washington, Jackson, and Pope—although at Lake Murphysboro, it is not uncommon in dry oak woods. SP. CIT.: Murphysboro, August 27, 1954, *M 4817;* Midland Hills, October 15, 1947, *BS 238*.

Coreopsis palmata Nutt. (3:361). This is a prairie species which has been collected in our area only along railroads where it is fairly common. SP. CIT.: seven miles north of Murphysboro, June 4, 1954, *VM 1576*.

Coreopsis lanceolata L. (3:358). Locally in sandy soil over most of Illinois is this species which has been found in Randolph, Jackson, and Pulaski counties. The Jackson County specimen belongs to var. *villosa* Michx. SP. CIT.: Southern Illinois University campus, June 6, 1940, *Mc 110*.

Coreopsis tinctoria Nutt. (3:363). Golden Coreopsis, native of western United States, sometimes escapes from cultivation in our area. SP. CIT.: near Murphysboro, September 17, 1954, *M 4241*.

Ambrosia L.

1. Leaves pinnatifid, petiolate . . . *Ambrosia artemisiifolia*
1. Leaves not pinnatifid 2
    2. Leaves opposite, petiolate, deeply 3- to 5-lobed, the lobes serrate; plants 3 to 12 feet tall . . *.Ambrosia trifida*
    2. Leaves alternate, sessile, lanceolate, somewhat cordate-clasping; plants 1 to 2½ feet tall . . *.Ambrosia bidentata*

Ambrosia artemisiifolia L. (3:375) [*Artemisia elatior* L.]. Common Ragweed is abundant throughout Illinois. Pollen from it and the following species is a principal cause of hay fever. SP. CIT.: Giant City State Park, July 5, 1953, *M 355*.

Ambrosia trifida L. (3:375). Giant Ragweed is one of our most rank and aggressive weeds. It should be destroyed wherever found. SP. CIT.: Giant City State Park, August 9, 1953, *M 522*.

Ambrosia bidentata Michx. (3:375). Texas Ragweed is known in Illinois only from the southern half of the state. It is common in dry waste ground. SP. CIT.: Giant City State Park, July 23, 1953, *M 443*.

Helenium L.

Helenium tenuifolium Nutt. (3:378). Bitterweed is known only in Illinois from southern counties where it is becoming widespread in fields and pastures. It is poisonous to cattle. The disk of flower-heads is yellow. The linear leaves are not decurrent along the stem. SP. CIT.: five miles north of Carbondale, August 14, 1940, *Mc 400;* one mile south of Ava, August 7, 1954, *M 4720*.

Helenium nudiflorum Nutt. (3:378). Sneezeweed is a southern plant which occurs in rather moist soil in all our counties. The disk of flower-heads is purplish and the narrowly lanceolate, entire leaves are decurrent along the stem. SP. CIT.: Giant City State Park, July 23, 1953, *M 1336;* one mile south of Carbondale, July 17, 1940, *Mc 294*.

Helenium autumnale L. (3:378). An occasional species along streams is this Sneezeweed which is apparently more common to the north. The disk is yellow, while the lanceolate, dentate leaves are decurrent along the stem. It occurs in Jackson County (see Jones *et al.*, 1955).

Erechtites RAF.

Erechtites hieracifolia (L.) Raf. (3:405). Fireweed, one of the first species to come into a recently burned clearing, also occurs in damp woodlands. It begins to flower the last of August. SP. CIT.: Giant City State Park, August 26, 1953, *M 541.*

Senecio L.

1. Basal leaves cordate-ovate or reniform, margins crenate or dentate
   .   .   .   . *Senecio aureus*
1. Basal leaves not cordate-ovate or reniform, margins pinnatifid or
   sinuate-dentate or entire 2
   2. Plants annual, glabrous .   .   .   .   . *Senecio glabellus*
   2. Plants perennial, slightly tomentose; leaves purplish on back
      side in early season .   .   .   . *Senecio plattensis*

Senecio aureus L. (3:403). Golden Ragwort is one of our most attractive wild flowers of low woodlands. The bright yellow flowers appear about mid-April. SP. CIT.: Midland Hills, April 29, 1948, *BS 386;* Lewis Creek, April 14, 1954, *M 1981.*

Senecio glabellus Poir. (3:399). Butterweed is our commonest species of *Senecio.* It is absent from the northern counties of Illinois. It flowers from the last of April through July. SP. CIT.: Giant City State Park, April 24, 1953, *M 739;* near Etherton, April 20, 1954, *M 2174.*

Senecio plattensis Nutt. (3:401). Unlike the preceding species, Prairie Ragwort occurs in dry soil. The only Southern Illinois records are from Jackson and Gallatin counties. SP. CIT.: Fountain Bluff, May 16, 1947, *BS 44.*

Parthenium L.

Parthenium integrifolium L. (3:371). The rough-leaved American Feverfew is a typical plant in prairie soil along railroads. SP. CIT.: seven miles north of Murphysboro, June 4, 1954, *VM 1579;* Giant City State Park, June 5, 1953, *M 203.*

Achillea L.

Achillea millefolium L. (3:384). Common Milfoil, a naturalized plant from Europe, is abundant in waste ground. A pink-flowered forma *roseum* Rand & Redf. occurs. SP. CIT.: typical: Carbondale, June 26, 1947, *BS 100;* Giant City State Park, May 9, 1953, *M 128.*

F. roseum: Scenic View, May 31, 1948, *BS.* A species with dense white "wool" on the stems is *Achillea lanulosa* Nutt. (3:384). This adventive from western United States has been found in Pope County.

Anthemis L.

Anthemis cotula L. (3:303). Dog-fennel is an ill-scented weed, particularly common in barnyards. The receptacle is chaffy only near the middle. The rays are neuter. SP. CIT.: Giant City State Park, June 25, 1953, *M 359.*

Anthemis arvensis L. (3:383). Field Chamomile, a native of Europe, is known only from two northern counties and Jackson County. The receptacle is chaffy throughout and the rays are pistillate. SP. CIT.: Makanda, July 18, 1917, *Cranwill.*

Matricaria L.

Matricaria chamomilla L. (3:389). Sweet False Chamomile is found locally in the southern half of Illinois. The rays are white. SP. CIT.: near West Point, 1955, *M 4841.*

Matricaria matricarioides (Less.) Porter (3:389). This plant has been recorded in Southern Illinois from near Kornthal Church in Union County but not in Jackson County. The pineapple odor possessed by this rayless flowered plant has caused it to be called Pineapple Weed.

Tanacetum L.

Tanacetum vulgare L. (3:387). Tansy is an infrequent escape from cultivation. It is aromatic. SP. CIT.: Murphysboro, 1956, *Mohlenbrock.*

Cacalia L.

Cacalia atriplicifolia L. (3:406). Indian Plantain is a large composite found in damp woods throughout most of Illinois. It flowers from August to October. The underside of the leaves is glaucous. SP. CIT.: Fountain Bluff, October 17, 1947, *BS 253;* near Sand Ridge, August 15, 1954, *M 4622.*

Cacalia muhlenbergii (Schultz-Bip.) Fern. (3:405). This is a local species known only from a few Illinois counties. The lower surface of the leaves is green. SP. CIT.: Giant City State Park, June 16, 1953, *M 498.*

Artemisia L.

1. Plants densely white pubescent; leaves entire . . . .
    . . . . *Artemisia gnaphalodes*
1. Plants glabrous; leaves pinnately dissected 2
    2. Leaves 2 to 6 inches long; heads in a paniculate inflorescence
        . . . . *Artemisia annua* °
    2. Leaves 1 to 3 inches long; heads in leafy spikes . . .
        . . . . *Artemisia biennis*

Artemisia gnaphalodes Nutt. (3:392). White Sage is an occasional adventive along railroads. SP. CIT.: Murphysboro, August 14, 1954, *M 4620*.

Artemisia annua L. (3:393). Annual Wormwood is found sometimes in waste ground in our area. It is known from Union, Hardin, Alexander, and Massac counties.

Artemisia biennis Willd. (3:392). The only record of Biennial Wormwood for Southern Illinois is based on a collection from Grand Tower (Jackson County) in 1871 by G. H. French.

Antennaria GAERTN.

Antennaria plantaginifolia (L.) Hook. (3:479). An early flowering species of wooded slopes is Pussy-toes. The pistillate heads are five to seven millimeters high. SP. CIT.: Giant City State Park, March 20, 1953, *M 631;* Little Grand Canyon, March 25, 1951, *Mohlenbrock*.

Antennaria fallax Greene (3:479). Not as common as the preceding species is this larger-flowered Pussy-toes. Pistillate heads are eight to ten millimeters high. SP. CIT.: Giant City State Park, April 8, 1953, *M 727;* near Etherton, April 19, 1954, *M 2177*.

Gnaphalium L.

Gnaphalium purpureum L. (3:483). Early Cudweed flowers from May to mid-June. It is abundant in woods and fields. SP. CIT.: Giant City State Park, May 6, 1953, *M 739;* Midland Hills, May 24, 1952, *Voigt*.

Gnaphalium obtusifolium L. (3:483). Sweet Everlasting occurs in usually dry open areas where it flowers from August to October. SP. CIT.: Giant City State Park, August 28, 1953, *M 224;* Thompson's Lake, October 2, 1940, *Mc 547*.

Solidago L.

16. Leaves serrate or dentate . . .
. . . . .*Solidago gigantea*
16. Leaves mostly entire or sparingly low
serrate . .*Solidago missouriensis*
15. Leaves with 1 principal vein . . . 17
17. Leaves mostly entire . . . .
. . . . .*Solidago speciosa*
17. Leaves serrate or dentate . . . 18
18. Racemes or panicle branches
few, arching . *Solidago ulmifolia*
18. Racemes or panicle branches numerous. 19
19. Heads in terminal panicles
or racemes and flowering in
July and August. . . 20
20. Leaves and panicle
branches glabrous or
nearly so . . .
. .*Solidago juncea*
20. Leaves scabrous and
panicle branches hir-
tellous . *Solidago
juncea* forma *scabrella*
19. Heads in axillary clusters or
racemes flowering from August until October . .
. . . *Solidago caesia*

Solidago petiolaris Ait. (3:425). Rather dry woods; not common. SP. CIT.: Giant City State Park, September 6, 1953, *M 540*.

Solidago hispida Muhl. (3:421). Dry woodlands; rare; only Illinois stations are in Jackson and Alexander counties. SP. CIT.: Little Grand Canyon, October 10, 1952, *BS 3168*.

Solidago rigida L. (3:435). Rigid Goldenrod; found along railroads and in hill prairies in our area; Jackson and Union counties. SP. CIT.: one mile south of Elkville, September 22, 1948, *BS 626*.

Solidago nemoralis Ait. (3:427). Field Goldenrod; common in open areas. SP. CIT.: Fountain Bluff, October 20, 1952, *BS 3218A;* Giant City State Park, September 1, 1953, *M 238*.

Solidago radula Nutt. (3:435). Rare in Illinois; southern counties include Jackson, Union, and Pope. SP. CIT.: Grassy Knob hill prairie, September 9, 1954, *M 4759*.

Solidago altissima L. (3:435). Tall Goldenrod is our most abundant species. SP. CIT.: Carbondale, September 19, 1947, *BS 188;* Fountain Bluff, September 14, 1940, *Mc 443.*

Solidago buckleyi Torr. and Gray (3:425). Usually around sandstone bluffs; limited in Illinois to southern counties. SP. CIT.: Giant City State Park, September 14, 1949, *BS 876.*

Solidago rugosa Mill. (3:431). Moist ground; known in Illinois from Lawrence and Randolph counties.

Solidago drummondii Torr. and Gray (3:429). Rare species of limestone areas; known only from southwestern Illinois; Jackson, Union, and Johnson counties in our area; one of the last species to flower. SP. CIT.: Fountain Bluff, October 20, 1952, *BS 3218B;* Grand Tower, August 24, 1948, *BS 590.*

Solidago latifolia L. (3:423) [*Solidago flexicaulis* L.]. Moist woods; occasional; probably in all our counties. SP. CIT.: Giant City State Park, September 20, 1953, *M 442;* Little Grand Canyon, August, 1954, *M 4026.*

Solidago graminifolia (L.) Salisb. (3:439). The typical variety is sometimes called *Solidago hirtella* (Greene) Bush. It is common in moist ground. Var. *nuttallii* (Greene) Fern. has a single station in Southern Illinois. Var. *media* (Greene) Harris [*Solidago media* (Greene) Bush] is fairly common in moist soil. SP. CIT.: typical: one mile north of Pomona, September 28, 1951, *V 996.* Var. *nuttallii:* moist soil, railroad prairie, Finney, August 26, 1954, *M 4664.* Var. *media:* north of the Big Muddy River, September 16, 1959, *BS 877.*

Solidago patula Muhl. (3:427). Rare species of very wet soil in our area; Jackson and Williamson counties. SP. CIT.: Murphysboro "marsh" (with *Dryopteris thelypteris*), 1953, *M 1702.*

Solidago gigantea Ait. (3:433). Late Goldenrod; occasional in moist ground; attributed to Jackson County by Jones *et al.* (1955).

Solidago missouriensis Nutt. (3:427) [*Solidago glaberrima* Martens]. Dry, often prairie areas; not common; Jackson, Union, and Williamson counties. Our specimens belong to var. *fasciculata* Holzinger. SP. CIT.: Giant City State Park, September 6, 1953, *M 538.*

Solidago speciosa Nutt. (3:421). Showy Goldenrod; dry open woods; fairly common. SP. CIT.: one mile south of Elkville, September 22, 1948, *BS 627.*

Solidago ulmifolia Muhl. (3:429). Elm-leaved Goldenrod; dry woods; common. SP. CIT.: Giant City State Park, August 20, 1953, *M 555.*

Solidago juncea Ait. (3:427). Along roads and in fields and woodlands; our earliest flowering goldenrod (June 25). Var. *scabrella* Gray is occasionally found. SP. CIT.: typical: Giant City State Park, June 25, 1953, *M 350;* Big Muddy River bottoms, July 16, 1940, *Mc 287.* Var. *scabrella:* Giant City State Park, June 25, 1953, *M 714.*

Solidago caesia L. (3:425). Blue-stemmed Goldenrod; near sandstone bluffs; occasional. SP. CIT.: Fountain Bluff, October 20, 1952, *BS 3215;* Giant City State Park, 1953, *M 500.*

The white-flowered *Solidago bicolor* L. (3:421) is reported from Union County, *Brendel.*

Chrysopsis NUTT.

Chrysopsis camporum Greene (3:411). Golden Aster occurs locally in sandy soil over the entire state. We have it from Randolph, Union, Alexander, and Pulaski counties.

Heterotheca CASS.

Heterotheca subaxillaris (Lam.) Britt. & Rusby (3:413). This waif has been collected twice in Illinois—in Henry County and in Union County.

Ratibida RAF.

Ratibida pinnata (Vent.) Barnh. (3:352). Drooping Coneflower (Fig. 67) is an attractive species along the edges of dry woods. It has been found from our area in Jackson, Randolph, Pope, and Hardin counties. SP. CIT.: two miles south of Elkville, July 9, 1951, *BS 1528.*

FIG. 67. *Drooping Coneflower* (Ratibida pinnata) *may be seen in most prairie situations and along country roadsides in summer.*

Rudbeckia L.

1. Leaves not cordate-clasping 2
   2. Leaves, at least the lower, pinnately lobed, pinnatifid, or 3-lobed 3
      3. Rays long (3 to 6 cm.), drooping; heads greenish-yellow . . . . *Rudbeckia laciniata*
      3. Rays shorter (1 to 3.5 cm.), not drooping; heads brownish 4
         4. Rays 8 to 12; lower leaves 3-lobed . . . . *Rudbeckia triloba*
         4. Rays 15 to 20; some lower leaves 3-lobed, tomentose below . . . . . *Rudbeckia subtomentosa*
   2. Leaves merely denticulate or entire 5
      5. Rays orange-yellow, 1 to 1.5 cm. long; pappus present, although sometimes minute 6
         6. Some of the leaves over 2 cm. broad; bracts glabrous or strigose . . . . . *Rudbeckia fulgida* [*]
         6. None of the leaves over 2 cm. broad; bracts hirsute . . . . . . *Rudbeckia missouriensis* [*]
      5. Rays light yellow; pappus absent 7
         7. Leaves 2.5 to 7 cm. broad; basal leaves ovate . . . . *Rudbeckia hirta*
         7. Leaves 1 to 3 cm. broad, linear-lanceolate; basal leaves oblanceolate 8
            8. Rays yellow throughout 9
               9. Rays 1.0 to 3.5 cm. long . . . . *Rudbeckia serotina*
               9. Rays 3.5 to 5.0 cm. long . . . . . . *Rudbeckia serotina* var. *lanceolata*
            8. Rays with a reddish spot near the base . . . . . *Rudbeckia serotina* f. *pulcherrima*
1. Leaves cordate-clasping . . . *Rudbeckia amplexicaulis*

Rudbeckia laciniata L. (3:349). Golden-glow is found occasionally in rich soil where it may become nearly ten feet tall. It is sometimes cultivated. SP. CIT.: Giant City State Park, August 2, 1953, *M 564*.

Rudbeckia triloba L. (3:347). Brown-eyed Susan is uncommon in dry woods. Unless it is carefully viewed it is easily confused with the following species. SP. CIT.: four and one-half miles north of Carbondale, August 15, 1940, *Mc 397*.

Rudbeckia subtomentosa Pursh (3:349). Fragrant Coneflower is more frequent in dry woods than the preceding. SP. CIT.: Giant

City State Park, August 27, 1954, *M 4657;* four miles south of Elkville, August 11, 1948, *BS 555.*

Rudbeckia fulgida Ait. (3:347). The only collection of this species in Illinois is from Herod, Pope County, by G. P. Clinton on July 29, 1898.

Rudbeckia missouriensis Engelm. (3:347). This characteristic species of limestone glades has been found on Illinois hill prairies in Monroe and Randolph counties by R. A. Evers.

Rudbeckia hirta L. (3:347). This native species of open woods and fields is local in our area. It is similar in appearance to the following species, both of which are called Black-eyed Susan. SP. CIT.: Carbon Lake, July 27, 1954, *M 4394.*

Rudbeckia serotina Nutt. This common Black-eyed Susan is abundant in fields and other open areas where it is adventive from western United States. An attractive form with blotched rays, forma *pulcherrima* (Farw.) Fern. and Schubert, has been found once in our area. Specimens in open woods and with rays three and one-half to five millimeters long belong to the apparently native var. *lanceolata* (Bisch.) Fern. and Schubert. SP. CIT.: typical: Giant City State Park, June 9, 1953, *M 296;* along the Big Muddy River, June 20, 1941, *Mc 832.* Var. *lanceolata:* Giant City State Park, June 25, 1947, *BS 99.* F. *pulcherrima:* near Lewis Creek, dry soil, *Mohlenbrock.*

Rudbeckia amplexicaulis Vahl (3:351) [*Dracopsis amplexicaulis* Cass.]. This adventive from southwest United States was found along a railroad in Murphysboro in 1951. It is also reported from Cook County. SP. CIT.: along railroad, Murphysboro, June 19, 1951, *Hardy.*

Chrysanthemum L.

Chrysanthemum leucanthemum L. (3:387). Ox-eye is very common in waste areas where it flowers from May to July. It is native to Europe. SP. CIT.: Giant City State Park, May 24, 1953, *M 179.*

Boltonia L'HER.

Boltonia interior (Fern. and Griscom) G. N. Jones (3:469) [*Boltonia diffusa* Ell. var. *interior* Fern. and Griscom]. This species, often mistaken by amateurs for an aster, is occasional in open areas. SP. CIT.: September 13, 1889, *French.*

Boltonia recognita (Fern. and Griscom) G. N. Jones (3:469) [*Boltonia latisquama* Gray var. *recognita* Fern. and Griscom; *Boltonia asteroides* (L.) L'Her. var. *recognita* (Fern. and Griscom) Cronq.]. Much more common than the preceding is this species which is found along rivers and streams. SP. CIT.: Carbondale Reservoir, October, 1953, *Bell;* Howardton cutoff, September 11, 1954, *M 4776.*

LITERATURE CITED

INDEX OF PLANT NAMES

LITERATURE CITED

Bailey, William M., and Julius R. Swayne. 1951. New Illinois plant records. Am. Midl. Nat. 46: 256.

———. 1952. Some Southern Illinois plant records. Trans. Ill. Acad. Sci. 44: 40–41.

Baldwin, J. T. 1942. Cytological basis for specific segregation in the *Sedum nevii* complex. Rhodora 44: 10–13.

Eggert, H. 1891. Catalogue of the phaenogamous and vascular cryptogamous plants in the vicinity of St. Louis, Missouri. Publ. by the author, St. Louis. 16 pp.

Eiten, George. 1955. The typification of the names "Oxalis corniculata L." and "Oxalis stricta L." Taxon IV, 5: 99–104.

Evers, R. A. 1956. Two plants new to the Illinois flora. Rhodora 58: 49.

Fernald, M. L. 1944. The geographic segregation of *Monarda fistulosa* and its var. *mollis*. Rhodora 46: 9–12.

———. 1950. Gray's manual of botany. Ed. 8. American Book Co., New York. 1632 pp.

Gleason, H. A. 1903. A second station for *Phacelia covillei*. Torreya 3: 89–90.

———. 1952. The new Britton and Brown illustrated flora of the northeastern United States and adjacent Canada. New York Botanical Garden, New York. 3 Vol. 1726 pp.

Hill, E. J. 1870. The fern flora of Illinois. Fern Bull. 20: 33–43, 73.

Jones, G. N. 1947. An enumeration of Illinois Pteridophyta. Amer. Midl. Nat. 38: 76–126.

———. 1950. Flora of Illinois. Ed. 2. Am. Midl. Nat. Monogr. 5: 1–368. University of Notre Dame Press, South Bend, Indiana.

———, G. D. Fuller, H. E. Ahles, G. S. Winterringer, and Alice Flynn. 1955. Vascular plants of Illinois. Museum Science Series, Vol. VI. The University of Illinois Press and the Illinois State Museum, Urbana and Springfield. 593 pp.

Mackenzie, K. K. 1931. *Carex* in N. Am. Fl. 18: 1–168.

———. 1935. *Carex* in N. Am. Fl. 18: 169–478.

Mohlenbrock, R. H. 1954. Some notes on the flora of Southern Illinois. Rhodora 56: 227–228.

———. 1955. Contributions to the flora of Southern Illinois. Rhodora 57: 319–322.

———. 1955–1956. The Pteridophytes of Jackson County, Illinois. Am. Fern Journ. 45: 143–150.

———. 1956. An unusual form of *Asplenium pinnatifidum*. Amer. Fern Journ. 46: 91–93.

Neill, J. 1950. *Isotes melanopoda* still grows in Illinois. Am. Midl. Nat. 44: 251.

Palmer, E. J. 1921. A botanical reconnaissance of Southern Illinois. Journ. Arn. Arb. 2: 129–153.

Patterson, H. N. 1876. Catalogue of phaenogamous and vascular cryptogamous plants of Illinois. Oquawka, Illinois. 54 pp.

Pfeiffer, Norma E. 1922. Monograph of the Isoetaceae. Ann. Missouri Bot. Gard. 9: 79–232.

Sargent, C. S. 1933. Manual of the trees of North America. Houghton, Boston. 93 pp.

Steagall, Mary M. 1927. Some Illinois Ozark ferns in relation to soil acidity. Trans. Ill. Acad. Sci. 19: 113–136.

Svenson, H. K. 1944. The New World species of Azolla. Amer. Fern Journ. 34: 69–84.

Tryon, R. M. 1939. The varieties of *Convolvulus spithamaeus* and *C. sepium*. Rhodora 41: 415–423.

Voigt, J. W. 1955. Southern Illinois flora: recent additions. Rhodora 57: 159–160.

———, and J. R. Swayne. 1955. French's shooting star in Southern Illinois. Rhodora 57: 325–332.

Winterringer, G. S. 1952. Flowering *Arundinaria gigantea* in Illinois. Rhodora 54: 82–83.

Woodson, R. E. 1954. The North American species of *Asclepias* L. Ann. Missouri Bot. Gard. 41: 1–212.

INDEX OF PLANT NAMES